Samira Spatzek
Unruly Narrative

American Frictions

Editors
Carsten Junker
Julia Roth
Darieck Scott

Editoral Board
Arjun Appadurai, New York University
Mita Banerjee, University of Mainz
Tomasz Basiuk, University of Warsaw
Isabel Caldeira, University of Coimbra

Volume 6

Samira Spatzek

Unruly Narrative

Private Property, Self-Making,
and Toni Morrison's *A Mercy*

DE GRUYTER

The open access publication of this title was supported by the publication fund of the Staats- und Universitätsbibliothek Bremen.

ISBN 978-3-11-163112-7
e-ISBN (PDF) 978-3-11-078057-4
e-ISBN (EPUB) 978-3-11-078066-6
DOI https://doi.org/10.1515/9783110780574

This work is licensed under the Creative Commons Attribution-NonCommercial-NoDerivatives 4.0 International License. For details go to http://creativecommons.org/licenses/by-nc-nd/4.0/.

Creative Commons license terms for re-use do not apply to any content (such as graphs, figures, photos, excerpts, etc.) that is not part of the Open Access publication. These may require obtaining further permission from the rights holder. The obligation to research and clear permission lies solely with the party re-using the material.

Library of Congress Control Number: 2022937392

Bibliographic information published by the Deutsche Nationalbibliothek
The Deutsche Nationalbibliothek lists this publication in the Deutsche Nationalbibliografie; detailed bibliographic data are available on the Internet at http://dnb.dnb.de.

© 2024 the author(s), published by Walter de Gruyter GmbH, Berlin/Boston
This volume is text- and page-identical with the hardback published in 2022.
The book is published with open access at www.degruyter.com

www.degruyter.com

Acknowledgements

Like most projects, this one owes to so many. The research for this monograph was generously funded through a dissertation fellowship by the Rosa-Luxemburg-Stiftung in Berlin. I also received financial support for research stays, summer school participation, and conference travels from the Rosa-Luxemburg-Stiftung, the University of Bremen, the German Academic Exchange Service, and the Postgraduate Forum of the German Association for American Studies. I thank the series editors Prof. Dr. Carsten Junker, Prof. Dr. Julia Roth, and Prof. Dr. Darieck Scott for trusting this project and for accepting the manuscript for publication in American Frictions. Dr. Julie Miess has been a wonderful interlocutor at DeGruyter, giving expert advice on all things related to the publishing process. Early versions and parts of chapters 3, 4.5 and 4.6 as well as the introduction were published in the following journals: *Black Studies Papers*, *Slavery & Post-Slavery / Esclavages & Post-esclavages*, and *Amerikastudien / American Studies —A Quarterly*.

Toni Morrison's *A Mercy* begins with three words uttered by an unnamed voice. Reassuringly, the voice tells the reader: "Don't be afraid." For the longest time, I was afraid that I would never finish this project. I managed to do so only thanks to the unflinching intellectual support, care, love, friendship, and stamina of those around me. Enormous thanks go to all of you – near and far – who have accompanied me over the past few years.

First of all, I thank my dissertation advisor Prof. Dr. Sabine Broeck. Had you not pushed me to do this, gently, I wouldn't have. Without your generous and ongoing intellectual support, feedback, criticism, questioning, rigor, time, and guidance, I would not have had the courage to think and write—and to think and write again, to revisit and develop my arguments. I thank Prof. Dr. Alan Rice, who joined this project at a later stage. I owe you a great deal and I am deeply grateful to you for years of invaluable critical conversation, intellectual exchange, and expert advice. I was very lucky that Prof. Dr. Peter Schneck agreed to become the third reader for my dissertation, doing so with as much intellectual curiosity as personal generosity.

During the years, I had the honor of working with Prof. Dr. Frank B. Wilderson, III. My thanks go to you for patiently engaging with all my questions, for encouraging me to "write with more muscle," and for generously sharing your thoughts. I am indebted to Cedric A. Essi and Paula von Gleich not only for (proof-)reading all drafts of the dissertation manuscript (night and day) but also for being my friends and colleagues, for challenging me, and for allowing me to grow in their presence. I thank Marius Henderson, Mariya Nikolova, Michel

Büch, and the other members of the doctoral students' network "Perspektiven in der Kulturanalyse: Black Diaspora, Dekolonialität und Transnationalität" (Bremen) as well as my fellow editors at *COPAS* for creating spaces to think and struggle together and for offering continued peer support. My thanks also go to my colleagues at the University of Bremen for lending their perspectives at all stages of this project. I thank Prof. Dr. Tracy D. Sharpley-Whiting, Prof. Dr. Houston A. Baker, Prof. Alice Randall, Prof. Dr. Tiffany R. Patterson, Prof. Dr. Claudine Taaffe, and the faculty at the Department of African American and Diaspora Studies for hosting me as a research fellow at Vanderbilt University in the fall of 2016.

I owe special thanks to my friends and family. I thank Julia Dück for years of coffee, complaint, and companionship. I thank Till Kadritzke for inspiring me to dig deeper into words and their meaning. I also owe thanks to Kalinka, Agnes Eilers, Gesine Leithäuser, Courtney N. Moffett-Bateau, Morten Paul, T. Mars McDougall, Sebastian Weier, Dominique Haensell, Moritz Altenried, Mustafa Fateh, Anne Potjans, Sarah Reimann, Julian Warner, Linda Krenz-Dewe, Takunda Matose, Nina Röttgers, Koen Potgieter, and Marie Rümelin for offering welcome distractions, love, and support at all times. When I struggled to drive this project over the finish line and needed just a little more time, Jan-Olaf Rodt encouraged me to go on, patient, arms wide open.

A big thank you goes to my father Tom van der Geld, my Spatzek family, and my U.K. family, Graham Beck and Lida Kindersley, who have tirelessly rooted for me over the years. I thank Edith Götze and Martin Vogl, who have given me more love and support than I could ever hope to return. Finally, I thank my mother Maria Spatzek—for everything, but above all for lovingly trusting me in the way that only you can.

Contents

1 **Introduction** —— 1
1.1 Forming Refusal, or Refusal as Form —— 1
1.2 Historical Contexts —— 8
1.3 Critical Contexts —— 12
1.4 Private Property as Cultural Metaphor and Literary Form —— 16
1.5 Situating the Study —— 18
1.6 Reading Methods —— 20
1.7 The Chapters —— 23

2 **Claims to Freedom: Private Property and the New World Liberal Subject** —— 26
2.1 Introduction —— 26
2.2 John Locke's *Two Treatises of Government* (1689): Critical Contexts —— 30
2.3 Post-Slavery Readings: The *Two Treatises* and "The Fundamental Constitutions of Carolina" (1669) —— 43
2.4 Whiteness as Property: "The Germantown Friends' Protest Against Slavery" (1688) —— 47
[Coda]: The Liberal Property Paradigm —— 52

3 **Interrogating Private Property: Black Studies and the Liberal Imagination** —— 54
3.1 Introduction —— 54
3.2 Interrogating the Property Paradigm —— 56
3.3 The Structure of Violence —— 63
3.4 Dispossession and Fungibility —— 69
3.5 Abjection and Abjectorship —— 75
3.6 Reproduction, Kinship —— 77
3.7 Anticipatory Wake —— 88
[Coda] —— 98

4 **Practicing Refusal: Narrative Interrogations of the Property Paradigm in *A Mercy*** —— 99
4.1 "The Chagrin of Being Both Misborn and Disowned": [Jacob Vaark], Freedom, and the Pursuit of Property —— 117
4.2 "I am Exile Here": [Lina], Self-Inventions, and Dispossession —— 137

4.3   "The Promise and Threat of Men": [Rebekka], Liberal Self-Making, and the Ruse of Solidarity —— 153
4.4   "My Name is Complete": [Sorrow], Anticipating Generations, and the New World Grammar of Property —— 174
4.5   "I Am a Thing Apart": [Florens] and the Ruse of Belonging —— 189
4.6   "There is No Protection": The [*Minha Mãe*], Slave Narratives, and the Sexual Economies of Atlantic Slavery —— 220

5   Coda —— 246
5.1   Refusing Private Property or, On Telling Impossible Stories —— 246

Works Cited —— 255

Index —— 279

1 Introduction

1.1 Forming Refusal, or Refusal as Form

> Freeing yourself was one thing, claiming ownership of that freed self was another.
> — Toni Morrison, *Beloved*

> The entanglements of bondage and liberty shaped the liberal imagination of freedom, fueled the emergence and expansion of capitalism, and spawned proprietorial conceptions of the self. [...] The longstanding and intimate affiliation of liberty and bondage made it impossible to envision freedom independent of constraint or personhood and autonomy separate from the sanctity of property and proprietorial notions of the self.
> — Saidiya V. Hartman, *Scenes of Subjection*

Within the first few pages of Toni Morrison's historical novel *A Mercy*, the Anglo-Dutch farmer, trader, and moneylender Jacob Vaark appears unannounced out of a thick and inscrutable mist on his way to his business partner, a Portuguese slave and tobacco trader. At this point, we as readers neither know his identity nor his business as we encounter him only as "[t]he man." As he happens on the novel's setting in seventeenth-century colonial North America,[1] he slowly moves "through the surf, stepping carefully over pebbles and sand to shore" until he arrives at "the ramshackle village that sleeps between two huge riverside plantations," where he buys a horse (Morrison, *A Mercy* 7, 8).[2] When "the man" signs a note as a means to finalize the act of sale, we finally learn his name: "Jacob Vaark" (*AM* 8). At this very moment of the purchase, Vaark not only signs himself into being, but *he signs himself into being as an owner*. Is it a coincidence that the coming-into-being of this character correlates with the sale of property? What is the relation between the sale of property, a signature, and the naming of this character? How can we examine this nexus, what does it mean? And what does it tell us about the making of liberal subjects?

The scene exposes the connections between the rise of liberal modernity and its subject and the questions it raises point to the importance of private property in this context. In this study of the positioning and formative powers of modern

---

[1] Most critics and readers seem to agree that the novel is set in colonial Virginia and Maryland even though some have suggested that it is set in the Northern colonies. In *Unruly Narrative*, I also consider the colonial Chesapeake and especially Virginia to be the novel's immediate setting while also suggesting more broadly that *A Mercy* allegorizes North American colonial beginnings as a whole.
[2] From here on, I will use *AM* whenever I quote from *A Mercy*.

liberal ideas of private property, I offer an intervention into hegemonic notions of the liberal Human. *Unruly Narrative: Property, Self-Making, and Toni Morrison's A Mercy* observes the importance of narrative for the liberal subject, which emerged through the formations of European liberalism, Atlantic slavery, and settler colonial expansion in the so-called New World. In questioning these connections, I turn to Morrison's *A Mercy* as a key literary text that generates a fundamental philosophical and political critique of the connections between self-making and private property as it interrogates liberal ideas of what it means to be a Human subject at its seventeenth-century New World scene.[3]

Unruly Narrative examines the complex social, cultural, political, and philosophical entanglements between power, race, and subjectivity that are so fundamental to U.S. society. The study scrutinizes this nexus, or grammar (Spillers, "Mama's Baby"), by way of examining the entanglements between self-making and private property within the realm of literary representation and in relation to what Saidiya Hartman has called "the liberal imagination of freedom[, the texture of which] is laden with [...] slavery" (*Scenes* 115–116). To this end, my study follows scholars of the racialized emergence of Western modernity, whose works show that, historically, liberal claims to individual freedom are intimately bound by the systems and practices of European settler colonialism and enslavement of African and African-descended human beings and the transatlantic trade in human flesh. Over the coming pages, I will use the terms Atlantic slavery, New World chattel slavery, chattel slavery, and slavery interchangeably in my discussion of colonial Virginia and its fictional representation in *A Mercy*. I do so in order to denote the global dimensions of transatlantic slavery as well as its local configuration in seventeenth-century Virginia. As Caribbean intellectual and feminist thinker Sylvia Wynter notes when she reminds us of the conjunction between the systems of Atlantic slavery and liberal claims to individual freedom:

> Western Europe's epochal shift was a product of the intellectual revolution of humanism [the effects of which were] the horrors that were inflicted by [...] the settlers upon the indig-

---

[3] In this study, I will use terms like subjectivity, subject, personhood as well as self and self-making interchangeably. Following Wilderson, I will use Human and Humanist with a capital H "to connote a paradigmatic entity that exists ontologically as a position of life in relation of the Black of Slave position, one of death" (*Red* 23). It is important to state clearly that my work does not position Black people as objects of analysis; I capitalize "Black" to denote a historical, socio-discursive construct and category of analysis, and a political position. I use "white" in small letters to show that "whiteness" also is a racialized (and not a universal) category and (analytical) position "that has been historically produced and that has real consequences" (Adusei-Poku 44).

enous peoples of the Caribbean and the Americas, as well as upon the African-descended Middle Passages and substitute slave labor force. ("1492" 13)

Today, dominant contemporary conceptions of the liberal subject as inherently free, white, and propertied relate back to these early moments of liberal subject formation described by Wynter.[4] It is this "white subject's universalist reign [which] keeps resurfacing [...] in much of the recent feuilleton and academic discourse of the legacy of Enlightenment as a haven of freedom, entitled subjectivity, and human rights" (Broeck, *Gender* 49). This white Human subject (of at least male and female if not all genders) and the discourse accompanying it have "endured not only in the face of hundreds of years of enslavism and colonialism, but also in our presence of white neoliberal capitalist expansion" (49).[5] *Unruly Narrative* interrogates the emergence and status of this subject and its claims to freedom on the literary level of representation and thinks through questions of self-making, personhood, and the meaning and positioning power of private property in relation to those who are barred from making such liberal claims at *A Mercy*'s early colonial scene.

---

4 A distinguished body of scholarship on the relation between *Race and the Enlightenment* (Eze; see also Buck-Morss; Hulme, "Spontaneous"; Ward and Lott) has shown that the Enlightenment's self-descriptive narrative of scientific and Human emancipation needs to be understood as "altogether bypass[ing] the historical experience of lively and angry early modern controversies around the slave trade, slavery, and issues of mastery, ownership, and oppression of human beings" (Broeck, "Never" 236). Historians, sociologists, and philosophers have also written extensively about the closely intertwined gendered and racialized histories of European economic expansion, (settler-) colonialism in the 'New World,' the transatlantic slave trade, and modern liberalism (e.g., Bhandar, *Colonial Lives*; Blackburn; Lowe; Grosfoguel; Dussel and Mendieta; Greene; Hartman, *Scenes*; McClintock; Mustakeem).

5 Sabine Broeck introduces the term "enslavism" to try and create a vocabulary for white people to address the "historical and ongoing practice of structural anti-Blackness" (*Gender* 5; see also "Legacies," "Lessons," "Abolish") and to "generate metacritical, epistemic potential" (*Gender* 47) to talk about these – white – practices. As such, it becomes a means to "critique the durable nexus between Euro-American transatlantic enslavement practices and (post-)modern discourses" as well as a means to think about "this continuity reaching into the future, in which anti-Blackness as violence, commodification and repression is contained as a kind of ongoing legacy in New World enslavement" (*Gender* 47, 3). With this term, then, Broeck seeks to take account of the "telling fact that humanist education, including recent so-called avant-garde theory, has so utterly abjected modern transatlantic enslavement from its purview" (*Gender* 47). Broeck seeks to create a vocabulary that helps white people to unlearn and "to dismantle [antiblackness's] conceptual, theoretical and epistemic hold and ubiquity, to move beyond prescriptive appeals to good behavior in ally-ship" (qtd. in Sirvent).

As a concept, private property has been subjected to much intellectual scrutiny at least since the seventeenth century and particularly within scholarly fields such as legal studies, political philosophy, and cultural studies. Historians and philosophers have thoroughly demonstrated that ideals of private property are central to the making of Western modernity and that private property has become "the basis for making claims of natural rights and political liberties" (Graeber 35). The multi-faceted concept of private property thus not only "references the things that are owned, as in common usage, but also a social system in which the right and ability to own are protected by the state" (Hong 180). Critical race and critical legal scholars have shown that private property also goes "beyond legal doctrine, extending to ideologies of the self, social interactions with others, concepts of law, and social concepts of gender roles and race relations" (Davies 2). Thus deeply woven into the social, political, and philosophical fabric of Western modernity, Whiteness has in this context gained "value as a property in itself, a value encoded in property law and social relations" (Bhandar, *Colonial Lives* 7; see also Bell; Harris, "Whiteness," "Afterlife"; Lipsitz). Black feminist thinkers' deep engagement with the entanglements between private property, racial capitalism, and gender has shown how the institution of slavery maintained and renewed itself through the calculated acquisition and reproduction of human property. As a post-slavery theoretical trajectory intervening into the discursive promises of universal liberty, Black feminist theorizing demonstrates that to reckon with the making of the white Human subject means to fundamentally engage with Western modernity's racialized "calculus" of private property (Hartman, *Lose*). I follow these insights in conceptualizing private property as a racialized cultural metaphor and abstract value determining various scales of existence. My study examines the complex entanglements of the regimes of private property and violence induced by Atlantic slavery with processes of liberal self-making and demonstrates how *A Mercy* becomes a critical lens with which we are able to interrogate ideas of the quintessential modern Human subject that characters like Jacob Vaark represent. Throughout, I conceptualize the connections between freedom, private property, and bondage as the *property paradigm*. I suggest that narrative form – specifically the text's strategies of characterization – becomes the means by which *A Mercy* allegorizes, criticizes, and ultimately rejects the property paradigm.

Published in 2008, *A Mercy* takes its readers back to late seventeenth-century North America and thus to a time when liberal ideas of individual rights, representational government, and political emancipation from feudal rule, as well as claims to individual liberty were first articulated through "metaphors of property" (Graeber 36). As David Graeber has it, "Where an earlier, hierarchical view assumed that people's identities (their properties, if you will) were defined by their place in society, the assumption was now that who one was was based on what one had, rather than the other way round" (36). Locating the action of the novel in colonial Virginia and Maryland around 1690, *A Mercy* begins *in medias res* and tells the story of the Black enslaved girl Florens. She is the legal property of Jacob Vaark (also called Sir) and lives on his farm with a group of women: Vaark's wife Rebekka (also called Mistress); an Indigenous woman servant called Lina; and a shipwrecked girl called Sorrow/Twin. Two white indentured servants, Willard Bond and Scully, whose services Vaark regularly makes use of, are also part of the farm life even though they belong to the household of neighbor. In addition, a blacksmith, whom one of the other characters describes as a "free African man" (*AM* 43), appears on the Vaark farm in the novel. Finally, the reader also encounters Florens' mother, the *minha mãe*. Florens' first-person text gives an account of her journey to the blacksmith, whom she is ordered to fetch to help cure Rebekka of the pox in the wake of Vaark's untimely death (37). Following Florens' six textual fragments are the respective sections of Jacob Vaark, Lina, Rebekka Vaark, Sorrow, and Willard Bond and Scully. Her fragmented auto-diegetic text thus takes turns with a third-person narrator who "provides the back-stories for Florens […] and the other characters who live or work on [Jacob Vaark's] burgeoning Virginia estate" (Jennings 646). The novel ends with the textual fragment of the *minha mãe*.

A Mercy urges its readers to critically revisit their Western liberal heritage as being shaped by slavery and settler colonialism. In doing so, the novel creates what I call a complex "character-scape." Critics have often read the novel's character-scape as representing seventeenth-century colonial Virginia's intricate social strata at a moment in the long history of Atlantic slavery "when the conflation of race and slavery was in its infancy" (Jennings 645). Take, for example, the section of Willard Bond and Scully, which is framed by Florens' texts and situated in the second half of the novel. In it, we learn that Willard and Scully have been subjected to different terms of indentured service. As a somewhat retrospective account of the novel's plot and action, this section of the novel exposes the reader to these characters' respective views of both the other characters and the events happening in the wake of Jacob Vaark's death. Focalized through Willard and Scully, the section paradigmatically speaks to colonial Virginia's said relative racial fluidity. In the novel's rendering, these two indentured servants are placed in a competitive relationship with the free African blacksmith, who is paid for the work that he performs on the Vaark farm. "The clink of silver was as unmistakable as its gleam. [Willard] knew Vaark was getting rich from rum investments, but learning the blacksmith was being paid for his work […] roiled Willard" (148). The fact that the blacksmith is not only able to "own things [and] sell his own labor" but that Vaark pays him for his work appears to suggest that his economic status at the New World colonial scene is elevated over that of white indentured servants like Willard and Scully (43). As we continue reading, however, we firstly learn that the blacksmith will be violently forced out of the novel's plot after Florens injures him in a fight; and, secondly, we learn that these two white men will in the wake of Vaark's death eventually also get paid for their help on the Vaark farm: "The shillings [Rebekka Vaark] offered was the first money they had ever been paid, raising their work from duty to dedication, from pity to profit. [...] Perhaps their wages were not as much as the blacksmith's, but for Scully and Mr. Bond it was enough to imagine a future" (142, 153). The text does not offer any clue as to what kind of future this will be. It seems to me, however, that the removal of the blacksmith from the narrative proper alludes to the notion that future social, political, and economic configurations of Virginia will no longer allow for the possibility of a free African man to conduct business with a white settler and, by extension, for racial lines to permit such cross-racial interaction in the first place.

What *A Mercy*'s motley cast of characters has in common, in other words, is that they attempt to negotiate and navigate their individual freedom, their servitude, and their enslavement, respectively, in the potentially dangerous environment of the New World colonial scene.[6] What separates and antagonizes these characters, as my study observes, are the ways formations of private property fundamentally position them in non/relation to one another. The idea that certain groups of human beings are positioned towards one another in some sort of structural relation and that others are positioned by the absence of such a relation stems from the trajectory of Afropessimism, which I discuss in more detail as part of Black Studies' post-slavery theoretical articulations in the third chapter of this study.[7] For now, suffice it to say that Afropessimism's claims about the non/relationality of subjects inform my study's core argument that *A Mercy* presents its readers with a critique of the liberal property paradigm by way of its strategies of characterization. I work with the premise that instead of relying on fully rounded and easily accessible fictional characters, the novel constructs its characters in the form of *allegorical figures* through which it interrogates the

---

[6] I use the terms "negotiate" and "negotiation" in the sense of Mary Louise Pratt's *contact zones*. Pratt describes these as referring to "social spaces where cultures meet, clash, and grapple with each other, often in contexts of highly asymmetrical relations of power, such as colonialism, slavery, or their aftermaths as they are lived out in many parts of the world today" (34).
[7] As I will show, Afropessimism has introduced a complex ensemble of concepts and terms in this context, which is geared towards accounting for the violence of Atlantic slavery as that which continues to position and structure Black life in(to) the present. Throughout, my study draws on Afropessimist vocabulary.

liberal property paradigm precisely. Indeed, reviewers have repeatedly hinted at the unease they felt when encountering *A Mercy*'s characters, stylizing the novel as a "wisp of a narrative [peopled with] *insubstantial* characters" (Mantel; emphasis mine). I take such notions of discomfort and lack with respect to *A Mercy*'s strategies of characterization as the study's point of entry to trace and unpack *A Mercy*'s fundamental philosophical and political critique of the property paradigm.

Throughout, I discuss *A Mercy*'s allegorical figures as a "refusal of narrativization." Under this rubric, I hope to account for a literary maneuver that needs to be understood as a refusal to restage and thus to partake in hegemonic, dominant discourses about North American beginnings and its liberal, possessing subjectivities. The *OED* defines the word "refusal" as the "action or an act of refusing; a denial or rejection of something requested, demanded, or offered"; and the "repudiation or renunciation of a contract, allegiance, obligation, etc." among many others. I think about and use the word "refusal" in all of these ways, but I am particularly interested in the notion of refusal as an action or an act, as I will demonstrate over the coming chapters. The challenge of *A Mercy*'s refusal/critique, in other words, lies in what Stephen Best calls the "agency of form[:] what form produces, what form generates" (*Fugitive* 21, 25). It lies in how form produces, enforces, and challenges connections – discursive and conceptual – between private property and self-making. This study draws extensively on recent deliberations on narrative stemming from post-slavery Black Studies, which have questioned narrative's ability to account for and emplot the violence that "wounds and positions" the enslaved (Wilderson, "Aporia" 134). My study follows this line of questioning and brings it to the study of fictional character (see Chapter 4). Following Wilderson, who writes that "for Blackness, there is no narrative moment prior to slavery" (*Red* 27), the enslaved are not part of the community of the Human. By extension, I suggest that Black Studies' questions about the "emplot-ability" of social death also concern the realm of fictional character. I claim that as long as cultural and literary critics cannot explain "how the Slave is of the world" (*Red* 11) and thus of (the structure of) narrative, any assumption or conceptualization of fictional character needs to be understood as being fraught with similar explanatory lacunae. In other words, I suggest that *A Mercy* draws an analogy between the making of liberal subjects (within the realm of the world) and the creation of fictional character (within the realm of literary narrative). If, following post-slavery interrogations of white Western modernity, to be the subject of property is to be a Human subject, then *A Mercy* suggests there can only be fictional characters if there is *subject form*. In dialoguing with Afropessimism's suggestion that social death "ruptures the assumptive logic of narrative writ large" (Wilderson, "Aporia" 135), I offer that *A Mercy* resorts to allegory in creating its characters as way to represent social death's explosion of narrative form. Put another way, *A Mercy*'s fundamental critique of liberal self-making is situated precisely in its form, in its strategies of allegorical figuration. *Unruly Narrative* ultimately will conceptualize *A Mercy* as unruly Black "anti-narration": as the practice and the site of an epistemic critique of modernity's calculus of private property that is ongoing—a critique that is both located in and constantly revisited, revised, and recalibrated with each of the text's allegorical figures.

With its analyses of *A Mercy*'s allegorical figures, this study will wrestle with a set of core questions that address the relationship between narrative form and/as epistemic critique. These are: How does narrative form problematize liberal conceptions of private property? How do *A Mercy*'s unruly strategies of characterization become the critical lever with which the novel advances its epistemic critique of the formative and positioning powers of private property? How do *A Mercy*'s strategies of characterization become a tool for interrogating and confronting, and perhaps also for redefining, the relation between property and personhood? How does one address, in a study that is concerned with literary narrative and/as epistemic critique, the notion that narrative itself is conscripted by the episteme in which it is produced? What is the relation between allegorical narration and the (un-)making of liberal, possessive subjectivities? How does one address the absence of narrative (social death) in what is, after all, a narrative text?

1.2 Historical Contexts

In order to understand how *A Mercy* stages its refusal of the liberal property paradigm, it is important to recall the historical context that the novel both critically revisits and allegorizes. I reconstruct this context here with an eye on what historians of slavery and of North American beginnings have often read as a moment of possibility in the long history of chattel slavery and of racial subjection on the North American mainland. In other words, historians have often deemed racial lines in colonial Virginia to be relatively permeable still at this point, which would, for instance, allow for enslaved and free African(-descended) people to coexist, trade with, or marry white settlers from all walks of life (Fields; E. Morgan; P. Morgan). Most often, this assessment of seventeenth-century Virginia is based on the fact that "the transition to slavery was slow, and free black men and women gained some autonomy and maneuverability over the course of the first fifty years of colonial settlement" (J. Morgan, "*Partus*" 3). Historians accordingly have traced a panorama of variously evolving subject and "non/subject"[8] positions, such as English settlers, colonizers, and gentry, white indentured servants, indigenous populations, both free and enslaved Africans and African-descended people, all of whom appear to navigate early Virginia's colonial space relatively peacefully as well as on relatively "equal" racial terms. This kind of emphasis on racial fluidity continues to serve "to distinguish Virginia from colonies such as Barbados, where slavery was in full force by the middle of the seventeenth century" (J. Morgan, "*Partus*" 3). As part of her research on reproduction, gender, and slavery in seventeenth-century Virginia, Black feminist historian Jennifer L. Morgan has recently both pushed back against and critically supplemented the historical narrative of Virginia's racially indeterminate and fluid social strata. Let me briefly trace this narrative for a moment here, before turning to her arguments in more detail.

In August 1619, a ship landed at the shores of what was then called Point Comfort (today's Hampton) in colonial Virginia, bearing a cargo of twenty to thirty enslaved Africans, who were sold to the Jamestown colonists. Bought from English pirates, who, in turn, "had stolen them from a Portuguese slave ship that had forcibly taken them from what is now the country of Angola" (Hannah-Jones, "Our Founding" 16), those enslaved Africans would mark the beginning of Atlantic slavery on the North American mainland. Although this was "not the first time Africans could be found in an English Atlantic colony" (Guasco), this event represents one crucial moment in which the interconnected concepts of freedom and individuality emerged and constituted themselves in parasitic relation to Atlantic slavery. Virginia was England's first successful, that is to say permanent, American colony on the North American mainland. The English first attempted to establish permanent settlements in the Chesapeake area in 1585, when the landed gentleman, adventurer, and writer Sir Walter Raleigh sponsored the colonial endeavor to settle at Roanoke. Famously, the colony was "lost" (Kelleter; Maier; L. Miller). Ultimately, what contributed to the permanent and economically successful settlement of Virginia was the growing of tobacco, which lead to a downright "tobacco boom" in the 1620s (E. Morgan 108–30). From their experiences at Roanoke, the English settlers and colonizers

---

[8] In *In the Wake: On Blackness and Being* (2016), Christina Sharpe uses the concepts of "non/being" and "non/status": "Living in/the wake of slavery is living 'the afterlife of property' and living in the afterlife of *partus sequitur ventrem* (that which is brought forth follows the womb), in which the Black child inherits the non/status, the non/being of the mother. That inheritance of a non/status is everywhere apparent *now* in the ongoing criminalization of Black women and children" (15).

knew that the Indigenous peoples "grew and smoked a kind of tobacco; and tobacco grown in the Spanish West Indies was already being imported to England," where it garnered high profits for the planters and traders (E. Morgan 90). As a "profit-seeking venture," Virginia's economic success was fueled by this boom, and it initially largely "rested on the backs of English indentured servants" (Fields 122). Only in the second half of the seventeenth century – at the end of which *A Mercy* is set – would this labor force come to be replaced by enslaved African and African-descended people large scale, whom the Virginia traders would initially buy from Barbados and other West Indian islands and, later, directly from the African continent (E. Morgan 295–315).

In general, then, economic success in colonial Virginia and, by extension, Virginia's emerging social makeup were fundamentally shaped by different forms of servitude that had already been partially in place before the tobacco boom but were significantly institutionalized by it. Apart from the more well off (socially as well as economically) English entrepreneurs and planters, the bulk of the white settlers who came to Virginia were subjected to various forms of indentured servitude (E. Morgan 115–116). White indentured servants "served longer terms in Virginia than their English counterparts and enjoyed less dignity and less protection in law and custom. They could be bought and sold like livestock, kidnapped, stolen, put up as stakes in card games, and awarded – even before their arrival in America – to the victors in lawsuits" (Fields 122). White English women, who were few in number throughout the seventeenth century and usually would come to the colony as 'mail-order-brides' (Zug), generally had greater economic advantages in Virginia than they would in England, and they would continue to enjoy these even as they were back on the marriage market after their husband's death (E. Morgan 165). Furthermore, the English initially envisioned 'their' land and future society along the lines of peaceful coexistence between the Indigenous as well as the English populations (E. Morgan 44). With recurring Indigenous resistance to white settlement and colonization, however, this one-sided utopian worldview would not materialize. While the white English population seems to have been in two minds about the issue of enslaving the Indigenous populations in the colony, Native Americans did in fact become increasingly subjected to slavery in the latter half of the seventeenth century (E. Morgan 316–337).

Historians have also observed how African and African-descended enslaved people "during the years between 1619 and 1661 enjoyed rights that, in the nineteenth century, not even free black people could claim" (Fields 126). In the wake of the tobacco boom, when white English settlers in colonial Virginia would slowly come to conceive of their environment in more permanent terms and the settler population would finally begin to grow, many of the previously indentured servants would become free and decide to try and make a living in the colony (E. Morgan 136). With the indentured servants gaining the status of free settlers, the labor force on the tobacco fields would slowly but steadily be replaced with African(-descended) slaves and the few rights that they enjoyed would be eradicated gradually (but nonetheless forcefully)—a development that was also fueled by the economical fact that prices for African and African-descended slaves dropped significantly in the latter years of the seventeenth century (Fields 126). Furthermore, free Black people continued to be part of the colony's social strata and they (together with free Indigenous people) went on to live in the colony even after "it was made plain to them and to the white population that their color rendered freedom inappropriate for them" through new sets of laws issued by the Virginia assembly from the 1660s. These laws were geared towards systematically racializing, subjugating, and abjecting enslaved African(-descended) people (E. Morgan 337). That is, colonial legislation came to regulate the spheres in which the enslaved and free populations would live, interact, marry, have children, or conduct business with one another while at the same time continuously inscribing and systematizing racist taxonomies, thus determining who would be considered free or someone else's property within Virginia's colonial fold. What such conventional historical assessments of seventeenth-century Virginia tell us, then, is that the colony and its emerging society need to be understood in terms of highly complicated, constantly changing social arrangements between Indigenous populations, white English settlers/colonizers, as well as African and African(-descended) people.

Jennifer Morgan's work has intervened in this historiographic narrative of colonial Virginia's racial fluidity and indeterminacy. In her discussion of the *Partus Sequitur Ventrem* law, which was passed in 1662 and which would be the first slave code of the English Atlantic to regulate slave status through maternal descent, J. Morgan argues that from the perspective of enslaved Black women, racial lines in colonial Virginia need to be understood as far from fluid in fact. Enslaved Black women's "maternal possibilities became a crucial vehicle by which racial meaning was concretized—and it did so long before legislators indexed such possibilities into law" ("*Partus*" 2). Throughout the sixteenth- and seventeenth-century Atlantic, slave owners "presumed that enslaved women's reproductive labor accompanied their manual labor in tobacco and sugar fields. Mobilizing the language of increase [...] was part and parcel of how nascent slaveowners shaped their newly racialized present and future" ("*Partus*" 2). If we follow J. Morgan, as I do throughout the study, colonial legislation made the connections between already existing assumptions about slavery, race, and heredity explicit rather than for those connections to suddenly materialize ("*Partus*" 2, 3). Put another way, only a few decades after the first arrival of enslaved

Africans in Jamestown, "white settlers demanded a new world defined by racial caste" (Stevenson 81). We already encountered this in the expectations of the indentured servants in *A Mercy*, who did not mind as much not to be paid while Vaark made money off them but felt uncomfortable knowing that the blacksmith was paid for his labor. The New World white settlers demanded was a world in which their claims to freedom would be made possible only through the enslavement of others. As Toni Morrison elsewhere reminds us, "The concept of freedom did not emerge in a vacuum. Nothing highlighted freedom – if it did not in fact create it – like slavery" (Morrison, *Playing* 38). At the same time that most of the white English settlers – be they part of the gentry and merchants classes or servants – were claiming their status as self-possessed New World liberal subjects, enslaved African and African-descended men, women, and children would already be positioned in antagonistic ways to this emergent Human subjectivity even though the laws cementing this antagonism would be introduced decades later.

1.3 Critical Contexts

For the majority of its critics and readers, *A Mercy*'s seventeenth-century setting and plot seems to neatly represent the various social positions outlined in the established historiographic narrative of racial indeterminacy in early colonial Virginia. La Vinia Delois Jennings' review of *A Mercy* in the journal *Callaloo*, for instance, explains how *A Mercy* invites

> twenty-first century readers to consider a sectarian America as its racial divide unfolded. It challenges us to historicize the racialized political momentum that ushered in perpetual servitude based on non-whiteness and to meditate on the analogous forms of early colonial servitude, formal and informal, that might have united rather than divided persons of disparate religions and nationalities, especially those of underclass status. (645)

As such, the novel's representation of American colonial beginnings in Virginia has functioned as a research catalyzer with most studies of the novel making this their analytical point of departure. Over the course of my study, I will return to and take up this line of inquiry periodically in each of my analytical chapters. These chapters will reconstruct the critical discourse's specific take on the character in question, thus providing a more detailed discussion of existing trends and strands within the research on the novel.

Of course, to write about Toni Morrison's fiction at this point in time is to sift through an enormous body of continuously growing scholarship. It is not in the scope of this study to comprehensively delineate this here.[9] In general, publications on *A Mercy* range from a huge number of reviews and interviews to a relatively small but steadily growing number of academic publications.[10] Scholarly articles on the novel have been published in major academic journals[11] and chapters on *A Mercy* have been published in monographs and edited collections.[12] The only book-length publication on *A Mercy* has been issued by Stave and Tally, whose collection offers readings of the novel from various disciplinary angles, such as the study of history, literary studies, sociology, and psychology. Thus published in various ways and forms, most of these readings usually start from the obvious: the novel's seventeenth-century plot, its motley cast of characters, as well as the strange and startling environment of the New World in which it is set. What astounds me is that this rich pool of literature often decen-

---

9 Roynon, *Cambridge Introduction*, and Raynor and Butler offer critical surveys of Morrison's work and its reception since the mid-1970s. Many of these early, often white-authored responses would assume "a white Euro-American subject position […] to which Morrison's concerns are always cast as relative" while Black-authored criticism already placed Morrison's writing as both emerging from and thriving within "a broader African-American tradition that was being documented and given scholarly validation at exactly this time" (Roynon, *Cambridge Introduction* 114, 115). With the rise, manifestation, and proliferation of the Black feminist and women's movements in the 1970s and 1980s, both within and outside the academy, critics would pay more attention to Morrison's equally growing literary output. At the time, that is, Black feminist scholars recognized Morrison as a Black female author alongside other Black women writers such as Toni Cade Bambara, Gwendolyn Brooks, Rita Mae Brown, Nikki Giovanni, Gayl Jones, Paule Marshall, Sonia Sanchez, and Alice Walker (DuCille). In this way, these scholars firmly established what is now understood, read, studied, and taught as a rich tradition of Black female/feminist literary writing that "defied known forms, invented new grammars, upset, inverted, and subverted traditional structures and narrative strategies[, creating and reconfiguring] the novel, unbound, blackened, feminized, repopulated, and unpunctuated" (DuCille 381; see also Hull, Scott, and Smith).
10 Reviews include, for example, Adams; Barbara Carey; Charles; Freeman; Gates; Grewal; Heltzel; Jennings; Mantel; C. Miller; C. Moore; Myers; Teele; Todaro; Updike. Most prominent examples of interviews are Morrison and Neary, and Norris and Siegel.
11 Articles on *A Mercy* have been published *Callaloo* (Jennings; Roye), *MELUS: Multi-Ethnic Literature of the U.S* (Babb; Cantiello; Morgenstern; Nicol and Terry; Sandy), *Early American Literature* (Bross; Curtis; Cillerai; Logan), *American Literary History* (Gustafson and Hutner), *Critique: Studies in Contemporary Fiction* (Strehle), *Black Women, Gender + Families* (Putnam), and *MFS Modern Fiction Studies* (Karavanta; Wyatt). For publications in graduate and postgraduate as well as online and open access journals, see, e.g. *Berkeley Undergraduate Journal* (Jimenez), *Teaching American Literature: A Journal of Theory and Practice* (Bartley), and *Black Studies Papers* (Andrès; Michlin; Müller, "Standing"; Raynaud; Spatzek).
12 For monographs, see, for example, Schreiber, "Echoes"; contributions in edited collections include Anolik; Carlacio; Conner, "Language and Landscapes," "Modernity"; Fultz; Mayberry; Montgomery, "Contested"; Waegner.

ters race in its reading of a novel that centers on the *very making of race* in North America. That is, the critical reception of *A Mercy* has shown a tendency to read the novel as a text in which *race* is not part of its narrative politics. Instead, it tends to argue that in writing *A Mercy*, Morrison created a novel that addresses our contemporary moment despite or, rather, *because of* its seventeenth-century plotting and cast of characters that allows Morrison "to present an expansive version of the prenational world, one that reveals the heterogeneity that characterized settlements then and the nation today" (Babb 149).

In other words, a significant portion of the novel's rich pool of critical readings situates itself within an environment of twenty-first century post-racial political discourses, in which Barack Obama's ascent to the White House is commonly read as a redemptive watershed in America's racial politics, and thus serves as proof that US-civil society has finally overcome its histories and its legacies of chattel slavery, on the one hand, and genocide of the Indigenous populations, on the other.[13] This kind of post-racial imperative figures prominently in the critical reception of *A Mercy*, not least since the novel was published one week after Obama was elected in 2008. This has led to these two events being coupled, making Obama and then-current politics "ubiquitous presences" in the discourse on the novel (Cantiello 165). If nothing else, this was a clever marketing stunt by Morrison's publisher, of course. With its analytical interest in the workings and powers of the property paradigm, my study critically revisits such post-racial readings of *A Mercy*, which typically follow the above established historical narratives stressing colonial Virginia's relative racial fluidity and indeterminacy. As previously suggested, Black feminist scholarship has complicated such widespread historical evaluations; it has fundamentally changed the ways in which we understand how Black women's interiority was violently bound to the market, as well as how the accumulation and proliferation of

---

[13] Here, I follow legal scholar Sumi Cho's theorization of "post-racialism" as an ideological formation in which "[r]ace-based affirmative action, race-based admissions or districting in school desegregation plans, and minority voting districts, as a few prominent examples, all come under scrutiny" (1594). In a post-racial world, Cho explains, "white normativity [is insulated] from criticism and opens the floodgates of white resentment when confronted with previously accepted and unquestioned civil rights inequities, [the effect of which is] the ultimate redemption of whiteness: a sociocultural process by which whiteness is restored to its full pre-civil-rights value" (1596). Within African American literary studies, we can observe a similar tendency to view our contemporary aesthetic moment along 'post-racial' lines. Conceptualizing African American literature as existing as specifically African American only in the historical period from the Jim Crow era until the 1960s Civil Rights Movement, Kenneth Warren has famously declared the end of African American letters (K. Warren; Touré; for a rebuttal of this, see e.g. Baker and Simmons).

the master's private property and economic increase was tethered to inheritable slave status as enshrined in colonial legislation. *Unruly Narrative* brings these insights to its study of *A Mercy*'s questioning and refusal of the liberal property paradigm.[14]

Another predominant trend in the academic study of *A Mercy* reads the novel as counterweight or counter-history to longstanding master narratives of North American beginnings, such as the myth of the promised land, the myth of discovery, and the myth of transatlantic love.[15] As critics variously speak of *A Mercy* as "rewriting" (Babb), "unwriting" (Strehle), "retelling" (Omry), or "counterwriting" (Karavanta) of American colonial beginnings, readers and critics variously read the novel's complicated narrative structure and "multi-voiced litany" as a smooth narrative of alternative American origins (Waegner 91; see also Montgomery, "Traveling Shoes") or as an alternative to civil religious representations of US-American origins (Strehle). While both these interpretive trends also reflect an interest in the novel's narrative structure and form, they do not make this their central concern. By contrast, Stephen Best's deep interest in the aesthetics, poetics, and politics of narrative form in this context becomes highly relevant to my own arguments. Both his essay "On Failing to Make the Past Present" (2012) and his monograph *None Like Us* (2018) are representative of recent scholarship that deals with the entanglements between ethics and aesthetics in Morrison's work (see also Baillie; Christiansë; Palladino; Morgenstern; V. Smith, *Writing*). In both these texts, to which I will return in detail in the fourth chapter of *Unruly Narrative*, Best positions *A Mercy* as a historical novel that reads as the paradigm of a "new" critical moment in which an ethical relation to the histories and the legacies of transatlantic slavery – as that which continues to structure the present moment and its modes of critical thinking and reading practices – should no longer fuel African American and African-diasporic theorizing ("Failing" 456–465). Best claims that *A Mercy*'s form abandons the reader "to a more baffled, cut-off, foreclosed position with regard to the slave past" (472), thereby creating a new relationship between the slave past (as historical event) and the ways in which Black Studies discourses "[apprehend] the black political present" (*None* 63). For Best, *A Mercy*'s form undoes the eth-

---

[14] In an interview with NPR's Michel Martin, Morrison herself has rejected this kind of post-racial rhetoric which "misreads the complexity of the racial relationships" examined in *A Mercy* (Cantiello 165) when she states: "I certainly don't like that word [post-racial] [...] [I]t seems to indicate something that I don't think is quite true, which is that we have erased racism" (Morrison and Martin).

[15] In *The Myths that Made America: An Introduction to American Studies* (2014), Paul offers a comprehensive overview and analysis of these myths.

ical imperative that *Beloved*'s poetics offered to generations of scholars—a poetics, in which the slave past and its afterlives in the present political and aesthetic moment continue to uniquely shape Black identities. If we follow Best, this also is a poetics, in which slavery's afterlives become common ground in most theoretical thinking about and attempts at creating Black (critical) community. By refuting such theoretical underpinnings for Black Studies and Black study (to paraphrase Christina Sharpe), Best argues that *A Mercy* fundamentally changes the ways in which we as readers should think about ethics and aesthetics in African American literature in the twenty-first century.[16]

1.4 Private Property as Cultural Metaphor and Literary Form

In different, while related, ways this study is also concerned with how *A Mercy* navigates the connections between ethics and aesthetics, thus joining the above Morrison scholarship showing an (renewed) interest in the relation between the two. Unlike Best (and others), however, I suggest that *A Mercy* not only deliberately returns its readers to the slave past, making this its very explicit narrative concern. I also argue that it does so in order to engage with how form both sheds light and becomes the means by which to investigate and to criticize the liberal property paradigm. In general, I understand property as an extremely flexible and mutable vehicle for the negotiation of social and cultural meaning as well as for the formation of power and of value systems (Bhandar, "Critical Legal Studies", "Disassembling"; Banner; Davies; Rose). In its Western liberal understanding, as legal scholar Margaret Davies notes, property is a

> multi-faceted, sometimes self-contradictory and internally irreconcilable notion which is variously manifested in plural (though inseparable) cultural discourses—economic, ethical,

---

[16] Much like Best, Yvette Christiansë in *Toni Morrison: An Ethical Poetics* (2013) turns to Morrison oeuvre as an African American writer whose work is situated in the "space of modern literature" (2). It is in this space and from "her own scholarship about and reading of modernism," Christiansë suggests, that "Morrison's fiction emerges from and depends upon" (2). At the center of her inquiry into Morrison's oeuvre is a set of questions on the relation between (the politics of) language, (African American) historical consciousness, racialization, and testimonial practices, among others (27). As amalgamation of close reading and critical theory, Christiansë's book establishes *A Mercy* as Morrison's single novel that is not concerned with "the enormity of America's slave past"—the legacies of which are otherwise "everywhere in her fiction and nonfiction" (21). Instead, Christiansë stylizes *A Mercy* as Morrison's attempt "to narrate a moment prior to that in which slavery had become codified and solidified by law, and naturalized through custom" (21).

legal, popular, religious. [...] [I]ts reach is not only material and political, but also cultural and symbolic. (3, 7)

In light of the varied meanings of property, I should point out that my understanding of property strongly resonates with Davies's observation that "property is also a powerful metaphor for *existence* in a liberal social framework" (7; emphasis mine; see also Radin). Discussing the property concept in the context of European settlement, colonization, and the implementation of slavery in North America, moreover, Critical Race theorists have shown how the conceptual conflation of property and personhood was racialized from the start. As mentioned already, whiteness became "the characteristic, the attribute, the property of free human beings" (Harris, "Whiteness" 1721). Thus marking whiteness and with it the promise of individual liberty as the most valuable property to be owned on the early American scene, such racially contingent forms of property remain protected in American law until today (1709; see also P. Williams).

My thinking about private property – as cultural metaphor and as abstract value determining and governing scales of existence – is also informed by scholarship on ownership and property in American and African American literature (Best, *Fugitive*; Clymer; Homestead; King, *Race, Theft*; Luck; Schneck) as well as by interventions from the scholarly field of law and literature (Coombe; Dolin). Flowering as a movement systematically investigating the correlation between the law and literature since the 1980s (Dolin 1), thinkers working in this context have since pointed to the "narrativity" of the law and its textual articulations (Brooks and Gerwitz; Rose) and have discussed this in relation to issues such as literary property rights, intellectual property rights, copyright, and constitutional rights (Buinicki; Ely; Hesse; Irr). Commenting on the importance of the politics of form for the domain of property and its various configurations in American law, Best in *The Fugitive's Properties: Law and the Poetics of Possession* suggests that the interplay between slave law and intellectual property law "help[s] redefine the very essence of property in nineteenth-century America" (16). Following Best, the conceptual correlations between property and personhood established under slavery continue to live "within the text of the law" and they do so within the specific frame of the emergence of intellectual property law towards the end of the nineteenth century (14). We read:

> The issues of personhood and property that slavery elaborates and the issues from the emerging law on intellectual property are part of a fundamental historical continuity in the life of the United States in which the idea of personhood is increasingly subject to the domain of property. Slavery is not simply an antebellum institution that the United States has surpassed but a particular historical *form* of an ongoing crisis involving the subjection of personhood to property. (*Fugitive* 16)

Best's analytical focus on slavery jurisprudence and intellectual property law configures a new ensemble of questions concerning "the social specificity of the person-property relation as the law tries to come to terms with new configurations of that relation and, in turn, generates new forms for that relation" (16). As mentioned earlier, it points to what Best has called the "agency of form." What follows is an attempt at taking up Best's concerns regarding form in my analyses of the relation between property and personhood and the critique of this relation as presented in *A Mercy*.

1.5 Situating the Study

Unruly Narrative is located at the interdisciplinary intersections of the scholarly fields of Early American Studies, African American Studies, U.S.-American Black (Diaspora) Studies, Black Feminist Criticism and Theorizing, as well as Critical Race Theory. It is a study that draws extensively on post-slavery theoretical trajectories (including recent US Black feminist articulations and Afropessimism)[17] invested in criticizing and dismantling white Western modernity's structural and epistemological histories and legacies of slavery and racial subjection. The study approaches Black Studies as an intellectual project equipped with an analytic lens that attempts to account for Blackness in an antiblack world, both in terms of structure and performativity (Sexton, "African American" 10). As such, Black Studies operates as a fundamental corrective and as an insurgent project of counter-epistemology to European and Western Enlightenment's narrative of universal subjectivity.[18] While Black Studies certainly can be understood as Black intramural critical conversation about the status and stakes of Black existence, articulation, and critique (Spillers, "Idea," *Black*), it also represents an important intervention into white knowledge productions and long-standing

---

17 Christina Sharpe defines the term 'post-slavery' as follows: "[W]hile all modern subjects are post-slavery subjects fully constituted by the discursive codes of slavery and post-slavery, post-slavery subjectivity is largely borne by and readable on the (New World) black subject" (Monstrous 3). I borrow the term here to delineate these articulations specifically (see also Chapter 3).
18 In general, Black Studies have labored to push a critical "transformation of the human into a heuristic model [over and against the idea of the human as] an ontological fait accompli" (Weheliye, "After Man" 322). Weheliye reminds us that Black Studies in North America has "existed since the eighteenth century as a set of intellectual traditions and liberation struggles that have borne witness to the production and maintenance of hierarchical distinctions between groups of humans" before becoming part of the US mainstream academy in the latter half of the twentieth-century (*Habeas Viscus* 3).

"historical racist sedimentation" (Yancy 233).[19] For someone who, like me, is positioned within as well as by the fold of the white Human – with the Human here "connot[ing] a paradigmatic entity that exists ontologically as a position of life in relation to the Black or Slave position, one of death" (Wilderson, *Red* 23) – engaging with and drawing from Black Studies necessarily comes with a persistent kind of tension or set of contradictions. That is, while I think about this study explicitly as a project of anti-racist critique and of dismantling white conceptualizations of the fashioning of a universal Human subject, Black thinkers and scholars of color, like Sara Ahmed, continue to remind me that "any project that aims to dismantle or challenge the categories that are made invisible through privilege is bound to participate in the object of its critique" ("Phenomenology" 150). That is also to say that while I hope to think, write, and speak from a position of being aware of the notion that "*whiteness* is a real category, that has been historically produced and that has real consequences" (Adusei-Poku 44) – a position confronted and challenged by Black (feminist) thinkers, novelists, and philosophers of color – this project inevitably runs the risk of being "ambushed" by white power formations (Yancy). It runs the risk of thus becoming a "[project] of critique [...] complicit with what [it] attempts to disrupt," including (but of course not limited to) the reproduction of epistemic violence (Applebaum 3).

With this comes the attendant problem and challenge of remove: Writing from a geographical location and disciplinary situatedness within the field of American Studies in Germany, this project aims to consciously reflect on the place from which it follows *A Mercy*'s critique. This place is that of the mostly white German university landscapes with which come the very substantial risk of imposing ventriloquist readings and offering "unbidden translation" of Black-authored novels like *A Mercy* (Broeck, *Gender* 11). This is another way of saying that my project of tracing and analyzing the interconnections between notions of private property, personhood, and (historical and contemporary) mechanisms of racialization and subjection within the literary realm of representation is heavily indebted to the intellectual labor of generations of (mainly US-American but also European and German) Black (feminist) thinkers. It is their intellectual work which has opened up critical epistemologies of white Western modernity and which continues to confront the modern white liberal subject. To say with Hartman's words quoted in the second epigraph to this chapter that *A Mercy*

---

19 In describing Black Studies as "a most difficult terrain" in the introduction to their *Companion to African-American Studies*, Gordon and Gordon also remind us of the various transformations that Black Studies as a discipline has undergone as locus of intense and dynamic debate.

challenges "the liberal imagination of freedom [and its attendant] proprietorial conceptions of the self" (*Scenes* 115), then, is to listen to and enter into an exchange with Black Studies post-slavery trajectories. This effort crucially needs to be informed by and committed to "critical vigilance" (Applebaum 3; Yancy).

In this vein, *Unruly Narrative* aims to contribute to the large body of scholarship on Toni Morrison's literary and critical oeuvre by offering the first in-depth single study of Morrison's ninth novel. Critics and readers have largely overlooked *A Mercy*'s profound engagement with and interrogations of the concept of private property as an integral part of the novel's re-telling and representation of American seventeenth-century beginnings in colonial Virginia. With its focus on private property as that which establishes, determines, and maintains scales of existence, the study seeks to critically supplement the existing critical discourse on the novel. By way of entering into an exchange with post-slavery interrogations of Atlantic slavery as the underside of white Western modernity's fashioning of freedom as self-authorizing teleological narrative of self-making, I furthermore hope to add to the existing discourse on post-slavery scholarship within American Studies in Germany and beyond. Engaging specifically with post-slavery thinking's questioning of the possibility of narrative emplotment of social death, this study pushes a set of questions on the discipline's "methodology and morals" as part of "the deconstruction of the anti-blackness [sic] structuring white western civil societies as well as large parts of their knowledge production" (Weier, "Consider" 430; cf. Essi et al.). As already suggested, my study also enters into conversation with existing concerns and conceptions on the making of fictional character as I bring Black Studies' questions about the "emplot-ability" of social death to narrative theory's vast archive of the study of fictional character.

1.6 Reading Methods

How does one read a historical novel published in the first decade of the twenty-first century against texts that come from this novel's historical setting and time frame of the seventeenth-century English Atlantic? And how does one read these in conjunction with twentieth and twenty-first century theoretical texts by Black theorists? To paraphrase Saidiya Harman, how does one tell impossible stories? Throughout the study, I employ the concepts of property, freedom, and subjectivity and I place them in relation to each other within the historical frame of white Western liberalism and modernity. As already suggested, my analytical focus will be on the concept of property and on the ways in which the concepts of freedom and liberal self-making connect to notions of ownership and possession. In the interconnected readings that *Unruly Narrative* pursues, I draw on the interdisciplinary methodology of *cultural analysis* as suggested by Dutch narratologist Mieke Bal.[20] Cultural analysis, which poses that concepts may "offer miniature theories" (*Travelling* 22), helps me embrace, combine, and connect the three analytical arenas of this project.[21] Following Bal, concepts are "intellectual tools, which determine how members of the academic community conceive of themes, approach objects and define relevant questions to be addressed" (Neumann and Nünning 3). For Bal, concepts are small theories in themselves —theories that are "flexible: each is part of a framework, a systematic set of distinctions, not oppositions, that can sometimes be bracketed or even ignored but never transgressed or contradicted without serious damage to the analysis at hand" ("Cultural Studies" 35–36). Thus contained, condensed, packed, and explicit, concepts will to a certain extent also always be normative, programmatic, as well as dated (39, 42). As mobile units of knowledge, concepts constantly commute through time and "through a nonlinear history" (*Travelling* 44, 40). As such, concepts are "not fixed […] between […] historical periods, and […] disciplines, their meaning, reach, and operational value differ" (24). In other words, the mobility of concepts always is "bound up with social and political

---

[20] Bal is not the first cultural critic who resorts to the idea of 'travelling concepts' (see also Said; Clifford). For an overview of how Bal's work relates to other approaches that make use of the metaphor of travel see Bachmann-Medick; Neumann and Nünning; Teller. Furthermore, it is noteworthy that Bal does not refer to the respective works by Said and Clifford, as a look at her bibliography shows.

[21] Bal's suggestions for cultural analysis and for 'travelling concepts' have been subjected to a good amount of critique (e.g., Nünning, "Kulturwissenschaft(en)"; Teller). German literary scholar Doris Bachmann-Medick, for instance, has observed that from a transnational, decolonial perspective, "the concept of 'travelling concepts' itself remains imprisoned in the tradition of a European history of travel, discovery, and expansion. This tradition has long been associated with concepts of mobility, flexibility, conquest, and expansionist ambition, which are not only eurocentric [sic] but also construed as middle-class and male dominated" (Bachmann-Medick 120, 121). Against this kind of "free-floating" of conceptual mobility (120), Bachmann-Medick calls for "more historical grounding and contextualization" of concepts (133). Her suggestion is to conceive of concepts as "concepts in translation," which would "allow for a more detailed exploration of exactly which social practices and social relations lie behind the specific concepts at issue, which intermediaries are active, and what obstacles and local resistances arise" (133). I seek to enrich my use of concepts and their mobility with Bachmann-Medick's suggestions for the incorporation and the recuperation of historical contexts in dealing with concepts (128–133). This study pursues this aim in that it – by following *A Mercy*'s allegories – identifies the conceptual nexus of private property and self-making in the historical frame of late seventeenth-century Western liberalism and European colonial expansion and then turns to recent critiques of this nexus articulated by post-slavery Black thinkers. In this two-pronged endeavor, I hope not to "[leave] universalizing assumptions unreflected" (Bachmann-Medick 129).

concerns" (Neumann and Nünning 8). Bal argues that concepts "can become a third partner in the otherwise totally unverifiable and symbiotic interaction between critic and object" (*Travelling* 23). This they can do only on the condition that "they are kept under scrutiny through a confrontation with, not application to, the cultural objects being examined" (24). Concepts thus need to be understood as "important arenas of debate" (27).[22] If concepts are not only "dynamic in themselves" but also "travel between ordinary words and condensed theories" (11, 29), it follows that property is by no means reducible to a single strand of meaning. That is to say, even though I for the purposes of this study conceptualize private property in relation to liberal self-making and against the backdrop of Atlantic slavery, other meanings of private property will also always resonate with this conceptualization. For while being present "in a given moment and a specific epistemological context, concepts also link that moment and that context to earlier moments, to earlier epistemological contexts" (Neumann and Nünning 4). Put another way, Bal's suggestions for cultural analysis help me account for the fact that the mutable concept of private property from its inception in the late seventeenth-century English Atlantic until today has traveled not only across centuries but also through the different analytical frames that this study puts in relation to one another. I also follow Bal in her suggestion to privilege the "close and detailed engagement with the object" of analysis that cultural analysis advocates in the form of close reading ("Cultural Studies" 38). I enrich the methodologies of cultural analysis and close reading with an additional reading device specifically developed for this project. Situated at the beginning of every individual chapter (excluding the introduction to and conclusion of the study) are single paragraphs marked with the subtitle "[Routing the Argument]." These single paragraphs not only succinctly summarize the arguments made in the respective chapter and trace how I develop, weave, and sharpen my overall argument with each step that I take. These paragraphs also index the movement of the property

---

[22] One needs to approach the project of cultural analysis, too, with "critical vigilance" (Applebaum 3). That is, one needs to be aware of the underlying assumptions of established methodologies within disciplines like American Studies or Cultural and Literary Studies more generally. To quote Bal again, concepts constitute "the backbone of the interdisciplinary study of culture primarily because of their potential *intersubjectivity*. Not because they mean the same thing for everyone, but because they don't" ("Cultural Studies" 35). To use a notion such as 'intersubjectivity' means to assume that all beings can be subjects or strive towards subject status. Post-slavery Black Studies trajectories have taught us, however, that this is not the case if by subject we mean critical theory's subject and its status "as a relational being" (Douglass and Wilderson 117). I draw on cultural analysis as a methodological framework for this project fully aware of such limitations while it is not in the scope of this study to examine the implications of these insights on cultural analysis.

concept as well as the critique of the property paradigm through the study's different analytical frames and from one discursive arena to the next. As such, the [Routing the Argument]-paragraphs function as a reading manual for the study.

1.7 The Chapters

This brings me to the study's overall structure. As should have become clear by now, *A Mercy* is this study's pivotal point, it is at the center of *Unruly Narrative*'s project of interrogating the property paradigm. In the second and next chapter of the study entitled "Claims to Freedom: Private Property and the New World Liberal Subject," I turn to a paradigmatic selection of texts from the late seventeenth-century English Atlantic to place my analysis of *A Mercy* on an historical footing. I examine these texts for the ways they stage, discuss, and navigate liberal ideas of what it means to be a Human subject with regard to private property. My focus will be on John Locke's *Two Treatises of Government* (1689), which, ever since its publication, has inspired dynamic intellectual debate on questions of government, sovereignty, political power, civil society, as well as the form and function(s) of private property (Laslett 1988). That is also to say that Locke's ideas on ownership and possession as put forth in the *Two Treatises* remain an important point of reference for many conceptualizations of (exclusive) private property as well as for critical discourses on such ideas until this day. In tandem with the *Two Treatises*, the chapter also discusses Locke's *The Fundamental Constitutions of Carolina* (1669) and "The Germantown Friends' Protest Against Slavery" (1688). As I work my way through these paradigmatic texts and the twentieth and twenty-first critical discourses on them, I argue that they constitute the New World's liberal Human subject through notions of ownership, metaphors of property, as well as "possessive investments" (Lipsitz) in white identity deliberations. My goal in engaging this paradigmatic selection of texts is to make visible an emerging conceptual nexus in seventeenth-century North America that connects notions of possession and ownership to questions of race and racialization.

In the third chapter – "Interrogating Property: Black Studies and the Liberal Imagination" – my study turns to Black Studies' post-slavery theoretical interventions into the discursive promises of universal liberty as interrogation of proprietorial conceptions of liberal selfhood. Bearing in mind the overall study's core questions on the relation between literary narrative and a fundamental theoretical critique of early modern liberal subjectivities, the chapter engages with Black Studies' interrogations of the liberal property paradigm, continuing my examination of the complex entanglements between self-making, slavery, and private property that I began in the previous chapter. From these post-slavery interventions, I extract a set of interrelated terms – *violence*; *dispossession* and *fungibility*; *abjection* and *abjectorship*; *reproduction* and *kinship*; and *anticipatory wake* – which not only provides the internal structure for the chapter itself but which, as analytical vocabulary, helps me address the ways Morrison's novel wrestles with the property paradigm within the representational realm of the literary. In other words, the aim of the chapter is to think with these Black Studies' post-slavery trajectories, whose interventions make it possible for me to address and examine the intricate connections between private property and self-making in *A Mercy*.

This, then, brings me to the fourth chapter of the study and to my analyses of *A Mercy*'s characters: "Practicing Refusal: Interrogating the Property Paradigm in *A Mercy*." This part consists of an introduction as well as six individual analytical subchapters, in which I examine the characters of Jacob Vaark, Lina, Rebekka Vaark, Sorrow, Florens, and the *minha mãe* separately. In the first introductory chapter of this part of the study, I turn to the vast field of narrative theory and to its various conceptualizations of literary character as I follow post-slavery Black theoretical trajectories' questioning of (white) narrative's ability to account for and emplot the slave/social death (Hartman, *Scenes*; Wilderson, "Aporia"). As I move towards examining the property paradigm within the realm of the literary in this way, I identify a fundamental tension at work in the ways in which *A Mercy* resorts to allegorical figuration when presenting its critique of the liberal property paradigm. On the one hand, this tension is caused by what I identify as *A Mercy*'s allegorical anti-narration. On the other hand, it relates to the fact that the vocabulary available to talk about form and narrative, such as the term "character," cannot account for social death. Throughout my close readings, I use square brackets as a way of connoting this tension, making it visible not only on an orthographical level but also to show how this tension is fundamental to my analyses of *A Mercy*. That is, I use "[character]" whenever I generally talk about *A Mercy*'s allegorical figurations and "[name of a character]," for example [Sorrow], whenever I talk about a specific [character] in *A Mercy*. Each of the [character] studies deals with how the [characters] under scrutiny tackle liberal conceptions of private property, ownership, and possession in relation to the making and unmaking of subjectivities at the New World colonial scene. Each of the [character] studies, too, establishes *A Mercy* as an epistemic critique that is ongoing. In refusing narrative and in creating Black anti-narrative, *A Mercy*'s critique is "always now" (Morrison, *Beloved* 248).[23]

---

[23] While I have briefly discussed the [characters] of [Willard] and [Scully] at the beginning of this introduction, these two [characters] are the only ones that I do not discuss in a separate chapter. I have chosen to do so because, in my reading of *A Mercy*, all the other [characters] play a much more central role for the novel's plot.

2 Claims to Freedom: Private Property and the New World Liberal Subject

[Routing the Argument] The chapter examines a paradigmatic selection of texts from the archive of the late seventeenth-century English Atlantic. These are: John Locke's *Two Treatises of Government* (1689), which will be the focus of my discussion and which I will read in tandem with "The Fundamental Constitutions of Carolina" (1669), which Locke co-wrote; and the "The Germantown Friends' Protest Against Slavery" (1688). By way of revisiting these texts, I seek to show how the New World's liberal subject is constituted in these paradigmatic writings through notions of ownership; how, in other words, concepts of freedom and individuality manifest themselves in these writings over and against the systems and practices of New World chattel slavery. The subject of these texts is the white liberal Human, who "establishes, maintains, and renews" (Wilderson, *Red* 11) its existence and its liberty in the modern Western world vis-á-vis the sentient being of the slave. Private property – here conceptualized as the ability to own one's own self as well as enslaved others – is that which creates the ground on which the liberal subject's claims to freedom in the New World are made. It also is the *structural* ground on which relationality between *subjects* is created. As I work my way through these paradigmatic texts and the twentieth- and twenty-first century critical discourses on them, I conceptualize the intricate connection between freedom and bondage as the property paradigm.

> The need to establish difference stemmed not only from the Old World but from a difference in the New. What was distinctive in the New was, first of all, its claim to freedom and, second, the presence of the unfree within the heart of the democratic experiment – the critical absence of democracy, its echo, shadow, and silent force in the political and intellectual activity of some not-Americans. The distinguishing features of the not-Americans were their slave-status, their social status – and their color.
> — Toni Morrison, *Playing in the Dark*

> Generally, property is divided into two major areas: realty and personalty. Realty is land, whereas personalty is possessions—for instance, jewelry, money, furniture, or (formerly) slaves.
> — qtd. in Patrice D. Douglass, "The Claim of Right to Property"

2.1 Introduction

Ideals of private property are central to the making of Western modernity. The multi-faceted concept of private property not only "references the things that are owned, as in common usage, but also a social system in which the right and ability to own are protected by the state" (Hong 180). Within a Western liberal context, that is, private property has become "the basis for making claims of natural rights and political liberties" (Graeber 35). Private property needs to be understood as a "powerful metaphor for existence in a liberal social framework [...] its reach is not only material and political, but also cultural and symbolic" and, as such, property affects "knowledge, social interactions, notions of law, and concepts of the self" (Davies 7, 24). Famously, John Locke "named property as a system that produces a subject defined through its ability [...] not simply to own but, first and foremost, to own itself" (Hong 181). As philosophers, political scientists, as well as legal scholars have shown with respect to the history of the United States, moreover, class, gender, sexual, and racial hierarchies have been elaborated and shaped by configurations of property from the very beginning (see e.g., Harris, "Whiteness"; Hong; J. Morgan, "*Partus*," "Archives").

The chapter revisits a paradigmatic selection of texts from the archive of the late seventeenth-century English Atlantic for the ways in which they employ private property as a means to constituting the liberal subject and its claim to freedom and/as (self-)possession. In doing so, I build towards the study's overall argument that Toni Morrison's novel *A Mercy* fundamentally interrogates the complex entanglements between individual liberty, slavery, and private property. In thus investigating the intricate connections between private property and (white) self-making, *A Mercy* breaks new critical ground on which to think about, confront, and dismantle those affiliations on the literary level of representation. The function of the chapter, in other words, is to delineate a conceptual frame in which Enlightenment tenets like freedom, subjectivity, and citizenship have taken center stage. As I will show in the third chapter of the study, *A Mercy* interrogates, decodes, and dismantles this frame as actually being specifically and exclusively about white freedom and self-making.

I will discuss John Locke's *Two Treatises of Government* (1689); "The Fundamental Constitutions of Carolina" (1669); and the "The Germantown Friends' Protest Against Slavery" (1688). The main emphasis in my discussion of these texts will be on Locke's *Two Treatises*, which I will discuss in a first step. Fundamentally, Locke's ideas on ownership and possession put forth in the *Two Treatises* remain an important point of reference for many conceptualizations of (exclusive) private property as well as for critical discourses on such ideas until this day (Graeber; Hong; Rose). Following political theorist Crawford Brough Macpherson, Locke was "the first to make a case for property *of unlimited amount* as a *natural* right of the individual, prior to governments and overriding them" (Macpherson, *Property* 15; see also Macpherson, *Political Theory*). He thus laid the conceptual foundation for the by now common definition of the liberal individual's inherent capacity for self-possession. My analysis of the *Two Treatises* is two-fold: First, I engage with the ways in which philosophers and political scientists, among others, have made sense of the intricate relationship between Locke's ideas of private property and his thoughts on slavery as configured in the *Two Treatises*. I then confront these discussions with a reading of the *Two Treatises* that critically looks at how Locke conceptualizes white liberal subjectivity in/for the New World. I supplement this reading with a brief discussion of "The Fundamental Constitutions of Carolina" (1669), the original draft of which Locke co-wrote (Hinshelwood 567). In a second step, I will turn to the third and final text to be examined in the chapter: the "Germantown Friends' Protest Against Slavery." Issued in 1688, this document presents an early moment in the development of Quaker antislavery discourse. My reading of the "Protest" is located at the fault line between, on the one hand, the text's antislavery impetus and, on the other, its situatedness in white practices of enslavement. As I hope to show, the text bears witness to an understanding of the value of whiteness at the New World colonial scene and thus speaks to the notion, pushed by Critical Race theorists and legal scholars such as Cheryl I. Harris and Derrick A. Bell, Jr., that whiteness and private property are in fact mutually constitutive categories (see also Hartman, *Scenes*; Hong, Lipsitz; Wilderson, *Red*).

My reading is geared towards showing how notions of private property provide the philosophical conditions for the white liberal subject to emerge in these seventeenth-century texts and that chattel slavery and the regimes of property subtended by it constitute the fertile economic, social, cultural, and political ground for this kind of coming into being. Scholars of slavery and of Western modernity have long argued that "[m]odernity's enabling fiction has been to see itself grounded in an advocation of universal freedom for humankind" that was defined against slavery (Broeck, "Never" 241). As Hong has it, if "property becomes the basis for freedom, defined as the ability to exercise one's will in the absence of the influence of others, this definition of freedom needs an antithesis—enslavement" (182). Inscribing this configuration "far beyond their own historical and geographical moment," moreover, the writings under scrutiny here need to be understood as part of a tradition of texts that would "[marginalize] the foundational function of slave trading/holding, mark[ing] it as a peripheral aside of state- and nation-building, and consequently instrumentaliz[ing] the notion of 'slavery' as an abstraction, as a useful foil" (Broeck, "Never" 241) for their own "claims to freedom" (Morrison, *Playing* 48). As one historian and philosopher aptly summarizes:

> By the eighteenth century, slavery had become the root metaphor of Western political philosophy, connoting everything that was evil about power relations. Freedom, its conceptual

antithesis, was considered by Enlightenment thinkers as the highest and universal political value. Yet this political metaphor began to take root at precisely the time that [...] slavery [...] was increasing quantitatively and intensifying qualitatively to the point that by the mid-eighteenth century it came to underwrite the entire economic system of the West[.] (Buck-Morss 821)

In recent Black Studies discourses, "the violence and dishonor and disaffiliation constitutive of enslavement and the radical breach introduced by the Middle Passage" continue to structure the status of freedom as ambiguous, past and present (Hartman, *Scenes* 72). By this route, in the post-Emancipation world "the roots of freedom [are] located in slavery and the meaning of freedom was[/is] ascertained by its negation" (172). In what follows, my thinking is influenced by and indebted to such post-slavery (as defined in the introduction to this study) scholarship on this inversion, which has opened critical epistemologies of white Western modernity and continues to confront its liberal subject. As one articulation of these trajectories, Afropessimism pushes any critical engagement of private property (and, by extension, of Western civil society more generally) towards an analysis on a structural plane.[23] In contrast to Marxist theorizations of private property that have tended to discuss property primarily in relation to rights or entitlements to something, which describe the social relationship between proprietors[24], Afropessimism focuses attention on a different kind of relationality, or non-relationality, that the property concept entails: namely, that the regimes of violence subtended by the Middle Passage create a relational void for the slave (Wilderson, *Red* 18). In *Red, White, and Black: Cinema and the Structure of U.S. Antagonisms*, Wilderson explains why a Marxist framework can neither account for the slave's status in the world nor for slavery's violent positioning power. Following Wilderson's critique of Marxism, the worker "labors in the market or sells [their] labor as commodity [and they] can claim to be alienated and exploited in the process of producing the commodity" (Weier, "Consider" 422). In contrast to the slave, who is "the commodity itself," the worker is able to "reclaim a reformation of the modes of production as a solution to [their] conflict with civil society" (422). In other words, the worker, while being exploited and alienated, is still able to claim rights to ownership, both of themselves and of other things and commodities. Wilderson goes on to tell us that the worker's "essential inca-

---

[23] I will discuss Afropessimism's premises and interrogation of white Western modernity in more detail in the next chapter.
[24] Wesley N. Hohfeld in "Some Fundamental Legal Conceptions as Applied in Judicial Reasoning" was first in pushing conceptualizations of private property towards the realm of the social at the beginning of the twentieth century. See also Macpherson, *Property, Political Theory*; Rose.

pacity (powers which cannot accrue to the worker, suffering as exploitation and alienation) is the essence of capacity, life itself, when looked at through the eyes of the Slave" (*Red* 8). For the enslaved, the violence of slavery fundamentally positions through the modalities of "accumulation and fungibility," which create an *absence* of relationality (59). We continue reading:

> It is sad, in a funny sort of way, to think of a worker standing in the same relationship to the sellers of goods as any other buyer, simply because his use-value can buy a loaf of bread just like the capitalist's capital can. But it is frightening to take this 'same relationship' in a direction that Marx does not take it: If workers can buy a loaf of bread, they can also buy a slave. (13)

The implications of this in thinking about private property are two-fold, at least: First, the above suggests that to think about private property as social relationality means to think about "a proletariat who 'stands in precisely the same relationship' to other members of civil society due to their intramural exchange in mutual, possessive possibilities, the ability to own either a piece of Black flesh or a loaf of white bread or both" (13). Second, the relational void created by slavery opens up precisely the white subject's 'possessive possibilities.' That is, the non-relationality of the Black 'sentient being' needs to be understood to create the relationality between different subjects of property in the first place (11, 41). In what follows, I conceptualize this nexus of proprietorial configurations of selfhood, freedom, and slavery – the intricate connections between freedom and bondage – as the property paradigm in my readings.

2.2 John Locke's *Two Treatises of Government* (1689): Critical Contexts

Ever since the first edition of Locke's *Two Treatises* was published at the end of the seventeenth century, the book has inspired dynamic intellectual debate on questions of government, sovereignty, political power, civil society, as well as the form and function(s) of private property (Laslett).[25] Since World War II,

---

[25] Peter Laslett furthermore notes in his introduction to the *Two Treatises*: "It has been printed over a hundred times since the 1st edition appeared with the date 1690 on the titlepage. It has been translated into French, German, Italian, Russian, Spanish, Swedish, Norwegian, Hebrew, Arabic, Japanese and Hindi: probably into other languages too. It is an established classic of political and social theory, perhaps not in the first flight of them all, but familiar to eight generations of students of politics all over the world, and the subject of a great body of critical literature" (3).

these debates have largely taken place across different fields and disciplines, such as political science, philosophy, the study of history, as well as, more recently, (comparative) literary studies.²⁶ In general, political theorists and philosophers have in the second half of the twentieth century been concerned with developing approaches to Locke's writings that focus both on the text itself as well as on its historical context (see e.g., Harpham; Kelly, *Reader's Guide*, "Reception"; Rogers; Stanton; Woolhouse and Stanton). Research on the concept of private property in Locke's *Two Treatises* encompasses numerous approaches, "show[ing] a confusing variety of often contradictory accounts of [the] genesis [of Locke's ideas on property, and their] meaning and status within his political and moral philosophical thought" (Stapelbroek 201). For instance, a dominant strand of the scholarship on the meaning of Locke's ideas of private property focuses on questions regarding the composition and dating, as well as the historical context of the text, for Locke scholars have often struggled to relate the fifth chapter of the *Second* Treatises, which is entitled "Of Property" and in which Locke elaborates on his views on private property, to the remainder of the *Treatises* (Laslett 59–66; Stapelbroek 201).

The chapter's focus on liberal individualism and its (propertied) subject as theorized in the *Two Treatises* does not permit any further thorough reconstruction of the reception history of Locke's work and the various co-existing interpretative frameworks within this history. Instead, in the context of my argument I restrict myself to re-visiting critical engagements of Locke's conceptualizations of private property with an eye to the reciprocal relationship between Western liberalism, colonialism, and slavery in the *Treatises*. That is, while twentieth-century Locke scholarship has often focused on reading Locke's *Two Treatises* in its domestic context (e.g., Woolhouse and Stanton), political scientists and philosophers have increasingly taken the historical context of English colonial and economic expansion into consideration in their readings since the 1990s (e.g., Arneil, "Trade"; Mishra).²⁷ At the heart of this line of inquiry is Locke's relationship with the American colonies, which James Tully brought to the scholarly scene with the publication of his study *An Approach to Political Philosophy: Locke in Contexts* in 1993.²⁸ More recently, the academic interest in questions

---

26 Political scientist Edward J. Harpham traces the main trajectories of twentieth-century scholarship on the *Two Treatises* in his field in *John Locke's* Two Treatises of Government: *New Interpretations*. For feminist readings of John Locke's writings, see e.g. Hirschmann and MacClure.
27 David Armitage in "John Locke, Carolina, and the "Two Treatises of Government" summarizes the main – and often opposing – positions of this research trajectory.
28 For a rebuttal of Tully's arguments and a perspective that sees "America belong[ing] only at the margins of [Locke's] main concerns in the *Two Treatises*," see Buckle (274).

about the alliance between Locke's writings and English colonial investments has shown in scholarly efforts to juxtapose conventional readings of the *Two Treatises* with interpretations of "The Fundamental Constitutions of Carolina" (1669). David Armitage, for instance, zooms in on Locke's active role in the drafting of the "Fundamental Constitutions" and argues that Locke was working on the fifth chapter of the *Second Treatises* at the same time that he was revising the "Fundamental Constitutions" in 1682 (602). For Armitage, this suggests that "there was an immediate and identifiable colonial context that contributed to [Locke's] distinctive theory of property" and that, subsequently, the arguments made in his famous fifth chapter were strongly influenced by "the hold the master-slave relationship had over his political imagination" (602, 619).

Marxist Critiques of John Locke's *Two Treatises*

Before continuing my discussion of this strand of Locke scholarship, I need to spend some time addressing C. B. Macpherson's influential study *The Political Theory of Possessive Individualism*, which was published in 1962.[29] As mentioned earlier, Macpherson suggested that the seventeenth-century political theories of John Locke, among others, largely attributed a "possessive quality" to the modern individual (*Political Theory* 3). While not explicitly interested in the affiliation of slavery and liberalism in Locke's work, I invoke Macpherson's Marxist perspective on modern political society, which he theorized as being fundamentally shaped by 'possessive individualism,' as one of the first, if not the most influential twentieth-century critique of Locke's *Two Treatises* from the academic, political Left. Although not all critical readings of Locke's views on modern individualism have overtly committed themselves to his Marxist agenda, Macpherson's arguments have "never been seriously challenged" (Graeber 36; cf. also Greeson, "American Enlightenment," "Prehistory"). As such, his work here not only stands as paradigmatic for Marxist analyses of Locke's book but also as a watershed in the critical evaluation of the function of property in Locke's *Two Treatises*.

In his study, Macpherson contended that the *Two Treatises* fundamentally were not only about exclusive private property but also about the claims and rights – most notably the right to one's own self – that the modern individual

---

[29] For critiques of Macpherson's arguments see generally, for instance, Broeck, "Never"; Pocock; Tully, *Approach, Discourse*.

gained precisely through the modalities of possession (*Political Theory* 1). This individual essentially becomes

> the proprietor of his own person or capacity, owing nothing to society for them. The individual was seen neither as a moral whole, nor as part of a larger social whole, but as an owner of himself. The relation of ownership, having become for more and more men the critically important relation determining their actual freedom and actual prospect of realizing their full potentialities, was read back into the nature of the individual. *The individual, it was thought, is free inasmuch as he is proprietor of his person and capacities.* The human essence is freedom from dependence on the wills of others, and freedom is a function of possession. (*Political Theory* 3; emphasis mine)

Macpherson thus asserts that Locke puts "a natural individual right to property [at the center of] his theory of civil society and government" (198). Drawing from this proposition, Locke's labor-based theory of private property and its appropriation spoke to emergent capitalist market relations (204–238). Macpherson argues that the notion of property in the *Two Treatises* and in Locke's version of political society is closely knit to questions concerning "class differentials" and that property of land or goods constitutes one of the fundamental assumptive logics of his thought (221–238). Only those individuals who own that kind of property will have the capacity to fully participate in Locke's version of political society: "Not every proprietor of land is necessarily a full member of the society but every full member is assumed to be a proprietor of land" (250). Macpherson ultimately claims that Locke's views on both modern individualism and civil society "[consist] of relations of exchange between proprietors. Political society becomes a calculated device for the protection of this property and for the maintenance of an orderly relation of exchange" (3).

However, while Macpherson's reading of Locke's text crucially points to the fact that it was the right to self-ownership and private property, which gave the modern individual its inherent freedom – with freedom becoming a "function of possession" – it fails to account for the notion that Locke did not refer to all human beings in the same way in thinking about the modern (self-)possessing individual. This comes to the fore if we consider Locke's conceptualizations of private property in relation to the systems of (settler) colonialism and slavery in the New World. Importantly, Broeck draws our attention to this when she argues that what is missing from Macpherson's otherwise crisp analysis of Locke's text is "the factor of New World slavery that constituted a particular group of humans as exterior to [the above] 'exchange of equals' in that they were turned, by force of the violence of European 'equals,' into the 'exchanged' objects of European equality" ("Never" 239). In other words, even though Macpherson's study critically re-visits Locke's *Two Treatises* within the historical context of English seventeenth-century political theory and political practice (1), his Marxist framework does account for Atlantic slavery as a fundamental precept of Western modernity and its possessing subject. (I will come back to this in my own reading of the *Two Treatises* below).

Two Treatises and Atlantic Slavery

I return now to the above debates on the intricate connections between liberalism, slavery, and colonialism in the *Two Treatises* in order to link those to the "possessive nature" (Macpherson, *Political Theory*) of Locke's liberal subject. As mentioned before, Locke scholars have become more and more interested in the manifestations of these conceptual conflations in Locke's thought—especially so in the wake of post-colonial criticism's entry into the academic landscapes of the global Northwest in the late 1980s and early 1990s. That is also to say that alongside this general interest in the colonial affiliations of Locke's writings, the debate has shifted to, and continues to revolve around, questions concerning the uses of the concept of slavery and the proclamation of universal individual liberty in Locke's thought (Bernasconi and Mann; Dunn; Farr, *Natural Law*, "So Vile"; Glausser; Hinshelwood; Welchman; Uzgalis, "Locke's Legacy"). There is, in other words, a keen and ongoing interest in the notion that Locke, as an historical figure, was actively involved not only in many of the English colonial affairs in the New World on an administrative level but also in the trading of African slaves, for example through his monetary investments in the Royal Africa Company (Bernasconi and Mann 89; Glausser 200–204; Welchman 71–74).[30] More often than not, critics and readers of Locke's work have struggled to reconcile these historical facts and Locke's views on slavery with his political arguments on universal liberty. That is, they have conceived of this seeming inconsistency between Locke's factual involvements in the transatlantic slave trade, on the one hand, and his groundbreaking conceptualizations of universal freedom, on the other, as an irresolvable paradox. Within the realm of philosophy, as Jennifer Welchman points out, the response to the questions raised by the 'incongruity' of Locke's conduct and his ideas has typically been that they are "of merely historical interest and that consequently, it is for historians, rather

---

[30] As early as the 1960s, Peter Laslett "connected Locke to new world slavery" when he published the definite edition of the *Two Treatises* (Farr, *Natural Law* 495–496; see also Arneil, "Trade," *John Locke and America*). Another important study from the 1960s is David Brion Davis's *The Problem of Slavery in Western Culture*, which was published in 1966 and in which Davis argues that Locke sought to justify chattel slavery in the *Two Treatises* (118–121).

than philosophers, to try to answer them. How as a matter of fact a philosopher comes [...] to advance the arguments he or she advances is entirely irrelevant to their critical evaluation" (68). Consequently, "the very facts that would seem to convict Locke of gross moral and/or philosophical turpitude," as Welchman continues to explain, "have been made the basis of ingenious reconstructions" of Locke's uses of slavery in the *Two Treatises* (69). In this context, Bernasconi and Mann explicitly point to the long history of the "attempt to reconcile Locke's involvement in the slave trade with his reputation as a philosopher of liberal freedom" (89). Because such attempts were already made in the early eighteenth century, they "cannot simply be dismissed as the product of the recent fashion for so-called political correctness, as some academics want to do" (Bernasconi and Mann 89). In other words, these reconstructions continue to provide a platform for lively debate, with the most prominent positions of this debate either claiming that Locke sought to justify chattel slavery in the New World or firmly rejecting the notion that Locke sought to defend slavery in his text and arguing instead that Locke's (rhetorical) use of the notion of slavery in his text to be geared towards opposing absolutist monarchical rule (Farr, *Natural Law*; Glausser). Another attempt at dealing with Locke's uses of the notion of slavery in the *Treatises* has recently been made by political theorist Brad Hinshelwood. In his "The Carolinian Context of John Locke's Theory of Slavery," Hinshelwood responds to these long-standing debates by claiming that the colony of "Carolina is in fact the focus of Locke's theory of slavery" (565). Reconstructing the *Treatises*' uses of the notion of slavery against the backdrop of Locke's active involvement in the drafting of the "Fundamental Constitutions," Hinshelwood suggests that the *Treatises* in fact discuss the enslavement of American Indians in this context (564). In this way, Hinshelwood follows in the steps of other political scientists walking down this explanatory route, like James Tully, who as early as 1993 speculated that Locke's theory and use of slavery in the *Treatises* "may also refer to Amerindian slavery" (*Approach* 143–144).[31]

---

[31] Another paradigmatic example is Barbara Arneil's study *John Locke and America: The Defence of English Colonialism*, which sets out to shed light on the "role of America and its aboriginal population in Locke's political theory which has been largely overlooked in previous scholarship on the *Two Treatises*" (2). Arneil's work shows painstaking attention to colonial detail, as she looks at the colonial discourses within which Locke wrote the *Two Treatises* and examines how these discourses manifest themselves in his ideas of private property (132–167; see also Arneil, "Trade"). However, despite this explicit focus on the substantiations of English colonial discourse in Locke's text, Arneil significantly does not contemplate Atlantic slavery in her study. For discussions on John Locke and Indigenous peoples in his thinking, see generally Miura and Squadrito.

While this research focus pinpoints the ways in which settler colonialism and Indigenous dispossession figure in Locke's text, however, this line of critical inquiry effectively subdues Atlantic slavery as a crucial element of Locke's configurations of private property and/as the basis of white liberal subjectivity. In this endeavor I follow Sabine Broeck, who in "'Never Shall We *Be* Slaves': Locke's Treatises, Slavery, and Early European Modernity" contextualizes the *Treatises* "as one of the discursive moments of early modernity, which *actually legitimize the slave trade*" (238, 243; emphasis mine).³² Both reflecting on and criticizing Locke criticism's previous engagements of the reciprocal relationship between Locke's modern individualism and English colonialism and Atlantic regimes of slavery, Broeck argues that it is the simultaneity – rather than the alleged contradiction – of chattel slavery and (bourgeois European) white liberty that needs to be foregrounded in any critical reading of Locke's text and its New World colonial context.³³ Broeck contends that Locke employs the notion of slavery in the *Treatises* not to reflect on chattel slavery in the New World but as a rhetorical tool with which to think about the bourgeois individual's emancipation from feudal rule: "Generations of critics have looked past the fact that when Locke speaks about 'slavery' – as the opposite of 'liberty' – he does not allude to actual New World chattel labor" (244). Instead, the notion of slavery becomes "in [Locke's] rhetorical repertoire [...] a signifier for the oppression of free gentlemen, and thus an indispensable move to define 'liberty'" (244).³⁴ In Locke's argument, that is, slavery functions "as the most effective signifier to attack what free Englishmen see as oppression of their rights" (237), or, to echo Buck-Morss again, a powerful metaphor that indexes "everything that was evil about power relations" (821). In Locke's formulation, then, the "the rebuttal of 'slavery' [...] had nothing to do with a universal rejection of slavery, but on the contrary, became a motor of the Atlantic slave trade *and* of early modern bourgeois emancipation in tandem" (Broeck, "Never" 237).

---

32 Similarly, Welchman has expanded on well-established interpretations of the *Two Treatises* by drawing attention to the transatlantic slave trade and the ownership of slaves as the essential context of the text's production. More recently, Shilliam has endeavored to make a similar argument as he shifts attention to a global vision of coloniality in Locke's *Two Treatises*.

33 Sibylle Fischer has recently made a similar argument in her essay "Atlantic Ontologies: On Violence and Being Human." Fischer notes, "Rather than seeing Locke as equivocating and disavowing knowledge he had (a critical reading that post hoc insulates his political theory against any infection from his view on slavery,) it makes more sense to think of Locke devising a political theory that actually responds to the realities of Atlantic slavery."

34 Farr rightly acknowledges this but fails to acknowledge the colonial context of transatlantic slavery as an important referential horizon of Locke's text (cf. *Natural Law*, "So Vile").

Before turning to my own re-reading of the *Two Treatises* in the next section, I want to address two interlocking points that Broeck makes, which will be crucial for my own project. One: Broeck connects Macpherson's arguments concerning possessive individualism to the histories of Atlantic slavery to argue that Locke's claims about private property need to be understood as "philosophical and political arguments [for the emerging enlightened liberal subject's claim to] *freedom as self-possession*" ("Never" 236). Locke did not in any way argue for freedom to "secure the gradual realization of a universal ethics, but to find the most effective rhetorical counterpoint to refute" his contemporaries' arguments in favor of feudal rule, most notably those of Robert Filmer (236). Following Broeck, Locke's arguments "propelled a European post-seventeenth-century discursive tradition in which 'freedom' became an object of negotiation, always already in relation to 'slavery' [...] This negotiation, however, becomes effective only in the abstract" (236). That is, the Enlightenment's enthusiasm for scientific and human emancipation needs to be understood as being condensed in a self-descriptive narrative that "altogether bypasses the historical experience of lively and angry early modern controversies around the slave trade, slavery, and issues of mastery, ownership, and oppression of human beings" (236).[35] Put another way, white European and Atlantic freedom needs to be understood as being parasitic, both on a material and an epistemic level, on slavery; or, as Toni Morrison has it in the second epigraph to the chapter, as parasitic on "the presence of the unfree within the heart of the democratic experiment."

Two: Locke essentially frames private property in terms of appropriation by means of labor in his *Treatises*. Broeck shifts critical attention to the question of slave labor as a fundamental yet unacknowledged presence within Locke's thinking when she writes that "African bodies and their labor capacity [...] function as the crucial absent presence, the invisible lever in [Locke's] argument about the legitimate accumulation of 'property'" ("Never" 242). What Broeck suggests is that Locke's conception of property as that which is accumulated through labor inextricably includes the "the purposeful ownership of chattel labor [...] [as] an *a priori element of property* deliberately built into the Lockean system" in the context of English colonial and economic expansion to the New World (242–243). That is, the forced labor done by the enslaved becomes the key means by which property is appropriated in the context of New World cultivation. This is also to say that it is *not* the labor of Locke's liberal subject by which property is appropriated in this context.

---

[35] On the relationship between *Race and the Enlightenment*, see generally e. g., Bernasconi and Cook; Eze; Uzgalis, "On Locke."

Two Treatises Re-Visited

So far, I have been trying to detail the myriad ways in which philosophers and political scientists have tried to think about John Locke's *Two Treatises* and his ideas about exclusive private property within a liberalist framework of English colonial and mercantile expansion. An important element of such endeavors is the attempt to make sense of the notion of slavery in Locke's writings from varied theoretical standpoints, most of which do not acknowledge chattel slavery as the conceptual foundation on which Locke builds his theory of individual freedom as self-possession in the *Treatises*. Drawing on Broeck's work, I have suggested that Locke criticism often avoids engaging with white practices of enslavement in the New World, thus ignoring the fundamental importance of these practices for early Enlightenment conceptions of political and human freedom on both sides of the Atlantic. In this section, I re-visit the interconnected concepts of *nature, land, labor,* and *cultivation* in Locke's *Two Treatises* as I trace the ways in which Locke employs these terms to conceptualize private property. At stake is the coming into being of Locke's liberal subject both in relation to these concepts and against slavery.

With *The Second Treatise of Government: An Essay Concerning the True Original, Extent, and End of Civil Government*, Locke announces that he seeks to provide an account of what he thinks society and the appropriate distribution of political power, civil government, and political representation should look like (Locke 267–268).[36] Following the rebuttal of Sir Robert Filmer's *Patriarcha* (1680) and his theses on authoritative, monarchical rule in the *First Treatise*, Locke begins the *Second Treatise* by defining political power as "*a Right* of making Laws with Penalties of Death, and consequently all less Penalties, for the Regulating and Preserving of Property" (Locke 268). Following from this, Locke argues that sovereign power should be devoted to securing the

> *Freedom of Men under Government* to have a standing Rule to live by, common to every one of that Society, and made by the Legislative Power erected in it; A Liberty to follow my own Will in all things, where the Rule prescribes not; and not to be subject to the inconstant, uncertain, unknown, Arbitrary Will of another Man [...] This *Freedom* from Absolute, Arbi-

---

[36] Throughout this section of this chapter, my reading will focus on the fifth chapter "Of Property" of the *Second Treatises* while also obtaining textual evidence for my arguments from other parts of the text, if necessary. For the sake of convenience, I have opted for quoting page numbers instead of section signs from Locke's text. All quotes are from Peter Laslett's 1988 edition of the *Two Treatises* and all emphases are John Locke's if not indicated otherwise. The spelling, too, is Locke's.

trary Power, is so necessary to, and closely joyned with a Man's Preservation, that he cannot part with it, but by what forfeits his Preservation and Life together. (284)

In other words, the main end of political rule is the preservation of private property in combination with a promise of individual freedom within the politico-legal framework of civil society. In the last few lines, the paragraph furthermore establishes an intimate connection between the liberal subject's freedom from the rule or will of others and its drive towards (self-)preservation. Arguing within the domain of natural law, Locke claims that it was God who gave the drive towards one's own preservation to the reasonable subject in the first place when he created the world (205). Anchoring a strong sense of self-preservation in this subject itself (286), Locke believes that the main reason why 'Man' would ultimately decide to enter civil society is to ensure self-preservation by way of the protection of one's property (209–210, 360; cf. also Euchner 97–98).

Land and the appropriation thereof are the building blocks of Locke's theorizations of private property and individual liberty, particularly as regards colonial rule in the so-called New World (Locke 290). Robbie Shilliam writes in this respect, "Locke cognitively works out the rights and obligations of primitive accumulation by reference to the specific rights and obligations of colonialism, and as part of the broader colonizing and proselytizing project that the English crown is embarking upon in earnest" (5). Starting from the premise that the world and all the divinely ordained land on it commonly belongs to 'humankind,' Locke argues that there must be some way in which individuals may be able to take a part of the commonly held land and its goods for their own subsistence:

> The Earth, and all that is therein, is given to Men [by God] for the Support and Comfort of their being [...] yet being given for the use of Men, there must of necessity be a means *to appropriate* them [land and its natural produce, such as fruit] some way or other before they can be of any use, or at all beneficial to any particular Man. (Locke 286–287)

While Locke here initially sets a limit on the amount of land provided by "the spontaneous hand of Nature" that any individual may appropriate (286),[37] he also clearly states that the notion of labor constitutes the primary way in which land and all its produce may be seized as private property: "*As much*

---

[37] Locke argues initially that any individual may appropriate as much as possible, provided that what has been harvested will be used completely in order for it not to spoil and that 'man' does not take so much that nothing will be left for other individuals (289–93; cf. also Euchner 88–97).

Land as a Man Tills, Improves, Cultivates, and can use the Product of, so much is his *Property*" (290). In Locke's architecture, that is, labor becomes a property-generating force (Euchner 90). Locke writes:

> Though the Earth, and all inferior Creatures be common to all Men, yet every Man has a *Property* in his own *Person*. This no Body has a right to but himself. The *Labour* of ["Man"'s] Body, and the *Work* of his hands, we may say, are properly his. Whatsoever then he removes out of the State that Nature hath provided, and left it in, he hath mixed his *Labour* with, and joyned to it something that is his own, and thereby makes it his *Property*. (Locke 287–288)

It is important to note that Locke's conceptualizations of labor and the appropriation of land/private property are intricately connected to his understanding of the liberal subject as a self-possessing individual. This entitlement to one's own self functions as a fundamental precept in all of Locke's subsequent theorizations of property. The above passage shows that it is this core tenet on which Locke grounds his ideas of a claim or right to private ownership, for the right to private property is built on the right one's own self. Locke illustrates this at numerous other places in the text, for example when he goes on to state, "And thus, I think it is very easie to conceive without any difficulty, *how Labour could at first being a title of Property* in the common things of Nature" (302).

Locke qualifies the concept of labor in two corresponding ways. Apart from being the principal way in which land may be appropriated as private property, firstly, labor is theorized as a function, which attaches value to any object seized. For Locke, it is "*Labour* indeed that *puts the difference of value* on every thing" (296). In the example he gives in order to prove this, Locke juxtaposes cultivated land with uncultivated land, arguing that the former holds much more value than the latter, precisely because it has been labored upon: "the difference is between an Acre of Land planted with Tobacco, or Sugar [...] and an Acre of the same land lying in common, without any husbandry upon it [...] the improvement of *labour makes* the far greater part of *the value*" (296). Thus, the notion of cultivation is paramount to Locke's labor-based theory of private property, for it is the cultivation of formerly common wasteland – or the "*value of Human industry*," as Locke has it (297) – which would ultimately provide prosperity based on possession. Secondly, Locke also defines this type of industriousness or "possessive accumulation" in terms of a division of labor (Shilliam 7). Locke writes that

> 'tis not barely the Plough-man's Pains, the Reaper's and Thresher's Toil, and the Bakers Sweat, is to be counted into the *Bread* we eat; the Labour of those who broke the Oxen, who digged and wrought the Iron and Stones, who felled and framed the Timber imployed

2.2 John Locke's *Two Treatises of Government* (1689): Critical Contexts — 41

> about the Plough, Mill, Oven, or any other Utensils, which are a vast Number, requisite to this Corn, from its being seed to be sown to its being made Bread, must all be *charged on the account of Labour*, and received as an effect of that: Nature and the Earth furnished only the almost worthless. (298)

This passage conveys a strong sense not only of the expenditure of human labor in the form of – to stay with Locke's example – the manifold steps to be taken to make bread; the trope of 'industriousness' also suggestively opens up a distinction between cultivated and uncultivated natures, respectively.

Cultivation thus takes center stage within this framework of the labor-based appropriation of land. While land constitutes "the *chief matter of Property*" in this framework (290), Locke makes a distinction between uncultivated land that lies 'waste' and cultivated land at several points in his text (e.g., 297, 299). The difference between the two, of course, lies in the extent to which the land has been labored on and, following from Locke's earlier logic, the extent to which value has been attached to the land. This merits closer examination. Locke writes:

> For I aske whether in the wild woods and uncultivated wast of America left to Nature, without any improvement, tillage, or husbandry, a thousand acres will yield the needy and wretched inhabitants as many conveniences of life as ten acres of equally fertile land doe in Devonshire where they are well cultivated? [...]
>
> There cannot be a clearer demonstration of any thing, than several Nations of the *Americans* are of this, who are rich in Land, and poor in all the Comforts of Life; whom Nature having furnished as liberally as any other people, with the materials of [...] fruitful Soil [...] yet for want of improving it by labour, have not one hundredth part of the Conveniences we enjoy [...]
>
> And as different degrees of Industry were apt to give Men Possessions in different Proportions, so this *Invention of Money* gave them the opportunity to continue and enlarge them. (294, 296–297, 301)

First, Locke connects the issue of cultivation to the geographical locations of 'America' and Devonshire in England, respectively, to claim that the 'needy and wretched American nations' will not benefit from their unimproved land as much as the industrious people in Devonshire who cultivate their land successfully and, therefore, enjoy 'all the Comforts of Life.' In Locke's architecture such a marked distinction between the "'success' of possessive accumulation at home, through enclosures" and ineffective and unprofitable agriculture abroad clearly helps Locke "[legitimize] the same procedures abroad, through colonialism [...] It is a global vision of colonial incorporation," as Shilliam holds in this respect (8, 7–8).

Second, the concept of cultivation functions as the lever with which Locke is able to differentiate between various groups of people, namely those who labor in order to cultivate their land and those who do not. In conjunction with the previous example regarding geographical location, cultivation here specifically signals the increase of the land's value, which only a particular group of people is able or willing to bring about. That is to say, the mastery of nature in the form of the cultivation of land becomes a distinguishing characteristic of Locke's liberal individual. Decidedly, property comes into play at this juncture: only those who are capable to own property will ultimately become successful cultivators/colonizers/subjects. As Broeck explains,

> According to Locke, everything a free man does to safeguard and accumulate his private property is legitimate, provided he does not encroach on another free man's property. This includes the appropriation and possession of formerly common 'waste' land beyond a free individual's own possibilities and needs to exhaust its riches as long as this person does not leave land uncultivated. Not to tolerate wasteland requires working it according to one's operational abilities, which are of course determined by one's property status. ("Never" 243)

Last but not least, the intimate affiliation of property, cultivation, and labor shows in conjuncture with the invention of money, as the above passage from the *Treatises* shows. In Locke's architecture, money functions as a means by which the abovementioned limitations to appropriation – spoilage and subsistence – may be circumvented. Money enables his liberal individual to possessively accumulate as much as possible, for it constitutes "some lasting thing that Men might keep without spoiling, and that by mutual consent of Men would take in exchange for the truly useful, but perishable Supports of Life" (Locke 300–301). With the invention of money, in other words, limitless appropriation is established as an incontrovertible fact in Locke's textual orbit. Importantly, Locke goes on to tell us that there are still parts of the world where the inhabitants have not consented to the use of money, which means that their land remains uncultivated and is thus not profitable in any way: "yet there are still *great Tracts of Ground* to be found, which (the Inhabitants thereof not having joyned with the rest of Mankind, in the consent of the Use of their common Money) *lie waste*" (299). Locke's solution to this, as Shilliam explains, is colonial expansion: Locke's "famous proclamation, 'in the beginning all the world was America,' is therefore not so much a reference to a primeval past as it is an invitation for freemen to exercise their natural right of possessive accumulation in the colonies with their expanding frontiers" (Shilliam 8).

In all of the critical discourse concerning (the uses of) slavery in Locke's *Treatises*, the question that is frequently bypassed is that of *who* labors on the vast uncultivated land masses that the New World provides for Locke's free individual (Broeck, "Never" 243).[38] Broeck explains that the accumulation of property

> includes the appropriation and possession of formerly common 'waste' land beyond a free individual's own possibilities and needs to exhaust its riches as long as this person does not leave land to lie uncultivated. Not to tolerate wasteland requires working it according to one's operational abilities, which are of course determined by one's property status. Because this status includes the potential ownership of chattel as *a matter of course*, working land productively with this chattel labor force becomes an advantage for the society's healthy development as a whole. ("Never" 243; emphasis mine)

In the colonial context of Locke's *Treatises*, the productive and profitable accumulation, appropriation, and cultivation of land "*by means of* slave labor is clearly implicated as the immediately available and even pressing option" (243). Against this backdrop, Locke's statement that "the Grass my Horse has bit; the Turfs my Servant has cut, and the Ore I have digg'd in any place where I have a right to them in common with others, become my *Property*" (Locke 289), needs to be reevaluated as one that *readily* assumes the easy availability of a slave labor force as part of the liberal individual's ability to exercise the 'natural right to possessive accumulation in the colonies.' Locke's narrative effectively conceals this, "presenting to readers a pastoral idyll of private wealth in harmony with social benefit, all produced originally by honest labor upgraded by the practical invention of money" (Broeck, "Never" 243).

2.3 Post-Slavery Readings: The *Two Treatises* and "The Fundamental Constitutions of Carolina" (1669)

Having looked at John Locke's theorizations of exclusive private property within the context of his late-seventeenth-century vision of civil society and appropriate political representation, it has become clear that these rely exclusively on a conceptualization of labor-based accumulation. At the center of the *Two Treatises of*

---

38 While the question of *who* labors is of utmost importance at this juncture, it also is necessary to note in this context that I do not understand enslavement/racial slavery to be merely defined as coerced labor. Instead, I follow Orlando Patterson's definition of slavery as social death (i.e. natal alienation, general dishonor, and openness to gratuitous violence), which post-slavery theoretical trajectories have taken up and expanded on by introducing terms/concepts such as *accumulation* and *fungibility* to describe the slave's violent positioning in the world. I will discuss this in more detail in the next chapter.

Government, Locke's meditations on ownership and possession, is the industrious individual, who acquires property by means of the cultivation of land. Locke fundamentally equips this individual with the right to a labor force built of slaves. At stake in Locke's uses of slavery, then, is not so much a project of universal liberation but the constitutive force that slave ownership generates for the liberal imagination of/and the self. Again, Wilderson reminds us that the social death of the slave (absence of relationality) and the fact that they can never be the subject of property opens up the master's possessive possibilities on a structural level. The slave is that "against which Humanity establishes, maintains, and renews its coherence, its corporeal integrity," to which (the claim of right to) private property is absolutely essential, as I have tried to show (Wilderson, *Red* 11). Again, freedom thus is a function of possession. The notion that ownership of chattel slaves and their labor capacity was indeed a matter of course and, by extension, a vital element of not only of European New World colonial and economic expansion but also of the emerging self-possessed liberal subject can be further traced and illustrated in Locke's other writings. Specifically, I am referring here to "The Fundamental Constitutions of Carolina" (1669). In turning to the "Fundamental Constitutions," I suggest that this set of colonial instructions, like the *Treatises*, needs to be read as another site of the liberal subject's emergence. First released on March 1, 1669, Locke was one of the principal authors of the "Fundamental Constitutions" (Armitage 607). Critics, historians, and political philosophers have labored considerably not only to show that Locke co-wrote the "Constitutions" but also to establish that he had been involved in later revisions to these colonial instructions in spite of the fact that Locke's "flight to France in November 1675 ended his tenure as secretary to the Proprietors" and, therefore, his immediate participation in their colonial affairs (Hinshelwood 573). That is, while scholars "long assumed that Locke's involvement with Carolina in France was virtually nonexistent, and that Locke did not pay serious attention to colonial affairs again until 1696, when he took up a post on the Board of Trade," it has by now been shown that Locke's involvement with Carolina continued while he was composing his *Treatises* (573; see also Farr, *Natural Law*). For example, both Hinshelwood and Armitage suggest that "Locke was in fact working on his sections on slavery" in the *Treatises* at about the same time that he participated in substantial revisions of the "Constitutions" in 1682 (Hinshelwood 574). Again, like other frequently cited and engaged (biographical) evidence that connects Locke to the systems and practices of Atlantic slavery, such as his material and monetary investments in the transatlantic slave trade, this goes to show that chattel slavery needs to be understood as a crucially important but *utterly disavowed* referential frame for Locke's early

Enlightenment thinking and fundamental conceptualization of liberty, subjectivity, and property.

With the "Fundamental Constitutions," the Lords Proprietors as the principal party interested in establishing a profitable colonial venture "between Virginia and the Spanish settlements in Florida" gave instructions as to the social and political organization of the nascent Carolina colony (Hinshelwood 567). Armitage explains that

> [t]hough frequently revised and just as often ignored by the settlers, the *Fundamental Constitutions* did formally provide the frame of government for the colony until they were overthrown by the settlers forty years after they had first been promulgated in 1669. They were repeatedly published in Locke's lifetime, both in manuscript copies for the settlers and in a variety of printed versions, from deluxe large-paper printings (presumably for the Proprietors) to abbreviated summaries designed to encourage emigrants. (607)

Initially a relatively small colony (Hinshelwood 568), the Lords Proprietors' envisioned Carolina to be ruled by landed aristocracy and they "hoped that property guarantees for lower-class migrants ('leet-men'), in addition to the provisions for religious freedom, would lure leet-men and 'attract and keep the weightier sort' to lead the settlement" (568). In other words, the aim was to create "a secure, agriculturally self-sufficient colony" (569). To this end, the "Fundamental Constitutions" "apportioned land and provided the legal and institutional framework for the infant colony" (Armitage 609). The "Constitutions" gave instructions on the organization of Carolinian colonial society along strict hierarchical lines, which, in turn, structured the colony according to claims of right to property of land. Thus, the "Constitutions" were

> explicitly designed to 'avoid erecting a numerous Democracy' and placed all authority perpetually in the hands of 'the true and absolute Lords and Proprietors of the province.' Beneath them would be the hereditary nobility composed of landgraves and caciques who would have jurisdictional authority over a further hereditary class of perpetual serfs or leet men. (Armitage 609)

Indeed, for many of the lower-class settlers/leet-men, the text's stipulations would give promises both of social and economic upward mobility by way of providing of land as well as a legal representation and a judicial system geared towards their adequate handling of their concerns (Hinshelwood 568). With the "Constitutions," the Lords Proprietors also "'consistently tried to balance their interests with Indian rights.' Indians were granted religious freedom and parcels of land within the baronies set up by the Proprietors, and the Proprietors reserved all rights to negotiate land purchases and other matters with the natives" (568). The Lords Proprietors hoped to avert any (potential) conflict about land rights between the European settlers and the indigenous peoples in the colony (568). From their very inception, then, the "Fundamental Constitutions" were written in order to secure and protect the settlers' property interests at the New World colonial scene.

The "Constitutions" refer to slavery in two out of the 120 provisions in total. In the 107th provision it is stated firstly that, like freemen, Carolinian law will allow 'slaves' to worship and that this shall not change "that civil dominion his master hath over him, but be in all things the same state and condition as he was in before" ("Fundamental Constitutions").[39] Secondly, section 110 reads: "Every freeman of Carolina shall have absolute power and authority over his negro slaves, of what opinion or religion whatsoever" ("Fundamental Constitutions"). This section of the "Constitutions" labors to achieve various points: For one, it represents part of the Lords Proprietors' efforts to attract a host of lower class settlers to secure a functioning, prosperous, as well as self-sufficient colony (Hinshelwood 568–569). That is to say, the Lords Proprietors hold out the prospect of a set of claims to ownership, "absolute power and authority over negro slaves" for these leet-men, which is something that they also maintain for themselves. Again, as a set of governmental instructions, the "Constitutions" delineated how the future colony of Carolina would be run along strict hierarchical, semi-absolutist lines, which put the Lords Proprietors, whose power and influence was based on their ownership of land (domestic and colonial) and would be maintained through heredity, at the top of Carolinian (political) society. Accompanying this set of claims is a promise of wealth and prosperity that is to be achieved, self-evidently, by means of the ownership, appropriation, and cultivation of land. Again, the Lords Proprietors conceived of the economic venture of Carolina as one that was to be headed by landed gentry at the same time that they "hoped that property guarantees for lower-class migrants" would help foster the development of the colony significantly (Armitage 568). What section 110 does, I suggest, is to create a relation between two different sets of English colonists across class differentials. For both groups, ultimately, the fact that they are the *subject* of property makes for their prosperity, economic and otherwise, both in the New World colony of Carolina and in the mother country.

---

[39] The full provision/section reads: "One hundred and seven. Since charity obliges us to wish well to the souls of all men, and religion ought to alter nothing in any man's civil estate or right, it shall be lawful for slaves, as well as others, to enter themselves, and be of what church or profession any of them shall think best, and, therefore, be as fully members as any freeman. But yet no slave shall hereby be exempted from that civil dominion his master hath over him. But be in all things in the same state and condition as he was before" ("Fundamental Constitutions").

2.4 Whiteness as Property: "The Germantown Friends' Protest Against Slavery" (1688)

My reading of the "Germantown Friends' Protest Against Slavery" needs to be framed in explicit ways: The "Protest" was issued as an early statement against slavery in 1688, five years after the founding of Germantown, a Pennsylvania village in the British colony of Philadelphia. Traditionally, historians have posed that the "Protest" had little to no influence on Quaker antislavery discourse until 1844, "at which point it was rediscovered by abolitionists, reprinted, and distributed more widely than its original authors could possibly have imagined" (B. Carey, *From Peace* 73). Brycchan Carey has recently contradicted this notion by arguing that the "Protest" was in fact "a seminal and connected moment in the development of Quaker antislavery discourse" (*From Peace* 73). While the "Protest" "raised the issue of slavery at a time when it was not a matter of general concern," as Carey notes (70 – 71), it needs to be pointed out explicitly that it was also issued at a time when "[c]olonial space and its mercantile and productive possibilities provided English gentlemen as a group with an experience of entitlement to being properly themselves and knowing/owning the world for themselves" (Broeck, "Never" 239). Like other actors on the New World colonial scene, that is, the Germantown Quakers depended on Philadelphia's flourishing economy and on the wealth produced by slave labor (B. Carey, "Inventing" 19). My reading is located at the fault line between the text's antislavery impetus, on the one hand, and its situatedness in white practices of enslavement, on the other. As I will show, the property of whiteness itself crucially determines conceptions of the New World subject in this early modern text. I argue that the Quakers' concerns for their reputation as concerns the meaning of slaveholding for their community (religious, economic) show that the "Germantown Friends' Protest" is in fact a highly self-conflicted text that configures the conceptual conflation of ownership of Black slaves and the making of white liberal selves.

As mentioned before, legal scholar Cheryl I. Harris clarifies that the conceptual conflation of private property and subjectivity was racialized from the start, marking whiteness and with it the promise of individual liberty as the most valuable property to be owned on the early American scene:

> Slavery as a system of property facilitated the merger of white identity and property. Because the system of slavery was contingent on and conflated with racial identity, it became crucial to be 'white,' to be identified as white, to have the property of being white. Whiteness was the characteristic, the attribute, the property of free human beings. ("Whiteness" 1721)

Harris's arguments on the ways in which the law would enshrine the intimate affiliation of race and property and that whiteness came to function as a "shield from slavery" ("Whiteness" 1718–1720) extend onto the idea that this property interest in whiteness complies with other conceptualizations of property as a metaphysical right in things that are intangible, such as social privileges or expectations (1725–1730). In this context, the law would recognize the notion of reputation, like whiteness, as a specific form of property. Following Harris, this ultimately was an 'ideological move" that recognized a *"reputational interest in being regarded as white"* and that would also construct the property of a white identity as a necessary tool for self-preservation (1734; emphasis mine).

Taking the relation between whiteness, property, and reputation as established by Harris into account, I argue that the protesters' concerns about their status in the eyes of the Quaker communities in Europe should be read as a concern about losing their reputational interest in being regarded as *white*. Losing the right to their property in whiteness in the dynamic and complex environment of the New World would mean losing valuable social and legal privileges on which they depended. This becomes somewhat clearer when reading the "Protest" in line with other paradigmatic texts from this period such as Locke's *Two Treatises*, published just one year after the "Protest" was issued. Again, Locke would break new ground with his arguments for *"freedom as self-possession* [that] strategically reject any absolutist voluntarism and boldly advance the rights and obligations of the emerging enlightened subject" (Broeck, "Never" 236). He would famously conceptualize these in his dictum of private property as the liberal subject's "life, liberty and estate," in which ownership came to be defined as both the free English gentlemen's right to own themselves *and* to own slaves (Locke 323–324, 350, 383). In this context, Locke's idea of self-ownership would become "particularly fertile ground for the idea that reputation [...] was similarly property" (Harris, "Whiteness" 1735). There is, however, another way in which reputation, as something that subjects can own, figures in this context. Recent post-slavery discourses have offered tools with which to account for the ways in which property, of reputation, like whiteness, becomes a function that establishes and delineates a relation between subjects (e.g., Hartman, *Scenes*; Sharpe, "Lose"; Wilderson, *Red*). That is to say, while Harris's claims are useful when it comes to describing or analyzing property formations on the level of performativity or identity as well as in relation to the realm of the law, I seek to supplement them here with a view on the structural operative dynamics of power subtended by the Middle Passage. This section, in other words, draws attention to the inextricable structural workings of private property, in which Blackness functions as the ground on which the liberal imagination and its subject unfold. This will become clearer in what comes next.

"Good Reports": Private Property, Slavery, Reputation

The "Germantown Friends' Protest" was written and signed by four men who were "part of a group of German and Dutch Pietists, Mennonites, and Quakers [...] who had arrived in Pennsylvania seeking religious freedom" (B. Carey, *From Peace* 72). Relating why these four men were against the slave trade, the document traveled through Quaker institutions and meetings, who each found it 'a thing of too great A wayt for this meeting to determine,'" before being ultimately dismissed by the Philadelphia Yearly Meeting in 1688 ("Protest"; B. Carey, "Inventing" 22; *From Peace* 72–73). The "Protest" is composed of a large section, made up of three paragraphs, in which the protesters advance their arguments against slaveholding in their colony; their signatures; and two smaller paragraphs, which demonstrate the document's travels through the Quaker hierarchy. The "Protest" begins with a statement against perpetual slavery – "These are the reasons why we are against the traffick of men-body" – and an invocation of the Quaker doctrine's Golden Rule[40] by asking whether there is "any that would be done or handled at this manner? viz. to be sold or made a slave for the time of his life?" ("Protest"; B. Carey, *From Peace* 77–78). My concern in this section is with the following excerpt from the "Protest," taken from the main body of the text:

> In Europe there are many oppressed for conscience sake; and here there are those oppressed ^{wh} are of a black colour [...] This makes an ill report in all those countries of Europe, where they hear off, that ^{ye} Quakers do here handel men as they handel there ^{ye} cattle. And for that reason some have no mind or inclination to come hither. ("Protest")

These lines show that the protesters are fully aware of the commercial objectives of their colonial enterprise in Pennsylvania. They know that the success of this venture would depend entirely on a "steady influx" of future colonists 'to come hither' (B. Carey, *From Peace* 80). Being themselves increasingly engaged in the transatlantic slave trade, the protesters are concerned about "the effect that slavery might have on the perception of potential immigrants that the colony was a place of universal toleration and brotherly love" (B. Carey, "Inventing" 25). Put another way, the notion of reputation plays an important role in the protesters' thoughts on the meaning of enslavement for their colony. In what fol-

---

[40] To briefly elaborate: The Quakers recognized the Golden Rule as their fundamental guiding principle. It instructed them to treat others in the same way that they would wish to be treated in similar circumstances and it would become the "final authority" in all of their arguments against slavery (B. Carey, "Inventing" 30).

lows, I suggest that the protesters' concerns about their reputation need to be approached with some hermeneutic suspicion: What implications may the concept of reputation have in this context, what may it also signify, and what would a loss of reputation mean to whom?

Unease about their reputation governs the main part of the "Protest" and is utilized throughout as the protesters refine their arguments against slavery, which become more radical as the document progresses (B. Carey, "Inventing" 25). That is, they contend that because Quaker doctrine condemns stealing, and because slaves essentially are "stolen things" ("Protest"), Pennsylvania Quakers should avoid purchasing slaves and put an end to capture and enslavement: "And we who profess that it is not lawful to steal, must, likewise, avoid to purchase such things that are stolen, but rather help to stop this robbing and stealing if possible" ("Protest"). As they continue to build their anti-slavery argument, the protesters again draw on the notion of reputation to support their claims. We read:

> And such men ought to be delivered out of ^{ye} hands of ^{ye} robbers [slave traders and slave holders], and set free as well as in Europe. Then is Pennsylvania to have a good report, instead it hath now a bad one for this sake in other countries. Especially whereas ^{ye} Europeans are desirous to know in what manner ^{ye} Quaker doe rule in their province; – and most of them doe look upon us with an envious eye. ("Protest")

The paragraph shows that the protesters believe that the colony of Pennsylvania can become a successful venture *only* if the Quakers do not engage in any activities concerning, or practices of, slaveholding. Not only do they argue in favor of ending all future slave trading activities, but they also go so far as to argue for the immediate release of all existing slaves in the colony (B. Carey, *From Peace* 80–81). Only then, they contend, will the Pennsylvania Quakers live up to the colony's reputation among the Quakers in Britain.

I contend that the transaction that takes place around the issue of reputation in fact needs to be understood not as a negotiation of sorts between masters and slaves, as suggested by the notion that they are willing to emancipate the enslaved in their colony, but as a negotiation between masters. That is, even though slaves and their proposed emancipation appear to be at the center of the Germantown Quakers' interests in their reputation, what the Quakers are in fact bargaining with is the *relation with their contemporaries* on both sides of the Atlantic. Slaves here function as barter in the transaction or exchange between subjects that is determined by, to echo Wilderson again, "the ability to own either a piece of Black flesh or a loaf of white bread or both" (*Red* 13). Indeed, linking their arguments about the colony's reputation with the call to emancipate all existing slaves probably was not a very successful rhetorical move by the protesters. As Brycchan Carey has argued, it would not have met with the Quaker community's enthusiasm, for it would have appeared as nothing less than a call to rid many colonists of their legally held property: "Neither English law, emerging colonial practice, nor Quaker notions about private property could countenance such a course of action" (*From Peace* 81). What this indicates is the precarious relationship between an emerging abolitionist discourse and the importance that the intersecting ideas of property, freedom, and white self-ownership would assume in shaping the liberal imagination in this period of the early Enlightenment. Arguing overtly along the lines of the notion of reputation, the protesters' aim appears to be a critique of the Quakers' engagement in the practices of slavery through a "system of trade with the British Caribbean colonies" developed in order to achieve economic prosperity for their colony (B. Carey, "Inventing" 19). What they do in effect, however, is a weighing of the gains and losses of slaveholding for the Quaker community in Pennsylvania. They negotiate their reputation as 'good' Quakers among their communities in Britain *versus* their reputation with the emerging white community of slaveholding propertied men in the colony.

That the possible property in slaves and the possible property in their whiteness seem to collide for the Germantown Quakers at this particular historical moment also shows in the last two paragraphs of the "Protest." These paragraphs are, in the original, handwritten notes that have later been added to the main text, recounting its travels and progress from one Quaker Meeting to the next (B. Carey, *From Peace* 76). Both notes state that the subject of the "Protest" was of "too great a weight" to make a final decision on it ("Protest"). And that, precisely, is what congregation would do—*not* make a decision against Quaker involvement in slavery (B. Carey, *From Peace* 76). This failure to reach a decision against their involvement in slavery shows that in weighing the potential gains and losses of slaveholding for their reputation, the Quaker community willfully invests in and relies on slaveholding as a means of securing their property in whiteness, as well as the attached privileges. Whiteness thus becomes and remains their crucial property to possess, a property that becomes more important than their religious sense of good and evil as Quakers, and their need to be assessed by Quaker standards of virtue as morally righteous in Europe. The protesters' call to emancipate all existing slaves accordingly posed too big a risk for all Quakers in Pennsylvania—a risk of losing their whiteness as property, to which reputation served as a necessary asset in that it was constitutive of the process of being regarded as white and therefore as having proper status in society (Harris, "Whiteness" 1734). In this sense Carey is right in stating that "alienating the property of colonists would have sent a far worse message to potential colonists in Europe than would the presumably widespread knowledge that Africans were being sold into slavery there" (*From Peace* 81). The Germantown Quakers would not have parted easily with their private human property or, for that matter, with their whiteness.

Finally, in this text from the archive of the English Atlantic, which has been read as a document against slavery, whiteness is conceptualized as reputation, assuming crucial importance in terms of the Pennsylvania Quakers' reliance on the system of chattel slavery for the success of their colonial project, specifically with respect to their wish to establish an economically and politically successful venture that would continue to draw more Friends from Europe. It thus shows that the protesters' anti-slavery agenda was an aborted effort. A close look at the conflict between its declaration of anti-slavery, on the one hand, and its simultaneously continued investment in slavery to maintain white identity, on the other, shows how the "Protest" prefigures the social, legal, and epistemological conflation of private property as self-ownership, possession of slaves, and whiteness. Crucially, it sheds light on how white investments in their "possessive possibilities" (Wilderson, *Red* 13) are established over and against the enslaved. It sheds light on how to be able to think and have conflict about the presence of the enslaved in their colony, and not be subjected to divine rule or feudal orders in this process, quintessentially comes within the emerging modern liberal subject's purview. Intra-white deliberation, in other words, becomes a white possessive investment to begin with and it becomes a means to make and maintain a status quo characteristic of the evolving liberal subject and its claims to freedom.

[Coda]: The Liberal Property Paradigm

Proprietorial conceptions of liberal selfhood are at the center of these early Atlantic texts and their negotiations of freedom at the New World colonial scene. As I have tried to show, what is being established as well as constantly recreated and nourished are the possessive possibilities of the liberal subject. Whether as a matter of course (as in the *Two Treatises* and "The Fundamental Constitutions of Carolina") or as the ambiguously tackled subject of intra-white negotiations and conflict, I have tried to elaborate on the ways private property functions as foundation for white self-making in this paradigmatic selection of texts. Drawing on Black Studies' post-slavery interrogations of the discursive promises of universal liberty, I suggest conceptualizing this constant (re-)negotiation as the liberal property paradigm. The theoretical interventions into the complex entanglements between individual, slavery, and private property

– or the "sanctity of property" (Hartman, *Scenes* 122) – as pushed by post-slavery Black Studies' discourses in recent years are the focus of the next chapter.

3 Interrogating Private Property: Black Studies and the Liberal Imagination

[Routing the Argument] The chapter reads Black Studies' post-slavery theoretical interventions into the discursive promises of universal liberty as interrogation of proprietorial conceptions of liberal selfhood. Bearing in mind the overall study's core questions on the relation between literary narrative and a fundamental theoretical critique of early modern of liberal subjectivities, the chapter engages with Black Studies' post-slavery interrogations of the liberal property paradigm. It thus continues the study's examination of the complex entanglements between individual liberty, slavery, and private property that I began in the previous chapter in my reading of late seventeenth-century liberal narratives of individual freedom. The aim of the chapter is to think with Black Studies' post-slavery trajectories, whose theoretical interventions make it possible for me to address and examine the intricate connections between property and self-making in the narrative orbit of *A Mercy* in the next part of the study. As for its structure, the chapter is organized around a set of interconnected terms that I have isolated from these post-slavery trajectories, and which shall serve as points of entry for my analyses of *A Mercy*. These are: *violence*; *dispossession* and *fungibility*; *abjection* and *abjectorship*; *reproduction* and *kinship*; and, finally, *anticipatory wake*.

> Black studies will have to disinvest our axiological commitments from humanism and invest *elsewhere*. Continuing to keep hope that freedom will occur, that one day the world will apologize for its antiblack brutality and accept us with open arms, is a devastating fantasy.
> — Calvin L. Warren, *Ontological Terror*

> A focus on violence should be at the center of this project because violence not only makes thought possible, but it makes black metaphysical being and black relationality impossible, while simultaneously giving rise to the philosophical contemplation of meta-physics and the thick description of human relations. Without violence, critical theory and pure philosophy would be impossible.
> — Patrice D. Douglass and Frank B. Wilderson, "The Violence of Presence"

3.1 Introduction

The chapter reads Black Studies' post-slavery theoretical interventions into the discursive promises of universal liberty as interrogation of proprietorial conceptions of liberal selfhood. As we have seen in the previous chapter, private property emerges as an extremely flexible and mutable vehicle for the negotiation of social and cultural meaning as well as for the formation of power and of value systems from at least the middle of the seventeenth century (see e.g., Bhandar, "Critical Legal Studies"; Davies). As a paradigm, that is, private property functions to keep power and knowledge systems in place and it provides fertile political, social, cultural, and philosophical ground for the ideal of the white liberal subject to emerge (cf. Quijano; Wynter, "Unsettling"). Again, the precepts and the promises of liberal individualism, including its anti-feudal impetus and its commitment to the granting of equal rights to "all" individuals, have from their very inception been intricately connected to the practices and economies of slavery and (settler) colonialism. And they continue to structure post-Enlightenment critical thinking and its concomitant literary, cultural, philosophical narratives of political and individual emancipation (Broeck, *Gender*; Dussel and Mendieta; Lowe; Mills, *Black Rights*). That is, slavery and freedom have been complicit from the very beginning, and it is their "vexed genealogy" that continues to structure both the liberal imagination of personhood and enslavement's "afterlife" in the present moment (Hartman, *Scenes* 115; *Lose* 6; see also Bennett; Walcott, *On Property*).

Working through an ensemble of questions on slavery, subjectivity/subjection, Blackness, liberty, whiteness, and private property, a number of scholars have greatly increased our understanding of the intricate entanglements of individual liberty and private property (e.g., Broeck, "Abolish"; Harris, "Whiteness," "Finding," "Markets"; Hartman, *Scenes*, "Belly"; J. Morgan, *"Partus,"* "Archives"; Nyong'o, "Barack"; Lowe; Sharpe, *Wake*, *Monstrous*; Spillers, "Mama's Baby"; P. Williams). While I generally draw on Orlando Patterson's seminal definition of slavery as social death, my argument fundamentally builds on the work of Black feminist historians, thinkers, and scholars whose research has focused on the institution of slavery's reproductive calculus and "afterlife of property" (Sharpe, *Wake* 15).[41] Critically supplementing and expanding on Patterson's pivotal text *Slavery and Social Death* in this way, Black feminist historians and thinkers have strongly influenced how we think and write about, for instance, Black women's reproductive capacities under slavery. They have fundamentally changed the ways in which we understand how Black women's interiority was violently bound to the market as well as how the accumulation and proliferation of the master's private property and economic increase was tethered to inherit-

---

[41] To recall, according to Patterson, the constituent elements of slavery-as-social-death are natal alienation (i.e. the absence of kinship structures), general dishonor, and gratuitous (as opposed to contingent) violence (Patterson; see also Wilderson, *Red*).

able slave status as enshrined in colonial legislation. As Christina Sharpe reminds us with respect to the longue durée of these intricate connections:

> Reading together the Middle Passage, the coffle, and, I add to the argument, the birth canal, we can see how each has functioned separately and collectively over time to dis/figure Black maternity, to turn the womb into a factory producing blackness as abjection much like the slave ship's hold and the prison and turning the birth canal into another domestic Middle Passage with Black mothers, after the end of legal hypodescent, still ushering their children into their condition; their non/status, their non/being-ness. (*Wake* 749)

Following in these scholars' vein, I seek to investigate private property as a means to white (self-)possession and liberty, which fundamentally includes the active and willful unmaking of human subjects into socially dead Black "sentient beings" (Hartman, *Scenes* 93; Wilderson, *Red* 55, 57) under slavery. In this context, the following questions arise: Which concepts or terms may describe the status of the enslaved as both socially dead and somebody else's legal and inheritable property? Which terms may address the idea that human freedom entails the notion of a right to property? Relatedly, who can be considered as being dispossessed and why? In wrestling with these questions, I aim to do two things: First, by engaging with these questions and by drawing on Black Studies' post-slavery theoretical trajectories that focus on the complicity of notions of subjectivity, freedom, and property with the histories, practices, and legacies of chattel slavery, I seek to complicate the liberal narratives of individual freedom that I examined in the previous chapter. Second, in thinking in relation to post-slavery theoretical trajectories such as Afropessimism and Black Feminist theorizing, I hope to address the various representations of the intricate connections between property and subjectivity that *A Mercy* navigates. That is also to say that without those theoretical interventions, it would not be possible for me, as someone who is positioned within the white/Human fold (Wilderson, *Red*), to unpack what I consider *A Mercy*'s fundamental critique of "the sanctity of property and proprietorial notions of the self" as fundamental building blocks of modern conceptions of freedom and subjectivity in the next part of the study (Hartman, *Scenes* 115).

3.2 Interrogating the Property Paradigm

In general, Black Studies have labored to push a critical "transformation of the human into a heuristic model [over and against the idea of the human as] an ontological fait accompli" (Weheliye, "After Man" 322; see also Weheliye, *Habeas*). It is not in the scope of the chapter to provide a detailed account of the development of the various critical trajectories of Black Studies in the United States ever since its inception as a twentieth-century academic field that developed out of 1960s Black social activism and political organizing during the Civil Rights and Black Liberation movements as well as the Black Arts and Aesthetic Movements.[42] Even before becoming an "institutional and disciplinary formation" (Weheliye, *Habeas* 3), Black Studies has always been the locus of intense and dynamic debate—debate about its intellectual, political as well as methodological stakes (Gates and Burton LII; Gordon and Gordon xxi; K. Wright). There is an extensive historiography on (the development of) Black Studies' various genealogies (see e.g., Bobo, Hudley, and Michel; Gates and Burton; Gordon and Gordon; Marable; Norment; Rojas). Recently, debates within Black Studies have continued to thrive under the guidance of media outlets such as #BlackTwitter, blogs, and other "hashtag politics," specifically in relation to social activism and political organizing in the era of BlackLivesMatter (Neal; Wadud).

I use the term "Black Studies" not only fully aware of these manifold transformations and controversies internal to the field but also aware of the fact that "Black Studies" focuses on different sets of questions depending on who asks these questions as well as on where this inquiry is located.[43] If the term "African American Studies" appears in this study, I use it interchangeably with "Black Studies" throughout. In doing so, I generally follow Jared Sexton, who discusses African American Studies as a twenty-first-century academic field stratified along "two general poles of inquiry and organization regarding the status" of nation and the United States ("African American" 211), with "neonationalist ideological tendencies under the heading of Afrocentrism, Afrocology, or Africology

---

[42] Weheliye stresses that Black Studies has "existed since the eighteenth century as a set of intellectual traditions and liberation struggles that have borne witness to the production and maintenance of hierarchical distinctions between groups of humans" before becoming part of the U.S. mainstream academy in the latter half of the twentieth century (*Habeas* 3). In describing Black Studies as "a most difficult terrain" in the introduction to their *Companion to African-American Studies* (2008), Gordon and Gordon likewise remind us of the various transformations that Black Studies as a disciplinary transformation has undergone.

[43] In Germany, for example, Black German feminist activist-scholar Peggy Piesche has recently called for the implementation of "Black (German) Studies," thereby designating the global dimension of Black and PoC European knowledge productions both within and outside of the academy. Black (German) Studies, according to Piesche, need to be understood as "an attempt to open an intellectual space for an interdisciplinary and international scholarly and artistic engagement with 'Black German Studies' that brings to the field insights beyond those demanded within the nationally inflected, colonial model of area studies" (Piesche). For a general overview of the development of African American studies as an academic discipline in West Germany see Boesenberg.

(and some variants of Africana and African Diaspora thought)," on the one hand, and "a postnational ideological tendency under the heading of [...] 'critical Black Studies,'" on the other (211, 212). Pointing out that the many debates happening in this context do so across this "general interpolation" and that they follow numerous methods and develop different approaches, Sexton argues that twenty-first century African American or Black Studies is at a crossroads at which questions concerning antiblackness and a sustained critique of Black suffering should take center stage ("African American" 213 ff.). In thus charting the political, theoretical, aesthetic, and philosophical stakes of Black Studies in the first decade of the twenty-first century, Sexton writes:

> The problem of speaking from the standpoint of the slave in a slave society, or, *pace* Gordon, as a black in an anti-black world, has structured black critical discourse from its earliest moments of articulation—primarily in aesthetic production (from music and dance to visual arts and literature), but also in political rhetoric and philosophical and theoretical writing as well [...] This is also to say that African American Studies, an academic project catalyzed in the political ferment and crisis of the mid-century social movements, inherits this problem; *it is the hard kernel around which it continues to grow.* ("African American" 211; emphasis mine)

Sexton's words here delineate Black Studies as a radical intellectual project equipped with an analytic lens that attempts to account for Blackness in an antiblack world, both in terms of structure and performativity. On the one hand, Sexton here discusses Black Studies as an "intramural" project in Hortense Spillers' sense, that is, as a project "interested in articulating the complex networks – historical, cultural, literary networks – that have shaped, and continue to shape, the contours of" Black life (Woubshet 925; see also Spillers, *Black*, "Idea"). On the other hand, and this is particularly important for my study, his words also show that Black Studies represents an important intervention into white knowledge productions that helped create the modern Western world. Black Studies' interventions as exemplified by Sexton's above words remind us that to reckon with the making of the white Human subject is to fundamentally engage with modernity's calculus of property and practices of "propertization" (Broeck, "Abolish" 212).[44] As Wilderson puts it, "The race of Humanism [...] could not

---

[44] I borrow the term "propertization" from Broeck, who uses it to describe the "scandal of reducing a human being to property" ("Abolish" 212). It is part of a repertoire of terms and concepts that Broeck has coined in her work on (early) European modernity, transatlantic slavery, and structural antiblackness (*Gender*), and which draws on Black post-slavery thinking's interrogations of white Western modernity. As part of this effort, I understand the term "propertization" to denote the white economic, political, philosophical, and epistemic effort to not only "re-

have produced itself without the simultaneous production of that walking destruction which became known as the Black" (*Red* 20).

In order to understand how a novel like *A Mercy* interrogates the complicity of slavery and freedom in/as the liberal imagination, it is important to remind ourselves of the groundbreaking work Black feminists have done in addressing the ways in which concepts of kinship, reproduction, and family have historically and epistemically been enmeshed with enslavement, the market, and racial capitalism in the United States. Intensely focusing on "questions of race and maternity" (J. Morgan, "*Partus*" 2), Black feminist thinkers have shown how a conceptual bind between property, slavery, and reproduction was and continues to be central to conventional understandings of Human subjectivity. Ever since Angela Davis' work on resistance, family, and women in slave communities put slavery on the agenda of Black feminist theorizing at the beginning of the 1970s ("Reflections"), that is, Black feminist thinkers in the twentieth and twenty-first centuries have examined and continue to interrogate what Adrienne Davis has described as the "sexual economy" of slavery[45] (see "Don't," "Slavery," "Private"; see also the work of, for example, Hazel V. Carby; Marisa J. Fuentes; Sharon Harley and Rosalyn Terborg-Penn; Cheryl I. Harris ("Finding," "Markets"), Darlene Clark Hine ("Female," "Rape"); Jennifer L. Morgan ("*Partus*, "Archive," *Laboring*), Deborah Gray White; Hortense J. Spillers (*Black*, "Mama's Baby"), and Patricia J. Williams, respectively). By this route, Black feminist thinking critically supplements this study's arguments on the socio-cultural, epistemic, and structural connections between private property and subjectivity in that it establishes at least three interlocking sites of critical inquiry: First, the connections between property and/as reproduction, the master's willful, violent, and targeted use of Black enslaved women's reproductive capacity to guarantee his economic increase. Hartman has recently described this as the "theft, regulation and destruction of black women's sexual and reproductive capacities [...] In North America, the future of slavery depended upon black women's reproductive capacity as it did on the slave market. The reproduction of human property and the social relations of racial slavery were predicated upon the belly" ("Belly" 166, 168). Second, the impossibility of *recognized* kinship formations for the enslaved because kinship "*can be invaded at any given and arbitrary moment by the property relations*" structuring racial slavery (Spillers, "Mama's Baby" 74). Third,

---

duce a human being to property" but also to establish and maintain whiteness's trajectories of individual liberty and self-possession.

45 In ""Don't Let Nobody Bother Yo' Principle": The Sexual Economy of American Slavery," Adrienne Davis describes the "interplay of sex and markets" and the law on the New World plantation as a "sexual political economy" (105).

the regulation of maternal descent through colonial law, which made racial slavery hereditary (J. Morgan, "*Partus*," "Archives"). In this context, scholars like Patrice D. Douglass ("Claim"), Saidiya V. Hartman ("Belly," *Lose*, *Scenes*), and Christina E. Sharpe ("Lose," *Monstrous*), among others, have also pointed to the ways in which the sexualized and racialized landscapes of the institution of slavery continue to animate hegemonic discourses on Black family formations in the present. As Hartman writes, for example, African and African-descended enslaved women's reproductive labor "not only guaranteed slavery as an institutional process and secured the status of the enslaved, but it inaugurated a regime of racialized sexuality that continues to place black bodies at risk for sexual exploitations and abuse, gratuitous violence, incarceration, property, premature death, and state-sanctioned murder" ("Belly" 169). Put another way, reading and studying Black feminist historians and theorists' work has taught me that "slavery and freedom presuppose one another, not only as modes of production and discipline or through contiguous forms of subjection but as founding narratives of the liberal subject" (Hartman, *Scenes* 116); and that engaging with the position of Black enslaved women on the New World plantation is key not only to further understanding and complicating this white, self-possessing liberal subject as the current hegemonic formation or "genre" (Sylvia Wynter) of the Human but also for the kind of reading of Toni Morrison's *A Mercy* that this study pursues.

In the first chapter of *Scenes of Subjection*, Hartman discusses the relationship between white narration, empathy, and enslavement in turning to the antislavery writing of white abolitionist John Rankin. Rankin – intent on documenting the injustices of slavery and on bringing those closer to his readers in order to win them over for the abolitionist cause – literally imagined himself and his family in the position of the enslaved in an "epistle to his brother" (17). Hartman points to the "difficulty and slipperiness of empathy" in the context of a narrative practice, exemplified by Rankin's writing, that bases its rhetorical gist on reiterating as well as identifying with black suffering (18). Because "empathy is a projection of oneself into another in order to better understand the other," white narrative emphatic identification here becomes an instance of obliterating the enslaved from discourse, even though it is designed to create such things as visibility, solidarity, or human common ground (19). Hartman writes:

> [B]y exploiting the vulnerability of the captive body as a vessel for the uses, thoughts, and feelings of others, the humanity extended to the slave inadvertently confirms the expectations and desires definitive of the relations of chattel slavery. In other words, the ease of Rankin's empathic identification is as much due to his good intentions and heartfelt opposition to slavery as to the fungibility of the captive body. (*Scenes* 19)

Hartman's concerns about the (im)possibility of white emphatic narration/identification with Black suffering strongly resonate with Frank Wilderson's questioning of narrative and whether narrative can "account for the violence that wounds and positions Black people" ("Aporia" 134). Wilderson raises these questions in the context of and as key proponent of the expanding theoretical trajectory of Afropessimism, which, as I will discuss in more detail below, "elaborates a paradigmatic critique of the Human that reckons [sic] civil society's perverse and parasitic relation to the hydraulics of anti-Black violence" (134).[46] As "both an epistemological and an ethical project" (Sexton, "Afro-Pessimism" par. 15), that is, Afropessimism thinks Blackness as coterminous with and inextricably bound by Slaveness (Wilderson, personal conversation. 9 December 2018) in its analysis of "how anti-black fantasies attain objective value in the political and economic life of society and in the psychic life of culture as well" (Sexton, "Afro-Pessimism" par. 7). In "Social Death and Narrative Aporia in *12 Years a Slave*," Wilderson argues that "narrative strategies that try to account for the violence of Black life," regardless of their purpose, will inevitably fail to achieve what they set out to do because they lack the "requisite explanatory power to make sense of violent context and performances that are prelogical and, as such, beyond the grasp of narration" ("Aporia" 134). That is, the violence that slavery gratuitously inflicts on the enslaved (and by which slaves are ontologically positioned as socially dead) cancels out any "narrative moment prior to slavery" on a structural level (*Red* 29). Following Wilderson, narrative is always imbued with both a temporal and a spatial dimension—dimensions to which the slave does not have access:

> Social death bars the slave from access to narrative, at the level of temporality; but it also does so at the level of spatiality. […] just as there is no time for the slave, there is also no *place* of the slave. The slave's reference to his or her quarters as home does not change the fact that it is a spatial extension of the master's dominion. ("Aporia" 136)

Any attempt of emplotting the slave within narrative, then, ultimately results in their eradication, "regardless of whether it [is] a leftist narrative of political agency […] or whether it was about being able to unveil the slave's humanity by actually finding oneself in that position," as in Rankin's case (Hartman and Wilderson 184). Wilderson explains that the slave's social death "when storied, should

---

[46] Wilderson has since published his philosophical and lyrical memoir/theory book *Afropessimism* (Liveright 2020), in which he continues to elaborate on Afropessimism's core ideas and arguments. I follow his spelling whenever I refer to or draw on Afropessimist thought in this study.

not be seen as producing a logical impasse or contradiction within narrative but, rather, social death is the very meta-aporia that interrogates narrative as a form" ("Aporia" 135). In other words, rather than being a means or a structure that can account for the slave, narrative needs to be understood as being within the purview of the Human. As Rankin's emphatic narrative imagining so aptly illustrates, narrative needs to be understood as being part of the liberal Human subject's repertoire of being/becoming (as examined in the previous chapter).

If narrative plays such a fundamental role in the making of Human subjectivities, as Hartman and Wilderson suggest, then what does this mean for a project that intends to dismantle white conceptualizations of subjectivity and liberal self-making bound by various conceptions of ownership at the same time that it is heavily indebted to Black knowledge productions? Hartman's and Wilderson's fundamental questioning of narrative situate not only their own work but also, more generally, Black Studies' inquiry of "speaking from the standpoint of the slave in a slave society" (Sexton, "African American") as critical projects of anti-narration—if by narrative we mean a spatiotemporal structure, which is part of the Human fold and for which the social death of the slave presents a fundamental impasse. In this respect, I need to explicitly frame my reading of *A Mercy* as a project with its own theoretical interest. That is, reading *A Mercy* as a literary intervention into the discursive promises of universal liberty as/ and white knowledge production produces a theoretical argument that both contrasts with the existing discourse on this novel and gets into conversation with the current debates about the legacies of slavery within post-slavery critical thinking. Hartman's questioning of white narrative's potential to tell stories that do not reproduce the "violence of identification" (*Scenes* 20) pushes me to conceptualize this project as a critical reckoning with whiteness and its imaginations and narratives of liberal selfhood. Hartman's and Wilderson's respective works help me raise the following core questions for this study: (How) Does a novel like *A Mercy*, which is set in the seventeenth-century North American colonial mainland, and which offers representations of slavery, take up Hartman's and Wilderson's respective concerns about narrative as a form and structure? How does a novel like *A Mercy* navigate its own incapacity (*qua* its narrative form) to account for the slave while at the same time mounting a fundamental critique of early modern liberal subjectivities on the aesthetic level of representation? With which narrative strategies does the novel attempt to rupture self-possessed subjectivities while simultaneously exposing its own inability to fully contain fictional beings that are enslaved? How does *A Mercy* "disnarrate" (G. Prince) self-possessed liberal subjectivities by creating Black fictional beings that resist and refuse modernity's grammar of property?

With these questions in mind, the task of the following pages is to delineate and to think about a specific set of interconnected terms stemming from recent Black Studies' post-slavery trajectories that tackle the complex political, philosophical, social, and cultural connections between transatlantic enslavement, liberal individualism, and the making of modern western subjectivities. These terms will help me structure the study's coming chapters and shall serve as points of entry for my analyses of *A Mercy*. They are: *violence*; *dispossession* and *fungibility*; *abjection* and *abjectorship*; *reproduction* and *kinship*. While related, the final term discussed in the chapter – *anticipatory wake* – somewhat differs from the former ones in that I construct it from my reading of Black feminist theorizing. The task here is to discern and to work through the ways in which these terms confront the conceptual conflation of individual liberty and/as self-possession put forth on the seventeenth-century New World colonial scene in political philosophical discourse á la John Locke and which continue to be produced and reproduced in(to) the present. It is important to note at this point that these terms do not neatly fall into chronology. Rather than examining them separately, I have grouped them into thematic clusters. As I elaborate on these terms as theoretical sites of inquiry (following Bal, *Travelling* 44, "Cultural Studies" 35 – 36) there will be some conceptual overlap. I will go through them in the order that they are listed above. Throughout, I will refer to and give examples stemming from *A Mercy* to make visible the connections between the novel and the epistemic, political formations that the terms under scrutiny here describe.

3.3 The Structure of Violence

Let me turn, then, to the first term to be examined: *violence*.[47] Within the first few lines of Toni Morrison's *A Mercy* we as readers encounter the slave girl Florens and her "telling" about what the reader will later come to learn has been a vio-

---

[47] Calvin Warren has expanded on this by introducing the concept of "ontological terror" in his eponymous study *Ontological Terror: Blackness, Nihilism, and Emancipation* situated at the intersections of African American Studies and Philosophy. Warren uses this term in the context of what he has coined as the philosophical trajectory of Black Nihilism to think about "the ontological crisis blackness presents to an antiblack world" (ix). Against this backdrop, he meditates on the "(non)relation between blackness and Being by arguing that black being incarnates metaphysical nothing, the terror of metaphysics, in an antiblack world. Blacks, then, have function but not Being—the function of black(ness) is to give form to a terrifying formlessness (nothing). Being claims function as its property (all functions rely on Being, according to this logic, for philosophical presentation), but the aim of black nihilism is to expose the unbridgeable rift between Being and function for blackness" (5 – 6).

lent fight between her and the blacksmith. We read: "Don't be afraid. My telling can't hurt you in spite of what I have done and I promise to lie quietly in the dark – weeping perhaps or occasionally seeing the blood once more – but I will never again unfold my limbs to rise up and bare teeth" (*AM* 1). These two sentences are saturated with words like 'blood,' 'weeping,' 'hurt,' and 'teeth' and they evoke notions of pain, fear, and violence. Not only do they proleptically set the novel's plotting for 'the fight-to-come' between Florens and the blacksmith but they also establish violence as one of *A Mercy*'s core concerns. Indeed, there are various representations of violence in the novel, ranging from social "disorder" and mayhem fueled by religious conflict in England (73)[48] and the forceful and unwarranted extension of indenture (146–147) and peonage (52)[49]; over the brutal uprooting violence of settler colonialism and the annihilation of Native American tribes in the New World (42, 44–45, 47)[50]; sexual violence and subjection (91–92, 117–118)[51] as well as the entanglements between sexuality, slavery, and the market as in the "breeding"/"increase" of slave property through sex (163–164)[52]; the violence of racialization (108–113)[53]; and the brutality and

---

[48] "Rebekka was ashamed of her early fears and pretended she'd never had them. Now, lying in bed, her hands wrapped and bound against self-mutilation, her lips drawn back from her teeth, she turned her fate over to others and became prey to scenes of past disorder. The first hangings she saw in the square amid a happy crowd attending. She was probably two years old, and the death faces would have frightened her if the crowd had not mocked and enjoyed them so. [...] She did not know what a Fifth Monarchist was, then or now, but it was clear in her household that execution was a festivity as exciting as king's parade" (*AM* 73).

[49] "Sold for seven years to a Virginia planter, young Willard Bond expected to be freed at age twenty-one. But three years were added onto his term for infractions – theft and assault – and he was re-leased to a wheat farmer up north" (*AM* 146).

[50] "Once, long ago, had Lina been older or tutored in healing, she might have eased the pain of her family and all the others dying around her: on mats of rush, lapping at the lake's shore, curled in paths within the village and in the forest beyond, but most tearing at blankets they could neither abide nor abandon. Infants fell silent first, and even as their mothers heaped earth over their bones, they too were pouring sweat and limp as maize hair" (*AM* 44).

[51] "The housewife told her it was monthly blood; that all females suffered it and Sorrow believed her until the next month and the next and the next when it did not return. [...] [W]hether it was instead the result of the goings that took place behind the stack of clapboard, both brothers attending, instead of what the housewife said. Because the pain was outside between her legs, not inside where the housewife said was natural" (*AM* 118–119).

[52] "But the first mating, the taking of me [*minha mãe*, enslaved mother of Florens] and Bess and one other to the curing shed. Afterwards, the men who were told to break we in apologized. Later an overseer gave each of us an orange" (*AM* 163–164).

[53] "Eyes that do not recognize me [the slave girl] Florens, eyes that examine me for a tail, an extra teat, a man's whip between my legs. Wondering eyes that stare and decide if my navel is in the right place if my knees bend backward like the forelegs of a dogs. They want to see if my

loss at play when being orphaned (30)[54] to the unspeakable act of child murder (121).[55] [56] Rather than to think about this list simply as a panorama of various representations of violence addressed in the novel, however, I suggest that it is indicative of how violence operates on a different level in the text—namely, how violence defines and positions *A Mercy*'s characters both within and towards one another in the text.

In order to understand how *A Mercy* navigates this, it is important to keep in mind how Black post-slavery thinking has made a focus on the "forms and functions of violence" its central analytical objective (Sexton, "Afro-Pessimism" par. 33). For example, Hartman in *Scenes* examines forms of domination and violence that manifest themselves in seemingly "innocent" scenes of the everyday in the aftermath of the American Civil War and the legal abolition of slavery. Explicitly, her work needs to be understood as "a mediation on metaphysical violence that asks first under what conditions of existence can injury become legible" (Douglass and Wilderson 119). In a similar vein, Afropessimism has made violence and its positioning power on the level of structure one of its core interests. As pushed in the respective works by Frank B. Wilderson, III., and Jared Sexton, Afropessimism examines

> the hidden structure of violence that underwrites so many violent acts, whether spectacular or mundane. [...] [I]n its formulation of power, and particularly of the nature and role of violence, Afro-Pessimism does not only describe the operations of systems, structures and institutions, but also, and perhaps more importantly, the fantasies of murderous hatred and unlimited destruction, of sexual consumption and social availability that animate the realization of such violence. (Sexton, "Afro-Pessimism" par. 6, par. 7)

---

tongue is split like a snake's or if my teeth are filing to points to chew them up. To know if I can spring out of the darkness and bite. Inside I am shrinking" (*AM* 112–113).

54 "From his own childhood he [Jacob Vaark, the master] knew there was no good place in the world for waifs and whelps other than the generosity of strangers. Even if bartered, given away, apprenticed, sold, swapped, seduced, tricked for food, labored for shelter or stolen, they were less doomed under adult control. Even if they mattered less than a milch cow to a parent or master, without an adult they were more likely to freeze to death on stone steps, float facedown in canals, or wash up on banks and shoals" (*AM* 30).

55 "Sorrow's birthing came too soon, Lina told her, for the infant to survive, but Mistress delivered a fat boy who cheered everybody up—for six months anyway. [...] Although Sorrow thought she saw her own newborn yawn, Lina wrapped it in a piece of sacking and set it a-sail in the widest part of the steam and far beyond the beaver's dam. [...] [I]t took years for Sorrow's steady thoughts of her baby breathing water under Lina's palm to recede" (*AM* 121).

56 This is, of course, a recurrent theme in Morrison's oeuvre, for example in *Beloved* when "Sethe commits the horrible act of killing her child in order to save her from a certain emotional and possibly physical death at the hands of slave-holders" (Raynor and Butler 181) or when Eva, in *Sula*, pours kerosene over the sleeping, dilapidated, heroin-addicted grandson-veteran Plum.

Afropessimism brings to the scene of critical inquiry a set of questions, which opens analytical ground for a radical re-focusing on violence as that which structures U.S. civil society, as that which has positioning power on the level of ontology. In *Red, White, and Black: Cinema and the Structure of U.S. Antagonisms*, Wilderson pushes for a structural analysis of civil society in the United States that goes beyond the emancipatory narratives that Leftist scholarship in the form of, e.g., Marxism, white feminism, or postcolonial studies has crafted in the wake of the liberation movements of the 1960s and 1970s in that it argues for a radical return "to think the vagaries of power through the generic positions within a *structure of power relations*" (Wilderson, *Red* x, 6; emphasis mine). [57] Wilderson argues that there are "the three structuring positions of the United States (Whites, Indians, Blacks)," all of which are situated in the world by different "grammars of ontological suffering," and that these three grammars are "predicated on fundamental, though fundamentally different, relationships to violence" (*Red* 29). Global in scale (2–3), this analysis calls for a paradigmatic framework which can elaborate on how violence structurally determines who is part of civil society and who is not. Following Wilderson, the Black positionality is cast *outside* the realm of civil society whereas the white and Red positionalities are fundamentally part of it. Always-already positioned as Slave (7), the Black positionality is constituted historically and ontologically by the excessive violence incited by the transatlantic slave trade and chattel slavery, which turned human bodies into (tradable) "flesh" (11; Wilderson borrows this term from Spillers, "Mama's Baby"). This grammar of suffering makes the Black position incommensurable not only with the white positionality but also with the structural po-

---

[57] Afropessimism largely builds on the work of such critics and intellectuals as Frantz Fanon ("Black Skin," "Wretched"), David Marriott, Saidiya Hartman (*Scenes*), Hortense Spillers ("Mama's Baby"), and Orlando Patterson. As a "contemporary phenomenon, some may even scoff that it is trendy, [Afropessimism's] political and intellectual evolution is considerably longer and its ethical bearings much broader than one might expect, and there is work yet to be done regarding a genealogy of its orientation and sensibility" (Sexton, "Afro-Pessimism" par. 1). This work is ongoing, as, for instance, in a 2016 issue entitled "Black Holes: Afro-Pessimism, Blackness, and the Discourses of Modernity" of the journal *rhizomes* featuring essays and interviews by, for example, Jared Sexton, Selamawit Terrefe and Christina Sharpe, Jaye Austin Williams and Frank B. Wilderson, Sabine Broeck, and Parisa Vaziri; or in a 2018 special issue of the journal *Theory & Event* edited by Tiffany Willoughby-Herard and M. Shadee Malaklou. Weier ("Consider") provides an excellent overview of Afropessimism's core concerns and arguments, especially of Wilderson's work. He suggests that Afropessimism's radical interventions into post-racial discursive formations in the United States and beyond also critically supplement previous critiques of post-racialist discourse and white supremacist thinking offered by scholars working in the field of American Studies in Germany (e.g., Berg; Knopf).

sitions of marginalized groups such as white women or "colored immigrants," whom Wilderson controversially calls the "junior partners" of civil society (*Red* 38). In this framework, the Black/Slave's position is one of gratuitous violence and incoherence, a state of being that renders the Black/Slave an "anti-Human, a position against which Humanity establishes, maintains, and renews its coherence" (11). Put differently, the relation between the structural positionalities of the white/Human, the Red/"Savage," and the Black/Slave is based on an ontological split or antagonism induced by the violence that the modern Western histories of enslavement and its economies of subjection brought about: "[T]his violence is peculiar in that, whereas some groups of people might be the recipients of violence, after they have been constituted as people, violence is a structural necessity to the constitution of blacks" (Douglass and Wilderson 117).

From an Afropessimist perspective, there is a difference not only between how positionality determines one's relationship to violence but also between two different forms of violence. Violence needs to be understood as being either *contingent* or *gratuitous*. While violence is *contingent* when it "happens because people transgress the unethical rules of civil society," *gratuitous* violence, by contrast, is that which "produces the 'inside-outside' of civil society" (Wilderson, Spatzek, and von Gleich 14, 15). *Gratuitous* violence is one of the three constituent elements of slavery; it is that which allows for the other two elements of social death, namely natal alienation and general dishonor (16). *Gratuitous* violence

> turns a body into flesh, ripped apart literally and imaginatively, destroys the possibility of ontology because it positions the Black in an infinite and indeterminately horrifying and open vulnerability, an object made available (which is to say fungible) for any subject. (Wilderson, *Red* 38)

It is this violence, in other words, which determines the status of the Black as a sentient nonbeing or as "existential negation" (Jackson, "Waking" 358). Wilderson goes on to tell us that *gratuitous* violence, as that which the slave receives, is "a kind of violence that is necessary not to produce a certain kind of behavior, but to give the other people who are not receiving this gratuitous violence a sense of stability in their own lives" (Wilderson, Spatzek, and von Gleich 14). Thus, while the Black is always "open gratuitous violence" and thus marked by it ontologically, the white positionality precisely is elaborated by "the freedom from violence's gratuitousness" (Wilderson, *Red* 31, 25, 20). Thirdly, the Red positionality is elaborated by its liminal status in civil society's political economy as being in between the white and the Black positions in that it "shuttles between the incapacity of a genocided object and the capacity of a sovereign subject[.] [...] [T]he Indian comes into being and is positioned by an a priori violence of genocide" (*Red* 49).[58] However, even though the violence of genocide separates Redness from the fold of whiteness, it does not fully isolate Redness from civil society on a structural level; the Red position can still submit claims to sovereignty (49–51). Working through the ways in which these different forms of violence (*gratuitous* v. *contingent*) position Blackness as well as Redness and whiteness against as well as alongside one another in political economy, violence's function(s) become apparent. That is, the relation between these positions – the Black v. the Red and the white positionalities – "demarcates antagonisms and not conflicts because [...] they are the embodiments of opposing and irreconcilable principles or forces that hold out no hope for dialectical synthesis" (29).

This chasm – a non-relation of antagonism versus a relation of conflict in Afropessimist terms – also plays itself out in the relationships between the characters in *A Mercy*, as I argue. Even though most characters in the novel's representation of the New World appear to be dispossessed, oppressed, or subjugated in similar ways, an analytical focus on violence, as Douglass and Wilderson have it in the third epigraph to the chapter, makes visible the ways in which this is in fact not the case. Consider, for example, Rebekka Vaark, the mail-order-bride of the settler and trader Jacob Vaark, as well as the Indigenous woman Lina and the enslaved girl Florens. The plot appears to suggest that all of these female figures suffer from early modern hetero-patriarchal power relations at the scene of Jacob Vaark's New World farm in very similar ways. However, a fundamentally different impression conveys itself if we examine their existence in the text with a focus on the hidden metaphysical and material structures, strategies, and systems produced by the violence of European economic and colonial expansion (as represented by Vaark). While Rebekka escapes her old life in England, which holds out few prospects to her other than "servant, prostitute, [or] wife," to begin a new life as Vaark's wife, Lina needs to build herself anew as the Vaarks' humble servant after having experienced the genocide of her tribe (*AM* 75–76, 43–47). A close relationship develops between Rebekka and Lina through their shared experience of trying to survive and run the Vaark farm in the wilderness of colonial Virginia (71–73). Florens, in turn, enters the household as the currency of one of Vaark's business transactions. While Vaark initial-

---

[58] Wilderson has since elaborated on this when he writes that "[i]n some ways, American Indians are a liminal category, and in other ways they are more profoundly on the side of 'junior partners' and antagonistic to Blacks" (Wilderson, Spatzek, and von Gleich 14).

ly considers her to be some sort of a replacement for one of their deceased children (24); and while Lina "adopts" her as something like her surrogate daughter/sister/friend who "assuaged the tiny yet eternal yearning for the home Lina once knew where everyone had anything and no one had everything" (58), the text clearly delineates Florens' place within the household as that of a slave. When the household collapses in the wake of Jacob Vaark's sudden death, Rebekka becomes the sole mistress of the other women on the Vaark farm, and the friendship between her and Lina dissolves. While Lina stays on to nurse her sickened mistress back to health, Rebekka sends Florens on an errand to fetch the blacksmith for help, equipping her with a letter in which she makes her claims of rightful ownership of Florens clear. Upon her recovery, Rebekka is free—free to marry again, for example. Her behavior towards Lina fundamentally changes: "She requires her company on the way to church but sits her by the road in all weather because she cannot enter. Lina can no longer bathe in the river and must cultivate alone" (158). She also makes plans to sell Florens (157). Far from being subjugated or dispossessed in similar ways, then, an analytical focus on violence will show how the relationships between these women are in fact structured by this violence even though they initially believe themselves to be, in Lina's words, a "tight-knit family" (56). As theorized in post-slavery discourses, then, violence can be deployed as an analytical tool that speaks to Atlantic slavery's fundamental, world-creating and world-destructing (to paraphrase Hartman, "Belly") powers. It introduces a relation of capitalism to African slaves, perpetuated inheritable property, white self-making, and conceptions of freedom and emancipation that encompass some but not others. What is at stake, in other words, is a reckoning with these structures, strategies, and systems of terror, violence, and death, on the one hand, and self-making, emancipation, and universal freedom, on the other. As will be discussed in the next part of the study, such an analytical focus also raises questions as to what this violence does to a literary text's strategies of characterization.

3.4 Dispossession and Fungibility

Dispossession and *fungibility* are terms relevant to my analyses of *A Mercy* because they delineate fundamentally different states of being and existence in the world. As I hope to show, it is their fundamentally different "assumptive logics" (Wilderson, *Red*) that will help me unpack and conceptualize the novel's strategies of characterization and, thus, its critique of the property paradigm. In its conventional definition, *dispossession* refers to the "action of dispossessing or fact of being dispossessed; deprivation of or ejection from a possession." This definition also includes meanings of (dis-)inheritance, birthright, and damage to reputation or personal injury caused by dispossession ("dispossession, n."). *Dispossession* is also defined as "the condition of those who have lost land, citizenship, property, and a broader belonging to the world" (see back blurb of Butler and Athanasiou's book). Within current academic, post-1960s social justice movements and political left-leaning discourses equipped with anti-capitalist and emancipatory rhetoric, the term *dispossession* largely functions as a theoretical trope. As such, it refers to a subject that is exposed to various forms of injustice and that is subjected to vulnerability in the context of neoliberal capitalism's global market economies (Butler and Athanasiou 2; cf. Bhandar and Bhandar). For example, in their book-length conversation on *Dispossession: The Performative in the Political* (2013), Judith Butler and Athena Athanasiou discuss *dispossession* as a useful tool with which to think about how "human bodies become materialized and dematerialized through histories of slavery, colonization, apartheid, capitalist alienation, immigration and asylum politics, postcolonial liberal multiculturalism, gender and sexual normativity, securitarian governmentality, and humanitarian reason" (10). For Butler and Athanasiou, *dispossession* signals "the contemporary production of social discourses, modes of power, and subjects [and it] works as an authoritative and often paternalistic apparatus of controlling and appropriating the spatiality, mobility, affectivity, potentiality, and relationality of (neo-)colonized subjects" (6, 11).

What these meanings share is the assumption of a *subject* that "has been deprived of something that rightfully belongs to them" (Butler and Athanasiou 6). That is, the above leftist political discourse on dispossession is structured around a set of questions concerned with Human subjects and "their capacities: powers subjects have or lack" (Wilderson, *Red* 8). As Black post-slavery thinkers, philosophers, and scholars have shown, however, the subject of dispossession is cast in a fundamentally different light, if examined against the backdrop of the histories and afterlives of transatlantic slavery and antiblackness in its longue durée (e. g., Sexton, "Social Life"; Walcott). *Dispossession* (as a term and as a theoretical trope) cannot account for the ways in which social death determines and structures the lives of the slaves as well as of their descendants, past and present (Sexton, "Afro-Pessimism"; Wilderson, *Red*). In drawing on this body of scholarship, Broeck in *Gender and the Abjection of Blackness* (2018) reads Butler and Athanasiou's use of *dispossession* in the context of their "post-Marxist, post-feminist meditation on the possibilities and exigencies of struggle against late neoliberalism's global production of permanent crisis for human life" (202) against the theoretical and literary interventions of Black feminism in general and, in particular, against Black feminist critique of gender (e.g., Douglass, "Black Feminist"; Hartman, *Scenes*; Spillers, "Mama's Baby"). Such interventions have aptly shown that to be a slave is to be absented from the purview of the Human and thus from "the possibility of subjectivity altogether" (Broeck, *Gender* 204).[59] Broeck writes:

> Being property cancels out dispossession; thus, if one speaks of dispossession as a general condition, one cannot speak of the (post-)slave. Where there has been no possession of self, let alone of property, of land, or other things, structurally speaking, but instead, propertied thingness of being, there can be no dispossession [...] the prefix tellingly assumes, at the very least, a metaphysical disposition to see the human [...] in possession, of his life, of herself, of a cartography of humanness, however embattled, if not altogether in possession of other property. (*Gender* 204, 207)

In light of this, *dispossession* as used by Butler and Athanasiou and, by extension, other intellectual projects similarly invested in accounting for "embattled humanness" needs to be understood as a term that can *only* account for the status of *subjects*. *Dispossession* cannot account for those who are positioned outside of the realm of the law and rights discourse and their epistemologies or, more generally, outside of civil society; it cannot account for the grammar of suffering of the enslaved. If a term like *dispossession* cannot do so, which term or terms can?

The Dispossessed Body, or the Fungibility of the Commodity

Let me turn to the Black feminist work of Saidiya Hartman and specifically to her seminal 1997 monograph *Scenes of Subjection: Terror, Slavery, and Self-Making in Nineteenth-Century America*. *Scenes* is Hartman's "provocative [...] exploration of racial subjugation during slavery and its aftermath" in the Reconstruction era, in which she focuses on the entanglements of slavery, subject-making, pleasure, and terror in the mundane scenes of the everyday (back blurb of *Scenes*). As part of her analyses, Hartman introduces a new term with the help of which she attempts to adequately talk about the enslaved's position of social death, racial subjugation, domination, and commodification, both on an analytic and semantic level. Hartman writes:

> The relation between pleasure and the possession of slave property, in both the figurative and the literal senses, can be explained in part by *the fungibility of the slave* – that is, *the joy made possible by virtue of the replaceability and interchangeability endemic to the commod-*

---

[59] This is part of the critical reckoning with and the anti-racist critique of the field of white gender studies that her book pushes towards.

> ity – and by the extensive capacities of property—that is, the augmentation of the master subject through his embodiment in external objects and persons. [...] [T]he *fungibility of the commodity* makes the captive body an abstract and empty vessel vulnerable to the projection of others' feelings, ideas, desires, and values; and, as property, the *dispossessed body* of the enslaved is the surrogate for the master's body since it guaranteed his disembodied universality and acts as the sign of his power and dominion. Thus, while the beaten and mutilated body presumably establishes the brute materiality of existence, the materiality of suffering regularly eludes (re)cognition by virtue of the body's being replaced by other signs of value, as well as other bodies. (*Scenes* 21; emphasis mine)

Fungibility delineates the slave's status as the exchangeable possession of their master, the uses and enjoyments of which are "essential aspects of property" (Harris, "Whiteness" 1734). As the usable commodity of the master, the enslaved may be replaced or substituted at any given moment and it is this fungible status, which caters to the master's needs (economic, personal) and brings them "joy" (Hartman, *Scenes* 21). The fungibility of the slave commodity thus serves the purpose of establishing and maintaining, as well as expanding the master's subjective coherence, for it is "through chattel slavery [that] the world gave birth and coherence to both its joys of domesticity and to its struggles of political discontent" (Wilderson, *Red* 20). Hartman here offers a powerful counterpoint to the above uses of *dispossession* by inserting *fungibility* to our critical lexicons. *Fungibility* draws analytical attention on the *function* of Blackness within the white liberal imagination because it emphasizes the exchangeability and usability of the slave commodity for the master. As Hartman explains, "the value of blackness resided in its metaphorical aptitude, whether literally understood as the fungibility of the commodity or understood as the imaginative surface upon which the master and the nation came to understand themselves" (*Scenes* 7).

However, in the above quoted paragraph there also seems to be a tension at work between, on the one hand, the term *fungibility* and, on the other, the words *dispossessed body*, both of which Hartman uses when referring to the bodies of the enslaved as an extension of or proxy for the master's body. If *fungibility* connotes the enslaved's status as commodity and as an 'abstract and empty vessel vulnerable to the projection of others' feelings, ideas, desires, and values'; if *fungibility*, in other words, connotes the absence of subjectivity for the slave, then the words *dispossessed body* evoke a different theoretical register, namely a Human(ist) analytical lens. So why use these differently theoretically accentuated and dissonant terms in the same semantic environment? What I want to suggest is that, rather than simply being a dissonance in her argument, Hartman pushes terms and tropes like *dispossession* to a point where what she talks about is *not* the subject's deprivation of something but the violent absenting of subjectivity subtended by chattel slavery. Hartman does not use *dispossession* in the above sense of a Human subject which is subjugated, dominated, and deprived of selfhood but, rather, the words *dispossessed body* signify the human "flesh" that the master owns (Spillers). They therefore do *not* delineate a dispossessed subject but a "sentient being" (Hartman, *Scenes* 93). [60] On a slightly different note, this tension also addresses and is testimony to the fact that language will only ever be an approximation of, only ever be a substitute for that which it seeks to signify. Hartman elsewhere not only questions the possibility of unearthing something from the archive of slavery that has not been fashioned from the viewpoint of the masters, captors, and traders but she also explains that "writing is unable to exceed the limits of the sayable dictated by the archive" ("Venus" 12). And archive here not only refers to the archive of the history of slavery but also to the epistemologies that slavery brought about. The question Hartman brings up in this context – "[H]ow does one tell impossible stories?" ("Venus" 10) – thus feeds directly back into the tension at work between her uses of terms like *fungibility* and *dispossessed body*, which, as I have suggested, speak to different epistemic as well as analytical registers. Ultimately, I understand *fungibility* to emerge as an analytical tool for thinking about *who* can actually be dispossessed. It shows that the register of *dispossession* cannot account for the enslaved on the level of theory. *Fungibility* introduces and locates the slave's "brute materiality of existence"; by bringing slavery's calculus of property to the scene of theoretical inquiry, this term critically subverts most left-leaning critical discourse's implicit assumptions of subjectivity.

Accumulation and Fungibility

Afropessimism, as a theoretical trajectory that offers an analysis of the violence produced and subtended by transatlantic regimes and practices of slavery as

---

60 Hartman continues to criticize and to play with Human(ist) analytical registers throughout her work. In her more recently published article "The Belly of the World," a meditation on the histories of Black women's (reproductive) labors, Hartman continues to use both terms in her discussion of the connections between reproduction, kinship, and economic concerns both on the New World plantation and within the domestic realm in slavery's aftermath. Hartman writes, for instance, "To be a slave is to be 'excluded from the prerogatives of birth.' The mother's only claim—to transfer her *dispossession* to the child. [...] For the enslaved, reproduction does not ensure any future other than that of *dispossession* nor guarantee anything other than the replication of racialized and disposable persons or 'human increase' (expanded property-holdings) for the master" (166, 168; emphasis mine).

that which structurally "underwrites the modern world's capacity to think, act, and exist spatially and temporally" (Wilderson, *Red* 2), both takes up and adds to Hartman's respective uses of the terms *fungibility* and *dispossessed body*. Again, Afropessimism's unflinching analysis of antiblackness as U.S. civil society's fundamental structuring "arithmetic" (Hartman, *Lose* 6) proposes "a critique of (post-)modernity's theorization of the subject whose claims within civil society are based on a supposed possession of the self and right thereto" (Weier, "Consider 421).[61] Those claims, Weier reminds us, are "constitutionally opposed to the literal possession of the slave or prison inmate as commodity and chattel and the structural *de facto* (if not always *de jure*) exclusion of blacks from that same civil society" ("Consider" 421). Afropessimism argues that Western modernity's theorization of the subject falls short in the face of the "Black; a subject who is always already positioned as Slave" and "whose structure of dispossession (the constituent elements of his or her [sic] loss and suffering" cannot be accounted for through a discourse organized around claims of or rights to self-possession (Wilderson, *Red* 7). That is, Afropessimism contends that a Black subject does not exist, if by subject we mean critical theory's subject and its status "as a relational being" (Douglass and Wilderson 117).[62] This conceptualization of Blackness as Slaveness fundamentally draws on Hartman's term *fungibility*. Following Wilderson, slavery "is and connotes an ontological status for Blackness" (*Red* 14). In his endeavor to provide a paradigmatic *structural* analysis that explains the Black/Slave's position in the modern world, that is, Wilderson goes on to theorize the "constituent elements of slavery [as] [...] *accumulation and fungibility* [...] the condition of being owned and traded" (*Red* 14, emphasis mine).

---

[61] Again, in arguing for a shift from "a politics of culture(s)" towards "a culture of politics," Wilderson offers a paradigmatic analysis of civil society along a "triangulation of antagonisms" between the Red/Indigenous, the white/Human, and the Black/Slave positionalities (*Red* 26, 53). In thus shifting the analytical focus away from the level of performativity and experience, these structural positions constitute "the embodiments of opposing and irreconcilable principles or forces" inside and outside of civil society (*Red* 29; Wilderson, "Prison Slave" 20; Wilderson, Spatzek, and von Gleich).

[62] In the introduction to Oxford University Press's *A Dictionary of Critical Theory*, Ian Buchanan loosely describes critical theory as follows: "I suspect critical theory has leakier borders than most disciplines, not least because at its origins it is a hybrid of history, philosophy, psychoanalysis, and sociology" (vii). With Afropessimism's interventions in mind, I use critical theory here to delineate Marxist/materialist thought and the (post)structuralist currents that came to dominate most of cultural and literary studies after the 1970s, including white feminist and postcolonial thinking. Here, critical theory assumes its subject to live, stand, act, or suffer in relation to other subjects (Wilderson, *Afropessimism*, *Red*).

These thinkers thus make similar points when it comes to thinking about the status of Blackness, "the world-making and world-breaking capacities of racial slavery" (Hartman, "Belly" 166), and how this can be accounted for on the level of theory. They show that the explanatory power of theoretical tropes/terms like *dispossession* falters in the face of Blackness; it falters because such tropes/terms assume a subject/relational being that has been deprived of such things as civil rights, propriety, livelihood, or sovereignty. This also means that such terms assume a "before," a moment *prior to* the moment in which one becomes dispossessed. It assumes a moment in time during which there existed possession (of self and rights), as brief as that moment may have been. It also suggests that there is the possibility of a future in which possession – of self, of rights, of coherence – may be restored.[63] Interventions such as those by Hartman and Wilderson show, however, that there will never be such a restorative moment for Blackness (as Slaveness) because "there was never a prior meta-moment of plenitude, never Equilibrium: never a moment of social life" (Wilderson, "Aporia" 139). As analytical categories and terms, *dispossession* and *fungibility* offer me a way to think about how *A Mercy*'s characters are positioned towards one another and how the structure of their positionality actually undermines any sense of "[companionship] they had carved [...] out of isolation" at the colonial scene of the novel's setting in seventeenth-century Virginia (*AM* 154). Of course, this connects back to my earlier point that even though most of the novel's characters appear to be dispossessed or subjugated in similar ways – because they are someone else's servant or because they are subjugated by hetero-patriarchal power formations, for instance – they are in fact structurally positioned by different regimes of violence.

3.5 Abjection and Abjectorship

In order to speak about the slave not as a dispossessed subject but as fungible being, post-slavery thinkers have introduced terms like "flesh" (Spillers, "Mama's Baby"), "object of property" (P. Williams), "sentient being" (Wilderson, *Red*; Hartman, *Scenes*), or "equipment in human form" and "merchandise" (C. Warren). In the context of discussing post-slavery thinkers' interventions into the discursive promises of universal liberty within American Studies in Germany,

---

[63] I thank Taija Mars McDougall (UC Irvine) for drawing my attention to this. See also generally, for example, her piece "'The Water is Waiting': Water, Tidalectics, and Materiality."

Broeck has done work on the terms *abjection* and *abjectorship* (*Gender*).[64] I discuss these terms here because they help me further unpack how *A Mercy* stages different states of being and existence in the world in relation to slavery's positioning power. These terms address the white subject that creates itself over and against the enslaved. They delineate a "system of *black abjection* and of *white abjectorship*" as being central to the making of white Western modernity ("Abolish" 214). Alienating the term "abjection" from its psychoanalytic and white feminist deployments as "a category descriptive of individual subjectivity and its contours," as for instance in the work of Julia Kristeva, Broeck thus uses the term *abject* as a "theoretical concept to discuss the underside of [white] Western modernity's terms of human *sociability*" ("Abolish" 215). Broeck claims:

> I am not, however, interested in a quasi-ethnographical description of Blackness as abject; quite to the contrary, I mobilize the term abjection here – pace Kristeva, and encouraged by Hartman's historicizing of the term – to speak of abjection as a white practice of subjectivity, as that which renders Black being abject in order to be. Thus, my interest is in the work of abjecting, which remakes white supremacy, and anti-Blackness on a daily basis, individually and collectively. I argue that the splitting off of the enslaved Black and of enslavism itself from the symbolic order was an act of successful externalization of allowing the white Western subject to engage with internal objects-to-be-subjects (as in gender struggle) but literally leaving the abjected *outside itself* and its parameters of subjectivation. (*Gender* 84)

Broeck's conceptualization of *abjection* and *abjectorship* as white practice of self-making helps me address what Hartman demarcates as the "cleavage or sundering as object of property, pained flesh, and unlawful agent situat[ing] the enslaved in an indefinite and paradoxical relation to the normative category person" (*Scenes* 56). This is especially pertinent to my readings of Jacob Vaark and Rebekka in *A Mercy*, in which I suggest that both these white characters make their respective claims to freedom vis-à-vis their slave property. For them, regimes of ownership become part of 'abjection as a white practice of subjectivity, as that which renders Black being abject in order to be.' For Jacob Vaark, this means that his freedom to opt for and invest in the West Indian sugar economies comes as a necessary step in the creation of his liberal self. With Rebekka becoming the sole owner and mistress of the slave and servant women on the Vaark farm after her husband's untimely death, this means that her struggle for subjectivity and, ultimately, her claim to white female co-mastery at the New World colonial scene are likewise bound by property. Put another

---

[64] For a discussion and theorization of the relation between *abjection* and Blackness in African American literature which foregrounds often neglected depictions of the sexual exploitation and humiliation of men, see D. Scott.

way, (co-)ownership of land, servants, and slaves, as well as ownership of herself open up an avenue towards the fold of the liberal Human and, as I will elaborate in my reading of her character, establish for this woman capacity for choice —the choice to opt for a certain way of existence in the first place (see Chapter 4.3).

3.6 Reproduction, Kinship

> Slaveowners in the early English colonies depended upon and exploited African women. They required women's physical labors in order to reap the profits of the colonies and they required women's symbolic value in order to make sense of racial slavery. Women were enslaved in large numbers, they performed critical hard labor, and they served an essential ideological function. Slaveowners appropriated their reproductive lives by claiming children as property, by rewriting centuries-old European laws of descent, and by defining a biologically driven perpetual racial slavery through the real and imaginary reproductive potential of women whose 'blackness' was produced by and produced their enslavability.
> — Jennifer L. Morgan, *Laboring Women*

> *The slave ship is a womb/abyss.* The plantation is the belly of the world. *Partus sequitur ventrem*—the child follows the belly. The master dreams of future increase. The modern world follows the belly. Gestational language has been key to describing the world-making and world-breaking capacities of racial slavery.
> — Saidiya V. Hartman, "The Belly of the World"

In *A Mercy*, the white Anglo-Dutch settler, farmer, and moneylender Jacob Vaark considers expanding his business activities and consequently resolves to invest in rum. This decision is provoked by his encounter with one of his clients, a Portuguese slave trader called Senhor D'Ortega. Vaark marvels at this man's economic buoyancy as well as at the opulence of his estate. On his return from their meeting, Vaark enters into conversation with the rum investor Peter Downes, who "[b]urly, pock-faced [...] had the aura of a man who had been in exotic places and the eyes of someone unaccustomed to looking at things close to this face" (*AM* 27). While Downes entertains his company with "mesmerizing tales ending with a hilarious description of the size of the women's breasts in Barbados," Vaark slowly arrives at his decision to invest in "kill-devil"/rum (28, 27). When Vaark responds to Downes's tales by questioning the sustainability of the sugar business in the Caribbean, the latter responds, "They ship in more. Like firewood, what burns to ash is refueled. And don't forget, there are births. [...] As long as the fuel is replenished, vats simmer and money heaps. Kill-devil, sugar—there will never be enough. A trade for lifetimes to come" (28–29). Morrison's text here explicitly references slavery's "sexual economies"

(Adrienne Davis, "Don't"). On the one hand, phrases like 'as long as the fuel is replenished' and 'what burns to ash is refueled' evoke Barbados's thriving seventeenth-century sugar industries driven by constantly replaced slave labor forces (Beckles; E. Williams). On the other, they strongly resonate with the hyper-sexualized imagery of African and African-descended women produced by European colonial discourses (Hulme, "Spontaneous," *Colonial Encounters*; J. Morgan, *Laboring* 12–49; Mackenthun)—imagery that *A Mercy* evokes by way of the rum investor's tales about the size of Black women's breasts. Together with Downes's insistent reminder that 'there are births,' these phrases speak to the importance of reproduction for the emerging capitalist marketplace in the New World, both in the Caribbean and on the North American mainland (see also Fuentes). As Jennifer L. Morgan reminds us in this context: "Women's lives under slavery in the Americas *always included* the possibilities of their wombs. Whether laboring among sugar cane, coffee bushes, or rice swamps, the cost-benefit calculations of colonial slaveowners included the speculative value of a reproducing labor force" (*Laboring* 3, emphasis mine).

In this section, I turn to *reproduction* and *kinship* as the next set of terms relevant for my analyses of Morrison's novel. In my thinking about these terms, I generally follow Alys E. Weinbaum's conceptualization of the race/reproduction bind.[65] With this conceptual frame, Weinbaum delineates an ideological constellation emerging from "transatlantic modernity's central intellectual and political formations," in which "competing understandings of reproduction as a biological, sexual, and racialized processes became central to the organization of knowledge about nations, modern subjects, and the flow of capital, bodies, babies, and ideas within and across national borders" (Weinbaum, *Wayward* 6, 2; see also Weinbaum, *The Afterlife*). As previously suggested, Black feminist thinkers and scholars of slavery continue to research the connections between the capitalist market, property, race, and sexuality. In what follows, I draw on their invaluable work on slavery's sexual economies and the "uncertainty of descent, the negation of paternity, the interdiction regarding the master-father's name, and the ambiguous legacy of inheritance" and abjection that these economies engendered (Hartman, *Scenes* 76).

---

[65] I follow J. Morgan in this respect ("*Partus*"; "Considering").

Partus Sequitur Ventrem: Reproduction, Slavery, Property

> Whereas some doubts have arisen whether children got by any Englishman upon a negro woman shall be slave or free, Be it therefore enacted and declared by this present grand assembly, that all children borne in this country shall be held bond or free only according to the condition of the mother—*Partus Sequitur Ventrem*. And that if any Christian shall commit fornication with a negro man or woman, hee or shee doe offending shall pay double the fines imposed by the former act.
> — Laws of Virginia, 1662 ACT XII

The above epigraph constitutes the first act with which the colony of Virginia legally codified and enshrined the complex intersections between slavery, reproduction, and property on the North American colonial mainland (qtd. in J. Morgan, "*Partus*" 1). Passed in 1662, the *Partus Sequitur Ventrem* act would tether notions of reproduction to questions of race, status, heredity, and descent. As such, the "American 'innovation'" of *Partus* would change genealogies of recognized family and kinship structures in North America in the *longue durée*, putting forth still hegemonic normative conceptions of white kinship and family formations vis-á-vis the so-called "pathology of the black family, rather than the necropolitics of slave life" (Nyong'o, "Barack").[66] Following J. Morgan's work on the histories of racial capitalism, gender, and reproduction in colonial slavery ("*Partus*," "Archives," "Considering," *Laboring*), the *Partus* ruling needs to be understood as a paradigmatic articulation of the ways in which the institution of slavery both depended on and maintained itself through a "reproductive calculus" bound by "a notion of heritability" (Hartman, "Belly" 169; J. Morgan, "*Partus*" 1).[67] Atlantic slavery, Morgan explains,

> relied on a reproductive logic that was inseparable from the explanatory power of race. [...] Building a system of racial slavery on the notion of heritability [...] did require a clear un-

---

[66] I am here of course referring to the notorious Moynihan Report, which tried to "explain racial subjugation in America by means of the supposedly inverted gender hierarchy in African American culture produced by chattel slavery" (Nyong'o, "Barack"). See also Spillers's seminal critique of this report and the symbolic order/American grammar it purports in "Mama's Baby, Papa's Maybe: An American Grammar Book."

[67] On a more general note, it is important to recall the constraints many historians and scholars of slavery face when working with hegemonic archives. Morgan, like others before and after her, talks about her "relationship with the archive [as] always one of struggle and frustration" given those constraints (J. Morgan, "Archives" 154). For Morgan, the problem she faces in her work is "not to weed through an overabundance of sources but to endure the absences, erasures, and mischaracterizations of racialized subjects" ("Archives" 155). I will return to this and similar notions when I discuss *anticipatory wake* as my final term of this chapter.

derstanding that enslaved women gave birth to enslaved children. Resituating heritability was key in the practice of an enslavement that systematically alienated the enslaved from their kin and their lineage. ("*Partus*" 1)

Partus sequitur ventrem is "the first explicit English articulation of hereditary slavery" ("Archives" 158). At stake here are the ways in which the law inaugurated a new symbolic order of gender and race that would regulate descent, heritability, and status through *maternal* instead of paternal lineage ("*Partus*"; see also e.g. Spillers, "Mama's Baby"; Nyong'o, "Barack"). In general, "[l]aws concerning slavery in the English Atlantic were not transposed from England but were an amalgam of legal borrowing and commonly held assumptions about who would be enslaveable and from whence the legal right to property in persons originated" (J. Morgan, "*Partus*" 2). In case of the *Partus* ruling, colonial and slaveowning legislators and stakeholders went to great lengths to establish a legal connection between reproduction and property – and thus between Black women's reproductive capacity and the emerging capitalist marketplace – as they turned away from "bastardy laws" and instead drew on property law in order to regulate descent on the New World plantation. "Bastardy laws would have clarified the status of a child born of a nonmarried couple; but by echoing the language of property, the legislators weighed in on something larger than sexual mores" ("*Partus*" 5). In other words, rather than being situated within a legal framework of familial (read: male) lineage, the "*partus* ruling derived from concerns around animal husbandry and conflicts over impregnated cattle and cultivated land" ("Archives" 159). Morgan reminds us that the "issue on the table was not simply the matter of heritable kinship but rather the matter of heritable property" ("*Partus*" 5). In linking race and status as "heritable qualities" to be passed on through the *maternal* line (14), slaveowners as well as

> slaveowning legislators enacted the legal and material substitution of a thing for a child: no white man's *child* could be enslaved, while all black women's *issue* could. This happens as though it were common sense, when, in fact, it was a profound reversal of European notions of heredity in the service of a relatively new notion of difference and bondage. ("*Partus*" 5)

With the 1662 act and the ways it bound Black women's interiority to the market, *reproduction* thus becomes a means to delineate freedom and bondage along racial lines. As Morgan argues, the law not only "locked enslaved women into a productive relationship whereby everything that a body could do was harnessed to the capital accumulation of another"; it also detailed that "some women's children became indelibly marked with the inevitability of enslavement, [while] other women's children became inevitably free" ("*Partus*" 17, 12–13). Enslaved women's reproductive capacities, then, become an important pathway to power for the English settlers, traders, and planters, both male and female, as well as to Black abjection. Morgan writes: "As enslaved women were situated as the antithesis to the rights-bearing citizen-subject emerging in this period, they became placed outside of historical processes, except as indexes of suffering" (16).

In close conversation with Morgan's work, Saidiya Hartman writes that "[r]eproduction is tethered to the making of human commodities and in service of the marketplace. [...] Slavery conscripted the womb, deciding the fate of the unborn and reproducing slave property by making the mark of the mother a death sentence for her child" ("Belly" 168, 169). As my analyses will show, *A Mercy* both addresses and negotiates the interplay between reproduction, slavery, and the market throughout. Most explicitly, it does so in the textual fragment of the *minha mãe* (Florens' enslaved mother). Situated at the end of the novel as a kind of coda to the novel's other fragments, the *minha mãe* tells the reader as well as her daughter about how she was first shipped to Barbados and then sold to her new master in the colonial Chesapeake. All of this happens, the text suggests, as part of her master's efforts to increase his profit—a profit calculated according to the proliferation of slave property both bought at the auction block and reproduced on the plantation. We read:

> Barbados, I heard them say. [...] One by one we were made to jump high, to bend over, to open our mouths. [...] It was there that I learned how I was not a person from my country, nor from my families. [...] So it was as a black that I was purchased by Senhor, taken out of the cane and shipped north to his tobacco plants. A hope, then. But the first mating, the taking of me and Bess and one other in the curing shed. Afterwards, the men who were told to break we in apologized. Later an overseer gave each of us an orange. [...] [T]he results were you and your brother. (*AM* 163–164)

With the *minha mãe* and her experiences of transport from one colony to another as well as of sexual subjection, the novel situates its setting in the colonial Chesapeake, where chattel slavery is nascent then; within a broader frame of a fully implemented economic regime of chattel slavery, property, reproduction, and sexual violence in the colonial English Atlantic. That is also to say that the *minha mãe*'s experience of "being taken in the curing shed" here represents what Hartman elsewhere describes as "the instrumental deployment of sexuality [...] The particular investment in and exploitation of the captive body dissolved all networks of alliance and affiliation not defined by property ownership" (*Scenes* 100). Of course, the irony of a gift of an orange does not ameliorate the sexual brutality inflicted against the *minha mãe* and the other women but instead emphasizes the institution's reproductive calculus by making visible their master's "investment" in them. Furthermore, the fact that the *minha mãe* does not hesitate to ask Jacob Vaark to take her daughter instead of herself when Senhor D'Ortega suggests human flesh/property as partial payment of his debt to Vaark (*AM* 20–25) speaks to her fear when she notices both her master's and her mistress's intentions of making sexual use of the *minha mãe*'s daughter: "Neither one will want your brother. I know their tastes. Breasts provide the pleasure more than simpler things. [...] It was as though you were hurrying up your breasts and hurrying also the lips of an old married couple" (160). Offering her daughter to Jacob Vaark under these circumstances – in an attempt to protect her from her master and mistress and without knowing that Vaark (and his wife) will spare her daughter – represents the *minha mãe*'s impossible "choice" between two masters/mistresses against the backdrop of slavery's reproductive calculus.

On a slightly different note, Morgan's arguments about the "intersections between intimacy and property" ("*Partus*" 9) critically supplement a widespread argument made by many historians of slavery: namely, that racial lines in colonial Virginia were more fluid than in other English colonies in North America and the Caribbean (e.g., Fields, E. Morgan). Indeed, in seventeenth-century Virginia "the transition to slavery was slow, and free black men and women gained some autonomy and maneuverability over the course of the first fifty years of colonial settlement" (J. Morgan, "*Partus*" 3). These facts have lead many historians interested in the "origins of racial thinking" to employ notions of fluidity and indeterminacy to chart colonial Virginia as a place where events might have taken a turn for the better, a place where "an egalitarian future could have been foretold" ("*Partus*" 3; "Archives" 158). In such "arsenal[s] for dislodging racist inevitability," as Morgan has it, the *Partus* law often appears to be "simultaneously anomalous and exemplary" ("*Partus*" 3). While historians distinguish Virginia "from colonies such as Barbados, where slavery was in full force by the middle of the seventeenth century" in this way, Morgan stresses the fact that, with the *Partus* ruling, lawmakers actually "put into code the assumptions about racial inheritance that prevailed throughout the Atlantic, even as those elsewhere simply acted on those assumptions" ("*Partus*" 3, 2–3). Put another way, rather than being "a simple and necessary corollary to racial slavery and the logical outgrowth of a labor system rooted in slavery in an increasingly inflexible and racialized understanding of heritability," the *Partus* law both enshrined and made explicit assumptions about as well as practices concerning the reproductive capacity of enslaved women that slaveholders had pursued in service of their capitalist ventures across the English Atlantic long before Virginia legislators passed the act (3, 2). In this light, the act complicates and perhaps even unhinges many scholars' emphasis on Virginia's racial fluidity and indeterminacy. That is also to say that the "ability to render this colony as the locus of possibility is itself a kind of testimony, a set of interpretative practices that could be made from a very particular location—but not a location in which the lives of pregnant African-descended women are the starting point" ("Archives" 158–159).

My close readings of *A Mercy* will follow Jennifer Morgan's arguments. I suggest that Morrison's novel both tests and contradicts the possibility that the property paradigm will *not* take hold in colonial Virginia throughout. What I hope to show is that *A Mercy* brings the sexual economies of New World slavery to its plotting and that it engages in a productive exchange with Black feminist historians' archival work on Atlantic slavery's histories of the experiences of enslaved women, whose "maternal" possibilities on the New World plantation were overwritten by chattel slavery's regimes of property. As I will elaborate below, it is this focus on *reproduction*, property, and slavery that critics and readers most often have neglected in their analyses of *A Mercy*. Placing the novel in conversation with Black feminist scholarship on *reproduction* and racial slavery, my analyses will take this nexus as its point of departure. The novel's thematic focus on and plotting of *reproduction*, then, becomes part of an effort of aesthetically representing a historical past in which African and African-descended women and their reproductive capacities take center stage. This is especially the case with characters like Sorrow and the *minha mãe*, but it also echoes in the other characters' textual fragments. My readings, then, will open up a counternarrative in Jennifer Morgan's sense and position *A Mercy* as being "home to the counternarrative, or at least to its possibility [...] [offering] a new way of thinking about slavery, gender, and reproduction" ("Archives" 154, 158).

The Property/Kinless Constellation

> It seems clear, however, that 'Family,' as we practice and understand it 'in the West' – the vertical transfer of a bloodline, of a patronymic, of titles and entitlements, of real estate and the prerogatives of 'cold cash,' from *fathers* to *sons* and in the supposedly free exchange of affectional ties between a male and a female of *his* choice – becomes the mythically revered privilege of a free and freed community.
> — Hortense J. Spillers, "Mama's Baby"

> The theft, regulation and destruction of black women's sexual and reproductive capacities would also define the afterlife of slavery.
> — Saidiya V. Hartman, "The Belly of the World"

The inextricable connections between race, reproduction, and property, as Black feminist thinkers have shown, are nowhere as clear as in the histories of New World chattel slavery. These connections were elaborated by the law, which mapped new meanings of status and made explicit assumptions about the connections between birth and race (J. Morgan, "*Partus*," "Archives"). That is, "Euro-American legislators amalgamated property with a reproducible kinlessness" (J. Morgan, "*Partus*" 14). In what follows, I chart *kinship* as the second term of this section. The Middle Passage resituated Black women's reproductive capacity within the marketplace. It "interrupted hundreds of years of black African culture" and made it impossible for the enslaved "to refer to one site of origin" (Spillers, "Mama's Baby" 68; Hartman, *Scenes* 76). The captors' and slaveholders' disregard of relations, familial or otherwise, enforced this "impossibility of origin" (Hartman, *Scenes* 76). As Spillers has it, "[w]hen the field of captives [...] is divided among the spoilers, no heed is paid to relations, as fathers are separated from sons, husbands from wives, brothers from sisters and brothers, mothers from children—male and female" ("Mama's Baby" 70). With *kinship* as the second term under scrutiny here, I seek to deal with questions that Morrison's text raises with regard to the possibility of recognized kinship or family formations, or the absence thereof, for the characters in the novel. Spillers's 1987 seminal work on questions of gender and family relations under slavery is particularly instructive in this respect. Published at a time when gender studies would come to be part of the Western academy and began to "compel us more and more decidedly toward gender 'undecidability'" (66), her essay "Mama's Baby, Papa's Maybe: An American Grammar Book" focused critical attention on how such a notion lacks analytical rigor when it comes to interrogating the lives and positionalities of Black women, children, and men during and after slavery, and it introduced a new vocabulary with which to address these issues (Spillers et al. 302).

Following Spillers, the conditions of the Middle Passage and of New World slavery uniquely position the captives in the world, situating them as "neither female, nor male, as both subjects are taken into 'account' as *quantities*" ("Mama's Baby" 72). Spillers claims that the capturing of human beings and the making of them into property that could be shipped across the Atlantic constitute an epistemic, cultural, political, and philosophical moment in which a distinction between "captive and liberated subject-positions" becomes the core organizing principle of the modern world (68). Spillers writes:

> the socio-political order of the New World [...] with its human sequence written in blood, represents for its African and indigenous peoples a scene of *actual* mutilation, dismemberment, and exile. First of all, their New-World, diasporic plight marked a *theft* of the body—a willful and violent (and unimaginable from this distance) severing of the captive body from its motive will, its active desire. Under these conditions, we lose at least *gender* difference *in the outcome*, and the female body and the male body become a territory of cultural and political maneuver, not at all gender-related, gender-specific. ("Mama's Baby" 67)

Under these conditions of the cultural "unmaking" of millions of men, women, and children (72), the captives are being forced to enter into a status of "*being* for the captor" (67). Their bodies become the "'flesh,' that zero degree of social conceptualization," that traders and slave owners will both regard and use as tradable, fungible, and collectible possessions (67). In the context of nascent racial capitalism in the New World, the status of the enslaved as their master's usable "flesh" has important implications with respect to notions of family formation and *kinship*. First, the offspring of the enslaved become "the man/woman on the boundary, whose human and familial status, by the very nature of the case, *had yet to be defined*" (74; emphasis mine). And second, Spillers explains that

> 'motherhood' is not perceived in the prevailing social climate as a legitimate procedure of cultural inheritance. [...] In effect, under conditions of captivity, the offspring of the female does not 'belong' to the Mother, nor is s/he 'related' to the 'owner,' though the latter 'possesses' it, and in the African-American instance, often fathered it, *and*, as often, without whatever benefit of patrimony. ("Mama's Baby" 80, 74)

For the enslaved, *kinship* "loses meaning, *since it can be invaded at any given and arbitrary moment by the property relations*" (74). Slavery's grammar of property, then, not merely rewrites reproduction along economic terms but it also codifies *kinship* and family as being within the purview of those deemed to be free (see also epigraph to this subchapter). With this conception of "the property/kinless constellation" in mind (74), which also is the overall title to section of the chapter, I argue that *A Mercy* deconstructs Western ideas of kinship and family, revealing the ways in which these conceptions are deeply dependent upon the intricately connected notions of property and human reproduction under Atlantic slavery. I will address this, for instance, in my close reading of Sorrow's textual fragment in *A Mercy* (see Chapter 4.4). I will also think about this in relation to nascent formations and identity deliberations of whiteness as staged in the novel.

Property, Kinship, Whiteness

Five months after white terrorist Dylann Roof murdered nine African Americans during a prayer service and Bible study group meeting at the Emanuel African

Methodist Episcopal Church in Charleston, South Carolina,[68] as well as eight days after the election of Donald Trump as President of the United States on November 8th, 2016, Christina Sharpe published a short piece in the online journal *The New Inquiry* entitled "Lose Your Kin." In it, Sharpe lends her perspective on slavery's longue durée in relation to the entanglements between property and *kinship*. Sharpe writes that "transatlantic chattel slavery's constitution of domestic relations made kin in one direction, and in the other, property that could be passed between and among those kin" ("Lose"). Tracing these formations of white kinship over and against the making of slave property through the examples of two U.S. Senators, both of whom had "fathered" Black children, whom they would not claim as kin, Sharpe writes:

> The laws of U.S. chattel slavery and Jim Crow made white kinship (legally, familially, and politically). These modes of recognizing white kinship and refusing to recognize Black personhood endure into the present; they make and unmake persons and families, and assign human beings value in and of themselves, or not. ("Lose")

Of course, this configuration of kinship as it relates to property and racial formations in many ways is in conversation with Cheryl Harris's arguments about "whiteness as property," as put forward in her eponymous 1993 article. As Bhandar has noted recently, Harris's piece "remains unsurpassed in the novelty of the theoretical framework she developed for understanding how whiteness has come to have value as a property in itself, a value encoded in property law and social relations" (*Colonial Lives* 7). As previously discussed, Harris interrogates how in conjunction with the formation of chattel slavery and practices of settler colonialism in the seventeenth-century century "the concept of race interacted with conceptions of property, to 'establish and maintain racial and economic subordination'" (Bhandar, *Colonial Lives* 7). According to Harris, whiteness became

> a shield from slavery, a highly volatile and unstable form of property. [...] Because whites could not be enslaved or held as slaves, the racial line between white and Black was extremely critical; it became a line of protection and demarcation from the potential threat of commodification, and it determined the allocation of benefits and burdens of this form of property. White identity and whiteness were sources of privilege and protection; their absence meant being the object of property. ("Whiteness" 1720–1721)

---

[68] See, for example, the news coverage by *The Guardian* ("Staff and Agencies") or the *Washington Post* (Bever).

Addressing these connections between property and whiteness and linking them to notions of kin and family formations, Sharpe suggests that *kinship*, like whiteness, becomes and remains "a way of sorting oneself and others into categories of those who must be protected and those who are, or soon will be, expendable" ("Lose"). Furthermore, both whiteness and *kinship* need to be conceptualized as being tied to notions of purity and pure blood: White legal identity also manifested in the assumed "purity" of white blood.[69] In a similar vein, notions of reproduction, lineage, and heritability under slavery, too, refer us back to such things as blood and blood lines (see also J. Morgan, *Partus*; Nyong'o, "Barack"). What is elaborated through the conceptual nucleus of *kinship*, in short, is a set of practices, belief systems, or a grammar, which organizes and structures civil society the in(to) the present: To be kin "means all of those *recognized* by the self – in some fundamental, indelible way – as being like the self" (Sharpe, "Lose").

We can recognize these entanglements in *A Mercy* as well. For example, we encounter them in the textual fragment of Jacob Vaark. As the text suggests, the decision to accept Florens as partial payment in his business transaction with the slave trader D'Ortega is also influenced by the fact that all of his children have passed. The loss of their children, in turn, weighs heavily on his wife Rebekka: "Three dead infants in a row followed by the accidental death of Patrician, their five-year-old, had unleavened her. A kind of invisible ash had settled over her which vigils at the small graves in the meadow did nothing to wipe away" (*AM* 19). In contrast to his wife, however, Vaark is "confident she would bear more children and at least one, a boy, would live to thrive" (19). What the text indexes here, as I hope to show in over the coming pages, is not only the prospect but also the project of *white male lineage* on the North American colonial mainland and, by extension, in the English Atlantic.

---

[69] Harris writes, "In adjudicating who was 'white,' courts sometimes noted that, by physical characteristics, the individual whose racial identity was at issue appeared to be white and, in fact, had been regarded as white in the community. Yet if an individual's blood was tainted, she could not claim to be 'white' as the law understood, regardless of the fact that phenotypically she may have been completely indistinguishable from a white person, may have lived as a white person, and have descended from a family that has lived as whites. Although socially accepted as white, she could not *legally* be white. Blood as 'objective fact' dominated over appearance and social acceptance, which were socially fluid and subjective measures" ("Whiteness" 1739–40).

3.7 Anticipatory Wake

> I am become wilderness but I am also Florens. In full. Unforgiven. Unforgiving. No ruth, my love. None. Hear me? Slave. Free. I last. I will keep one sadness. That all this time I cannot know what my mother is telling me. Nor can she know what I am wanting to tell her. Mãe, you can have pleasure now because the soles of my feet are hard as cypress.
> — *A Mercy*

> I'm interested in ways of seeing and imagining responses to terror in the varied and various ways that our Black lives are lived under occupation; ways that attest to the modalities of Black life lived in, as, under, and despite Black death. And I want to think about what this imagining calls forth, to think through what it calls on 'us' to do, think, feel in the wake of slavery—which is to say in an ongoing present of subjection and resistance; which is to say wake work, wake theory.
> — Christina E. Sharpe, *In the Wake*

While the study of the aftereffects of chattel slavery in the United States is anything but new, Black Studies post-slavery theorizing has recently pushed analytical emphasis towards slavery's ongoing structural, ontological, and metaphysical positioning power for Blackness in a continuum from the plantation to the penitentiary. As suggested earlier in the chapter, theoretical trajectories such as Afropessimism (e.g., P. Douglass, Hartman, Sexton, Terrefe, Wilderson), as well as others following in their vein, have inserted into our critical lexicons a focus on analytical terms such as *fungibility, gratuitous violence,* and *abjection.* Such terms, I have argued, matter to critical thinking not merely because they offer new ways of accounting for Blackness in the modern Western world on a structural level—as well as, for that matter, for how whiteness's imaginary is "parasitic on the Middle Passage" (Wilderson, *Red* 11); but they also allow for an unflinching examination of the "vexed genealogy" of property and white liberal self-making as subtended by chattel slavery (Hartman, *Scenes* 115).

The last term that I introduce and discuss in the chapter is *anticipatory wake.* I combine this from my reading of Saidiya Hartman's ("Belly," "Venus," *Lose*) and Christina Sharpe's (*Wake,* "Black Studies") Black feminist theorizations of Atlantic slavery and its ongoing aftereffects. In their own ways, Hartman and Sharpe take up a set of questions that the trajectory of Afropessimism does not seem interested in pursuing.[70] That is also to say that Afropessimism's ensemble of questions generally has sparked some controversy among critics,

---

[70] Here, it is important to remember that Hartman's book *Scenes of Subjection* is often regarded as a "founding document" for Afropessimist thinking (Weier, "Consider" 425). To my knowledge at this point, Hartman herself has never referred to her work as Afropessimist. However, like Sharpe's, her scholarship appears to be in conversation with Afropessimism.

with arguments against this strand of Black critical thinking often being based on the observation that it does not offer any "way out" of the abyss of white supremacist antiblack modernity.[71] Afropessimism does indeed not offer a prescriptive "roadmap to freedom so extensive it would free us from the epistemic air we breathe" (Wilderson, *Red* 338). Several scholars have responded to this "impasse" and have in a more or less explicit manner set out to think about what Jared Sexton elsewhere calls "the social life of social death." Prominent among them, for instance, are the interventions into (some) Afropessimist precepts by Fred Moten ("Nothingness," "Case"; cf. also Moten and Harney), who, rather than dwell on the ontological impossibility of Black being, seeks to "think (not so much about, but rather) in social life, which is to say black social life, which is to say the social life of the alternative in a different way" (Moten and Harney 12–13).[72] While the politics of Moten's work are invested in thinking about "the possibility and the law of outlawed, impossible things" rather than

---

[71] Jared Sexton offers a somewhat comprehensive overview of the numerous arguments against Afropessimism's analytical lens when he writes: "Afro-Pessimism [...] is thought to be, in no particular order: a negative appraisal of the capabilities of black peoples, associating blackness with lack rather than tracing the machinations through which the association is drawn and enforced, even in the black psyche, across the *longue durée*; a myopic denial of overlapping and ongoing histories of struggle and a fatal misunderstanding of the operational dynamics of power, its general economy or micro-physics, reifying what should be historicized en route to analysis; a retrograde and isolationist nationalism, a masculinist and heteronormative enterprise, a destructive and sectarian ultra-leftism, and a chauvinist American exceptionalism; a reductive and morbid fixation on the depredations of slavery that superimposes the figure of the slave as an anachronism onto ostensibly post-slavery societies, and so on. The last assertion, which actually links together all of the others, evades the nagging burden of proof of abolition and, moreover, fails to acknowledge that one can account for historically varying instances of antiblackness while maintaining the claim that slavery is here and now. Most telling though is the leitmotif of *offense*, and the felt need among critics to defend themselves, their work, their principles and their politics against the perceived threat. In place of thoughtful commentary, we have distancing and disavowal. The grand pronouncement is offered, generally, without the impediment of sustained reading or attempted dialogue, let alone careful study of the relevant literature. The entire undertaking, the *movement* of thought it pursues, is apprehended instead as its lowest common denominator, indicted by proxy, and tried *in absentia* as caricature" ("Afro-Pessimism" par. 4–5).

[72] Here, I am of course referring to an intra-mural debate – often stylized as unfolding along the two contesting poles of Afropessimism and what has come to be called "Black Optimism" – that both stresses the importance and the necessity of Black studies now more than ever and grapples with the orientation/locus of Black (studies) critique. This debate can be traced through the following works, among others: Moten, "Case," "Nothingness," "Black Op"; Moten and Harney; Sexton, "African American," "People-of-Color-Blindness," "Unbearable Blackness," "Ante-Anti-Blackness," "Social Life"; Sharpe, "Response."

pursuing a Black theoretical project that thinks Blackness as social death (Moten, "Case" 178; "Nothingness" 738),[73] others have made Afropessimism's conceptualization of antiblackness and Black social death the locus of their thinking through "the vitality of the impossibility" (Jaye Austin Williams, private conversation).[74] For example, in "Response to Jared Sexton's 'Ante-Anti-Blackness: Afterthoughts,'" Sharpe writes that Afropessimism's explanatory power

> makes clear the existence of black social life in all of its modalities does not alter the fact of black social death. That black life is not recognized as life (or life lived) on the order of other lives [...] [Its trajectory enables Black Studies] to build a language that, despite the rewards and enticements to do otherwise, refuses to refuse blackness, that embraces 'without pathos' that which is constructed and defined as pathology. ("Ante-Anti")

This section charts conceptualizations of 'black social life in all of its modalities' in the face of social death in the work of Hartman and Sharpe, who theorize programmatically slavery's technologies of self-(un)-making as that which is being produced and reproduced into the present. In grappling with questions about the (im)possibility of redressing the positioning brutality of slavery, issues concerning the representation of lives lost and obliterated, and ways of dealing with or mourning the "interminable event" of slavery, both Hartman and Sharpe's respective thinking oscillates between "disaster and possibility" (Sharpe, *Wake* 19, 134). As such, their work also comments on the stakes of Black study as a "continued reckoning the longue durée of Atlantic chattel slavery, with black fungibility, antiblackness, and the gratuitous violence that structures black being, of accounting for the narrative, historical, structural, and other positions black

---

[73] Moten writes: "More to the point, if Afro-pessimism is the study of this impossibility, the thinking that I have to offer (and I think I'm as reticent about the term *black optimism* as Wilderson and Sexton are about *Afro-pessimism*, in spite of the fact that we make recourse to them) moves not in that impossibility's transcendence but rather in its exhaustion. Moreover, I want to consider exhaustion as a mode or form or way of life, which is to say sociality thereby marking a relation whose implications constitute, in my view, a fundamental theoretical reason not to believe, as it were, in social death. Like Curtis Mayfield, however, I do plan to stay a believer. This is to say, again like Mayfield, that I plan to stay a black motherfucker" ("Nothingness" 738).
[74] At a workshop on the connections between antiblackness, subjectivity/subjection, and (practices of) performance that I co-organized as a member of the doctoral students' network "Perspectives in Cultural Analysis: Black Diaspora, Decoloniality, Transnationality" at the University of Bremen, artist, scholar, teacher, writer, actor, and director Jaye Austin Williams described U.S.-Black radical studies' multiple critical inquiries of transatlantic slavery as regime of Western modernity as the study of the 'vitality of the impossibility.' The workshop took place in August 2016. For more information on the University of Bremen's doctoral students' networks see the University of Bremen's website ("Doc Netzwerke").

people are forced to occupy" and of Black Studies as the "continued imagining of the unimaginable: its continued theorizing from the 'position of the unthought'" (Sharpe, "Black Studies" 59). I reconstruct their respective works here in regard to the ways they pay attention to notions of grief, loss, the (im)possibility of Black narration/narrative, and the "afterlife of property" (Sharpe, *Wake*), and I do so as I work towards my own term, *anticipatory wake*. With this term I seek to account for the ways in which *A Mercy* again and again walks a tightrope between, on the one hand, taking its readers to colonial Virginia and Maryland at a time when chattel slavery was not yet fully implemented on the North American mainland and, on the other, doing so with the full knowledge that slavery will become full-scale only a few decades later, unfolding across the eighteenth and maintaining itself as a regime of violence well into the nineteenth century and with its legacies ongoing.

Living (in) Slavery's Afterlife

> She had discovered a way off the ship. It worried her that the ancestors might shun her, or the gods might be angry and punish her by bringing her back as a goat or a dog, or she would roam the earth directionless and never find her way beyond the sea, but she risked it anyway, it was the only path open. When the two boys plummeted into the sea, they had made leaving so easy [...] If the story ended there, I could feel a small measure of comfort. I could hold on to this instant of possibility. I could find a salutary lesson in the girl's suffering and pretend a story was enough to save her from oblivion.
> — Saidiya V. Hartman, *Lose Your Mother*

The above epigraph comes from Hartman's book *Lose Your Mother: A Journey Along the Atlantic Slave Route*. Part autobiography, part fiction, part travelogue, part history book, and part theory (Newman et al. 1), this book offers Hartman's personal, theoretical meditation on her attempt at grappling with a historical past that is ongoing. Hartman begins *Lose Your Mother* by narrating how she travels to Ghana in search of a way to belong in the world, wanting "to engage the past, knowing that its peril and dangers still threatened and that even now lives hung in the balance" (6). For Hartman, the lasting effects of slavery which continue to structure and define Black existence need to be understood as slavery's "afterlife." I echo her by now seminal formulation here:

> Slavery had established a measure of man and a ranking of life and worth that has yet to be undone. If slavery persists as an issue in the political life of black America, it is not because of an antiquarian obsession with bygone days or the burden of a too-long memory, but because black lives are still imperiled by a racial calculus and a political arithmetic that were

> entrenched centuries ago. This is the afterlife of slavery—skewed life chances, limited access to health and education, premature death, incarceration, and impoverishment. I, too, am the afterlife of slavery. (*Lose* 6)

Notions of longing and of loss are at the center of Hartman's epistemological project of interrogating slavery's afterlife and these notions are tied to questions of narrative and the writing of the past, present, and future of slavery's positioning power ("Venus," *Lose, Scenes*). From the very beginning, that is, *Lose Your Mother* wrestles with the loss of lives during and after the Middle Passage and Hartman questions how these lost lives can be mourned: "As both a professor conducting research on slavery and a descendant of the enslaved, I was desperate to reclaim the dead, that is, to reckon with the lives undone and obliterated in the making of human commodities" (6).

The above epigraph comes from the very end of a chapter in the book entitled "The Dead Book" and it is an example of how Hartman, on the one hand, tries to navigate the fact that the archive of slavery does not contain the stories of those whose lives have been erased during and after the Middle Passage. On the other hand, it is an example of how she negotiates the longing to "represent the lives of the nameless and the forgotten, to reckon with loss, and to respect the limits of what cannot be known" ("Venus" 4). In the chapter, Hartman goes to great length to tell the story of the murder of an enslaved girl aboard the slave ship *Recovery* and to represent the ensuing court case, in which The Committee for the Abolition of the Slave Trade sued the captain of the *Recovery*, John Kimber, for the murder of this girl and another slave girl. Variously focalized through, for example, the captain, the third mate, the enslaved girl, and the abolitionist William Wilberforce, Hartman's chapter is a multi-perspective narrative that tries to bring to the fore the slave girl's perspective in the face of obliteration, archival and other. Hartman writes: "On April 2, 1792, William Wilberforce immortalized the girl in a speech delivered before the House of Commons and the world lent its attention, at least for a few days. When the trial ended, so did any interest in the girl. No one has thought of her for at least two centuries, but her life still casts a shadow" (*Lose* 138). Thus writing (in) the shadow of the girl, the chapter becomes "an instance of possibility" (see epigraph).

In "Venus in Two Acts," an essay published in *small axe* as a companion piece to the book, Hartman returns to her writing this particular chapter as she discusses the connections between the archive of slavery, the writing of history, and the "impossibility of discovering anything about [the dead] that hasn't already been stated" (1). At the center of this piece is the second enslaved girl who died on the *Recovery*, who merely appears in the archive of slavery as a *dead girl* named in a legal indictment against a slave ship captain tried for the murder of two Negro girls. [...] We stumble upon her in exorbitant circumstances that yield no picture of the everyday life, no pathway to her thoughts, no glimpse of the vulnerability of her face or of what looking at such a face might demand. We only know what can be extrapolated from an analysis of the ledger or borrowed from the world of her captors and masters and applied to her. ("Venus" 1, 2)

What Hartman addresses in this essay is the difficulty of creating a narrative that will embody the dead girl and that, at the same time, does not pretend to "provide closure where there is none" ("Venus" 3, 8). While writing stories about the dead here can become a "form of compensation or even [reparation], perhaps the only kind we will ever receive," these stories remain confined to the archive and the longing to ameliorate the pain and to seek redress for something that cannot be remedied weighs heavily on those who attempt to write such stories (4).

Together, "The Dead Book" and "Venus in Two Acts" raise questions about narrative's ability to account for Blackness's material and social death during and after the Middle Passage. What they reissue is a Black critique of narrative strategies (including but not limited to white/Human narrative articulations of empathy for the slave) that strive "to account for the violence of Black life" (Wilderson, "Aporia," *Red*). For Hartman, a way to deal with this conundrum – the grief and the pain about lives lost, the longing to tell their stories so as to rescue them from oblivion, and the impossibility of narrative to actually account for the (living) dead – is a new Black writing practice that both "tell[s] an impossible story and [amplifies] the impossibility of its telling" ("Venus" 11). Hartman calls this new aesthetic mode/method, which attempts to account for Blackness's "narratively condemned status" (Wynter, "No Humans" 70), "critical fabulation." Employing this mode/method means to be "playing with and rearranging the basic elements of the story, by re-presenting the sequence of events in divergent stories and from contested points of view [so as to] jeopardize the status of the event, to displace the received or authorized account, and to imagine what might have happened or might have been said or might have been done" ("Venus" 11). The goal of this writing practice is to write a "history of the present" with which to "imagine a *free state*, not at the time before captivity or slavery, but rather as the anticipated future of this writing" (4). Ultimately, Hartman seeks to create a counternarrative with these texts, doing so in spite of the fact that everything that can actually be said or told "*take*[s] *for granted* the traffic between fact, fantasy, desire, and violence" (5). Violating the boundaries of the archive of slavery would turn that which is being told into ridicule instead of honoring the dead (10). And yet, this kind of writing also wants "to escape the slave hold with a vision of something other than the bodies of two girls settling on the floor of the Atlantic" (9) even though it is aware that this project is doomed to fail:

"It's hard to explain what propels a quixotic mission, or why you miss people you don't even know, or why skepticism doesn't lessen longing. The simplest answer is that I wanted to bring the past closer" (*Lose* 17). For Hartman, this notion of 'bringing the past closer' is what living or being *in* and *as* the afterlife of slavery means. We continue reading:

> I am the relic of an experience most preferred not to remember, as if the sheer will to forget could settle or decide the matter of history. I am a reminder that twelve million crossed the Atlantic Ocean and the past is not yet over. I am the progeny of the captives. I am the vestige of the dead. (*Lose* 17–18)

"Wake Work"

In Christina Sharpe's highly acclaimed *In the Wake: On Blackness and Being* (2016), Hartman's project of writing an "anticipated future" and a "history of the present" becomes Sharpe's critical interest in "plotting, mapping, and collecting the archives of the everyday of Black immanent and imminent death, and in tracking the ways we resist, rupture, and disrupt that immanence and imminence aesthetically and materially" (*Wake* 13). The book is divided into four chapters entitled "The Wake," "The Ship," "The Hold," and "The Weather," respectively. Sharpe deals with questions of Blackness and being in the afterlife of slavery as she interrogates multiple visual, literary, cinematic, and other aesthetic responses to the ongoing abjection and commodification of Blackness in slavery's afterlives, both in the United States and globally. However, rather than "seek to explain or resolve the question of [Black] exclusion in terms of assimilation, inclusion, or civil or human rights," Sharpe interrogates notions of survival and persistence in the face of Black social death, past and present (14).

To this end, Sharpe introduces the metaphor of the wake to "depict aesthetically the impossibility of such resolutions by representing the paradoxes of blackness within and after the legacies of slavery's denial of black humanity" (*Wake* 14). The metaphor of the wake delineates the paradox of "surviv[al in] this insistent Black exclusion, this ontological negation" (14). As she extracts and discusses the multiple layers of meaning of this metaphor – including the track left in the water behind a ship, keeping watch with the dead, commemorating the dead (1–13) – Sharpe argues that for Blackness to be "*in* the wake is to occupy and to be occupied by the continuous and changing present of slavery's as yet unresolved unfolding" (13–14). Explicitly, Sharpe here enters into conversation with Hartman (and, by extension, Black feminist historians like Jennifer Morgan, given her conceptual focus on property and reproduction) when she writes that

> Living in/the wake of slavery is living 'the afterlife of property' and living the afterlife of *partus sequitur ventrem* (that which is brought forth follows the womb), in which the Black child inherits the non/status, the non/being of the mother. That inheritance of a non/status is everywhere apparent *now* in the ongoing criminalization of Black women and children. (*Wake* 15)

With the metaphor of the wake, Sharpe thus seeks to examine "how we imagine ways of knowing that past [i.e. slavery], in excess of the fictions of the archive, but not only that. I am interested, too, in the ways we recognize the many manifestations of that fiction and that excess, that past not yet past, in the present" (13). And in thinking with Hartman, Sharpe's metaphor of the wake and her theorization of how antiblackness continues to shape Black life in the United States and beyond put analytical attention to notions of mourning and of memorialization (19–20). However, rather than to think about these notions in relation to redress, the questions that ultimately drive Sharpe's theorizations target the ways in which the "interminable event" of slavery (19) can be mourned, commemorated, remembered, respectively: "What, then, are the ongoing coordinates and effects of the wake, and what does it mean to *inhabit* that Fanonian 'zone of non-Being' within and after slavery's denial of Black humanity?" (20). The metaphor of the wake ultimately needs to be understood not as something that is employed in order to seek "resolution of blackness's ongoing and irresolvable abjection" but as way to describe and represent a "form of *consciousness*" (14). Following Sharpe, inhabiting such a

> blackened consciousness [would mean to] rupture the structural silences produced and facilitated by, and that produce and facilitate, Black social and physical death. For, if we are lucky, we live in the knowledge that the wake has positioned us as no-citizen. If we are lucky, the knowledge of this positioning avails us particular ways of re/seeing, re/inhabiting, and re/imagining the world. (*Wake* 22)

In dialogue with Hartman's critical fabulation, Sharpe thus positions "wake work" (17–22) as a "*mode* of inhabiting *and* rupturing this episteme with our known lived and un/imaginable lives," as a mode for Black thinkers and scholars, a way of "encountering a past that is not past" (*Wake* 18, 13; emphasis mine). As mode and as consciousness, the wake is "a theory and a praxis of Black being in diaspora" (19). The work that this theory and this praxis do is to find and create a language that is able to account for the manifold inhabitations of the wake and of Black being in the afterlife of slavery's sexual economies of property; following Sharpe, it is to chart "responses to the terror visited on Black life" and to possibly "imagine otherwise from what we know *now* in the wake of slavery"

(116, 18). This study positions Toni Morrison's *A Mercy* as an articulation *in* and *of* the wake.

Anticipatory Wake, or Anticipating the Afterlife of Property in *A Mercy*

"To anticipate" something means to expect or to foresee something, to forestall someone else's moves, or to await something ("anticipate"). The *Oxford English Dictionary* also defines "to anticipate" as "to take into consideration before the appropriate or due time" ("anticipate, v."). With these definitions in mind, I join the notion of anticipation to Sharpe's concept of *the wake* to create the last analytical term for my readings of *A Mercy*: *anticipatory wake*. My term *anticipatory wake* both situates and delineates *A Mercy*'s narrative as a space in which grief about the "interminable event" of slavery is navigated on the literary level of representation (Sharpe, *Wake* 19). It fundamentally speaks to the novel's complicated workings of time that are best delineated as the text's anticipation of a historical future yet to come. That is, Morrison's novel in the present moment – in the twenty-first century as well as in each reader's present/presence – tries in imaginary hindsight to represent and to show us what did not happen. It attempts to imagine and to represent a moment in the past, in which chattel slavery and, with it, the property paradigm would not be implemented full-scale on the North American mainland—and it does so with the full knowledge and anticipation that this will in fact happen. In other words, at the same time that *A Mercy*'s diegetic orbit asks, *What else could have happened? What could have been had history not taken us down the devastating path of the epistemic, philosophical, social, political, and cultural entanglements of slavery and freedom fueling the liberal imagination of self?*, it asks these questions while charting what Sharpe calls "slavery's yet unresolved unfolding" (*Wake* 14). As such, *A Mercy* becomes a writing praxis – Morrison's writing praxis – "of Black being in diaspora" (Sharpe, *Wake* 19).

On the level of diegesis, moreover, *A Mercy* becomes a history of the present in Hartman's sense, a history that "imagine[s] a *free state*, not at the time before captivity or slavery, but rather as the anticipated future of this writing" ("Venus" 4). However, the anticipated future that *A Mercy* presents its readers with does not strive towards resolution and/or redress. As I will suggest in the next part of this study, *A Mercy*'s form in fact fundamentally defies such notions. This is best illustrated with the slave girl Florens' "telling" (*AM* 1), which is broken up into six textual fragments that take turns with the other character's texts. The last of Florens' fragments ultimately connects back to the first, which is also the beginning of the novel, so that her "telling" will start all over again.

As I will argue in more detail below, her text thus takes a circular form, which points to the fact that her telling will never end. She will continue to tell her story without ever getting to a point of resolution. In the first epigraph to this section, we read that the narrating I is both "wilderness and Florens," both "slave and free," and that Florens will "keep one sadness." These lines constitute the very last words of Florens' textual fragments and they are situated towards the end of *A Mercy*. Like Hartman's grieving meditation on the connections between slavery and narrative, Florens' words are filled with sadness about the fact that she will never be able to talk to her mother, from whom she was separated when her Portuguese master decides to sell her in order to settle his debt with Vaark. Florens will never "know what my mother is telling me. Nor can she know what I am wanting to tell her." Her words "will talk to themselves" (*AM* 159). Florens here also defies categories of "freedom" and "slavery," which, incidentally, are categories along the lines of which the other characters (as well as critics of the novel) have read her. Instead, I suggest, Florens claims for herself a space in which she is both slave and free, in which she is "wilderness." Like Sharpe's "wake work," then, *A Mercy* aesthetically ruptures and disrupts the "immanence and imminence" of social death with Florens's character. Lasting as wilderness, she claims and occupies modernity's grammar of slavery and property, using it to "imagine existence otherwise" (Sharpe, *Wake* 18) and outside of the New World's grammar of property and its positioning power. As I will elaborate below, Sorrow's fragment offers another example of this notion of claiming one's ambiguous status as a "survivor of insistent Black exclusion [and] ontological negation" (Sharpe, *Wake* 14). Not only will she become a mother at the end of her text, she will also rename herself "Complete" (*AM* 114–132). Sorrow refuses the "non/status" that slavery's reproductive calculus forces on enslaved women by claiming her child in spite of *partus sequitur ventrem* (Sharpe, *Wake* 15).

With the term *anticipatory wake*, finally, I want to make visible that *A Mercy* needs to be understood as an aesthetic investment in an imagining elsewhere. As a post-slavery historical novel published in the twenty-first century United States, *A Mercy* performs wake work in Sharpe's sense. The novel makes the absence of resolution its primary concern. And with this refusal to offer narrative resolution, *A Mercy* brings to the fore "the paradox of survival" (Sharpe, *Wake*) *in spite of* the abjection of Blackness within and after the Middle Passage and its regimes of private property.

[Coda]

This chapter has isolated the following terms from recent post-slavery Black interventions into the discursive promises of universal liberty: *violence*; *dispossession* and *fungibility*; *abjection* and *abjectorship*; *reproduction* and *kinship*; as well as *anticipatory wake*. These terms offer me multiple avenues to interrogating the nexus between private property and subjectivity in the remainder of my study. These terms help me chart what I have previously described as *A Mercy*'s refusal to restage and to partake in hegemonic discourses about North American beginnings and its liberal, possessing subjectivities. This also is a refusal to (re-)produce resolution where there is none. As I hope to show in the chapters to come, this refusal lingers and sprouts in the novel's combination of theoretical intervention with narrative form. I will trace this refusal by way of closely examining how *A Mercy* interrogates the intricate connections between subjectivity, slavery, and private property with its allegorical figures. In thus examining the property paradigm within the realm of literary representation, I will also raise questions with respect to the wide array of conceptualizations of fictional character within the field of narrative theory and their ability to account for social death.

4 Practicing Refusal: Narrative Interrogations of the Property Paradigm in *A Mercy*

[Routing the Argument] The present chapter examines the property paradigm in Toni Morrison's novel *A Mercy*. It moves from the discursive field of the white liberal English Atlantic and from Black Studies' theoretical interventions into the discursive promises of universal liberty towards examining the complex entanglements between self-making and private property within the realm of the literary. The chapter's overarching aim is to examine how *A Mercy* takes up, allegorizes, confronts, criticizes, and ultimately rejects liberal ideas of what it means to be a Human subject on the literary level of representation. It understands these ideas to be inextricably bound by notions of ownership and (self-)possession. I argue that that the novel stages its critique of the property paradigm by way of its strategies of characterization. Across the vast field of narrative theory, scholars have developed a wide array of conceptualizations of fictional character. What these conceptualizations share, as I will show in drawing on post-slavery theoretical trajectories, is that they lack the explanatory power to account for and emplot the slave in narrative. In what follows, I use square brackets as a way of connoting this: I use "[character]" whenever I generally talk about *A Mercy*'s allegorical figurations and "[name of a character]," for example [Sorrow], whenever I talk about a specific [character]. Finally, the chapter delivers close readings of *A Mercy*'s [characters]. Core questions in this chapter are: How does one address the absence of narrative (social death) in what is, after all, a narrative text? How does one address, in a study that is concerned with literary narrative and/as epistemic critique, the notion that narrative itself is conscripted by the episteme in which it is produced (Wilderson, *Red* 27–28)? What is the relation between allegorical anti-narration and the (un-)making of liberal, possessive subjectivities?

> It is fair to say that characters do not exist.
> — Mieke Bal, *Narratology*

> It is impossible for narrative to enunciate from beyond the episteme in which it stands, not knowingly, at least.
> — Frank B. Wilderson, *Red, White and Black*

The Agency of Form

In a 2012 essay entitled "On Failing to Make the Past Present" and published in the *Modern Language Quarterly*, literary scholar Stephen M. Best positions Toni Morrison's novel *A Mercy* as a literary text and a historical novel that reads as the paradigm of a "new" critical moment. Here, an ethical relation to the histories and the legacies of transatlantic slavery as that which continues to structure the present moment and its modes of critical thinking should no longer fuel African American and African-diasporic theorizing ("Failing" 456–465). Best writes: "*A Mercy* opens the door to an appreciation of the slave past as it falls away, as *that* which falls away[.] The *form* of *A Mercy* thus undoes a crucial aspect of the historical ethics that *Beloved* played such a pivotal role in bringing about" (466; emphasis mine). The argument continues: While the publication of Morrison's Nobel Prize-winning novel *Beloved* in 1987 and its poetics "shaped the way a generation of scholars conceived of its ethical relation to the [slave] past" (459), both in literary studies and in the study of history, *A Mercy* makes

> abandonment itself a primary concern. [...] *A Mercy* conjures up a moment of pure possibility, before a decision has been made and history begins to rumble down the path that leads to us, and to get here, Morrison settles on a moment, not when things come together but when things fall apart. [...] *A Mercy* abandons us to a more baffled, cut-off, foreclosed position with regard to the slave past. ("Failing" 467, 472)

Best has recently elaborated on these claims in his book *None Like Us* (2018), which is located at the disciplinary intersections of African American Studies and Queer Studies. In it, he critically revisits the "unassailable truth [in Black Studies discourses] that the slave past provides an explanatory prism for apprehending the black political present" (*None* 63). Best overall concern in the book is with a critical impetus within Black Studies grounded in a

> communitarian impulse [which] announces itself in the assumption that in writing about the black past "we" discover "our" history; it is implied in the thesis that black identity is uniquely grounded in slavery and middle passage; it registers in the suggestion that what makes black people black is their continued navigation of an "afterlife of slavery," recursions of slavery and Jim Crow for which no one appears able to find the exit[.] (*None* 1)

Situating such 'communitarian tendencies' within the historiography of slavery, Best offers the term "melancholy historicism" (*None* 1–26) as a way to account for "the view that history consists in the *taking possession* of such grievous experience and archival loss" (15). As suggested above, *Beloved* would set the terms for this kind of witnessing of and accounting for the slave past and its losses within the realm of the literary so that with "Morrisonian poetics as a guide," as Best argues, "the black Atlantic provided a way to make history for those who had lost it and thus secured the recent rehabilitation of melancholy in cultural criticism" (68). Best confronts melancholy historicism and its "impulse" or "desire" to recover a sense of community from/in the history of chattel slavery by arguing that there is no such thing as community in the violent negation of Black subjectivity that slavery was. He writes: "[W]hatever blackness or black culture is, it cannot be indexed to a 'we'—or if it is, that 'we' can only be structured by and given in its own negation and refusal. There is no mutuality, no witnessing, no acknowledgement to be discovered in the archive" (132). There is, in other words, no way in which "abundant recompense" required by the loss that slavery generated can be given (16). These arguments directly engage with postslavery theoretical trajectories like Black Optimism and Afropessimism, both of which, if we follow Best, assume a "kind of lost black sociality [grounded] in horror," which establishes for both notions of community in the present (22). Fundamentally, Best questions such notions of a relationality between the loss of an assumed previous community and a newly constituted sense of community based precisely on this loss in the present. This questioning is part of his attempt at building "a new set of relations between contemporary criticism and the black past on the basis of aesthetic values and sensibilities that I espy in works of literature and art that [...] strive to forge critical possibilities by way of a kind of apocalypticism, or self-eclipse" (22).

A Mercy constitutes a paradigmatic example of the "disintegrative impulse" that Best's project is driven by (*None* 23). Rather than advocating a sense of mutuality in violence and (traumatic) loss (as in *Beloved*'s poetics), *A Mercy* is concerned with writing a "history of discontinuity" (24). Rather than to recover an "impossible community" and create a sense of belonging, Best suggests that *A Mercy* questions "the very condition of possibility, the origin, of that 'us' [and] renders it impossible" (9). In this context, the novel's form gains center stage. For example, Best writes that reading "*A Mercy* requires an attentiveness to who is speaking, and to whom, and through which medium, and in which genre, but then the novel evades capture by resetting all of these conditions of utterance with every turn of the page" (75). We continue reading that "Morrison's prose has often isolated readers by depriving them of the usual coordinates in time and space"; and that "*A Mercy* intensifies that aesthetic: the chapters oscillate, confusingly at first, between Florens' first-person narration and a third-person omniscience, with the apparent goal of isolating the book itself, leaving it, too, with no place in the world" ("Failing" 468). While I take my cue from Best's observations of *A Mercy*'s form and "abandonment aesthetic," I seek to engage with these towards a different end. While I share Best's concerns about the novel's form as the primary means by which *A Mercy* offers its criticism, my reading of the novel in this chapter ventures to make a different argument. That is, my argument precisely takes the slave past and its calculus of ownership and possession as critical paradigm. *A Mercy* opens the door not "to an appreciation of the slave past as that which falls away" (to echo Best) but instead to the notion that the grammar of the liberal property paradigm continues to structure the present political, cultural, and aesthetic moment. This also is to suggest that *A Mercy* opens a narrative window onto the (im)possibility of self-making beyond the modalities of possession.

To read *A Mercy* from a post-slavery point of view is to read it, contra Best, from a theoretical perspective that does not make any recuperative gestures. It is to read *A Mercy* with a focus on slavery's sexual and racialized economies and following a logic of structural antagonism.[75] In some ways, Best's project does not appear to be that far removed from this, for in dialoguing with post-slavery trajectories like Afropessimism, Best himself stresses a kind of antagonism or impossibility when he writes that "there is and can be no 'we' in or following from such a time and place, that what 'we' share is the open secret of 'our' impossibility." Whereas Best makes these arguments in the context of a "search for a selfhood that occurs in disaffiliation rather than in solidarity" (*None* 22), it is important to remember that Afropessimism does not assume Black subjectivity but instead argues that Black subjects do not exist, if by subject we mean critical theory's subject and its status "as a relational being" (Douglass and Wilderson 117). What is more, Afropessimism does not offer a recuperative narrative for Blackness; neither does it gesture towards "the germ of a new beginning if not a new world" (Wilderson, *Red* 337). Instead, Afropessimism advocates for antagonism, incommensurability, or "pyrotechnics" (*Red* 337). As someone who is positioned within the fold of the Human, my goal in this study absolutely cannot be (to attempt) to comment on notions/conceptions of Black community, as Best does. Instead, in attempting a post-slavery of *A Mercy*, I hope to enter into an exchange with theoretical projects like Best's, in which aesthetic concerns about literary form take center stage. The present chapter's overarching aim is to examine how *A Mercy* takes up, allegorizes, confronts, criticizes, and ultimately rejects liberal ideas of what it means to be a Human subject—ideas that are inextricably linked to notions of ownership and self-possession. The chapter thus takes a third step in the study's endeavor to interrogate the connections between private property, self-making, and Western liberalism that I have previously conceptualized as the property paradigm. In focusing on *A Mercy*, it establishes the realm of

---

75 Like most critics and readers of *A Mercy*, Best neglects to discuss Atlantic slavery's reproductive calculus. I will return to this in my readings of Florens and the *minha mãe*.

the literary as this study's main site of critical inquiry of this nexus. The chapter delivers close readings of each of *A Mercy*'s characters, suggesting that they resist interpretation that follows the hegemonic meta-narratives of *The Myths that Made America* (Paul). Set at an historical moment in which racial divisions unfolded on the North American mainland, my study suggests that *A Mercy* pushes its readers to consider and scrutinize the property paradigm. And while it speculates about whether it would have been possible for history to follow a different route, the novel portrays a world in which the property paradigm and the formation of racial slavery factually exert a positioning force on its characters at the New World colonial scene. As John Updike comments in this context, *A Mercy* "circles around a vision, both turgid and static, of a new world turning old, and poisoned from the start."

A Mercy advances its epistemic critique of the conceptual as well as philosophical conflations of subjectivity and ownership not only on the level of the plot but also through a vast array of aesthetic strategies, and specifically through its strategies of characterization. For instance, the text strategically invokes actual historical events while simultaneously making it difficult to decode them. A case in point is Bacon's Rebellion in colonial Virginia (1675–1676) at which the novel gestures in a single sentence that runs "Half a dozen years ago an army of blacks, natives, whites, mulattoes – freedmen, slaves and indentured – had waged a war against local gentry led by members of that very class" (*AM* 8; cf. e. g., "Bacon's Rebellion"; A. Taylor). A few pages later, the Anglo-Dutch businessman Jacob Vaark discusses his potential investments in Barbadian rum with an experienced investor who explains to Vaark that there is an ever self-reproducing work force on Barbados which keeps up the production of rum on the sugar plantations. We read: "And don't forget, there are births. The place is a stew of mulattoes, creoles, zambos, mestizos, lobos, chinos, coyotes" (28). In these sentences, designations like 'zambos,' 'mestizo,' or 'lobos' are actually site-specific designations that mainly refer to people of mixed racial heritage in different parts of colonial Latin America rather than specifically to Barbabos itself.[76] The text here strategically alters historical facts in order to submit its critique, emphasizing in this case not only how white businessmen fundamentally relied on enslaved labor forces for their fortune but also that the notion of white liberal subjectivity established itself across the English Atlantic (this will become clearer in my close reading of Jacob Vaark below). As I argue, the challenge of *A Mercy*'s critique is in what Best elsewhere calls the "agency of form" (*Fugitive* 21);

---

[76] I thank Dr. Alicia Monroe for pointing this out to me. Personal conversation at Vanderbilt University, 2 December 2016.

that is, the challenge is in "what form produces, what form generates" (*Fugitive* 25), in how form produces, enforces, and challenges connections, discursive and conceptual, between private property and self-making. I argue that the novel stages its critique of such early Enlightenment ideas as universal freedom, citizenship, and modern Western subjectivity by way of its strategies of characterization. Indeed, at least one critic has in this context hinted at the unease they felt when encountering *A Mercy*'s characters, stylizing the novel as a "wisp of a narrative [peopled with] *insubstantial* characters [whose] half-told tales leave cobweb trails in the mind, like the fragments of a nightmare" (Mantel; emphasis mine). I claim that the novel constructs its characters in the form of *allegorical figures* instead of relying on fully rounded and easily accessible fictional characters and that it interrogates the liberal property paradigm precisely through these allegories.

Allegorical Anti-Narration

At this juncture, it is important to recall that post-slavery theoretical trajectories have largely questioned (white) narrative's ability both to account for and emplot the slave, regardless of its purpose, and have thus unpacked the intricate connections between narrative (as a structure), meaning-making, and liberal/ Human self-making (Hartman *Scenes*; Hartman and Wilderson; Wilderson, *Red*, "Aporia"). Afropessimist thinking allows for an understanding of widely accepted literary criticism definitions of narrative as "hav[ing] a capacity for stasis and change, and, most importantly, for that stasis and change to be recognized and incorporated by human beings" ("Aporia" 136). Wilderson goes on to explain that "[n]arrative time is always historical (imbued with historicity): 'It marks stasis and change *within* a [human] paradigm, [but] it does not mark the time *of* the [human] paradigm, the time of time itself, the time by which the slave's dramatic clock is set[']" ("Aporia" 136). Narrative's temporal and spatial dimensions do not have any bearing on the Black/Slave. For them, if we follow Wilderson,

> historical "time" is not possible. Social death bars the slave from access to narrative, at the level of temporality; but it also does so at the level of spatiality. [...] [J]ust as there is no time for the slave, there is also no *place* of the slave. The slave's reference to his or her quarters as home does not change the fact that it is a spatial extension of the master's dominion. ("Aporia" 136)

From an Afropessimist point of view, then, what structures Blackness is the absence of a transformative promise in narrative. Blackness' "narrative arc [needs to be understood as] a flat line" ("Aporia" 139).[77]

If a transformative promise underlies meaning-making, narrative, and narrativization in the world and, by extension, in fictional story worlds, this means that this promise also pertains to fictional character in narrative.[78] By extension, this also means that fictional characters need to be understood and scrutinized as being part of the Human fold (Wilderson, "Aporia" 139). *A Mercy* engages with these theoretical premises as it pushes its own epistemic critique of white Western modernity and the liberal property paradigm on the literary level of representation. That is, the novel draws an analogy between the making of liberal subjects (within the realm of the world) and the creation of fictional character (within the realm of the literary narrative). If, following post-slavery interrogations of white Western modernity, to be the subject of property is to be a Human subject, then *A Mercy* suggests there can only be fictional characters if there is subject form. Building on the idea that social death "ruptures the assumptive logic of narrative writ large" (Wilderson, "Aporia" 135), I argue that *A Mercy* resorts to allegory in creating its characters and as way to represent social death's explosion of narrative form. That is, *A Mercy*'s fundamental critique of liberal self-making is situated precisely in its form, in its strategies of allegorical figuration.

In a conventional sense, allegory usually is understood to "occur whenever one text is doubled by another" so that the former's meaning is recast and mediated by the latter's terms (Owens, "Allegorical Impulse 1" 68; Hejinian 285). While I turn to the vast archive of literary criticism in my use of allegory, I will not give a detailed account of how critics, writers, and philosophers have debated allegory's aesthetic potential, its philosophical nature and function(s), or its role within psychoanalytic inquiry, to name only a few (Owens, "Allegorical

---

[77] To quote Wilderson in full: "This kind of change, this transformative promise belongs to White men and their junior partners in civil society, meaning non-Black immigrants, White and non-Black people who are queer, and non-Black women—but only in relation to each other. These fully vested citizens and not-so-fully vested citizens live through *intra-communal* narrative arcs of transformation; but where the Black is concerned, their collective unconscious calls upon Blacks as props, which they harness as necessary implements to help bring about their psychic and social transformation, and to vouchsafe the coherence of their own human subjectivity" ("Aporia" 139).

[78] For now, I generally consider fictional character as participants in story worlds or as an "effect" within a narrative (Margolin; Bal *Narratology*). I will discuss this in more detail below.

Impulse 1" 68).⁷⁹ Rather, I draw on a recent contribution to this archive by avant-garde poet and essayist Lyn Hejinian, who in "Wild Captioning" (2011) places allegory at the intersections of "creative work, political activism, and everyday life" (Hejinian 281). Allegory here emerges as the function of an aesthetic practice that seeks to bring together the notions of the creative, the political, and the quotidian (Hejinian 282). Bringing into play (post-)Marxist, poststructuralist as well as dialectic materialist approaches, Hejinian explains that allegory "depicts what has been undepicted in a depiction. [...] [Allegory] seeks to dig into time, to secure a place for what's gone and for what's not gone, the loss itself, per se. [...] [T]he allegory uses the raw materials of memory to restore time to an absence and to bind an absence to time" (285). Because allegory narrates "temporal stories" in this way it animates (past) matters as it brings them to the present moment (285, 286). As such, allegory stands in stark contrast to metaphor, which according to Hejinian describes connections between different places and thus tells "spatial stories" (285). If we follow Hejinian (who draws on Walter Benjamin's seminal *The Origin of German Tragic Drama*), allegory also needs to be distinguished from symbol: whereas the symbol is that which is complete by and in itself, allegorical representation is mobile and constantly in flux as it progresses over time (Hejinian 288). By relocating meaning across temporalities in this way, allegory as the "purveyor of a known and purportedly well-understood code" also produces contradictions, which offer "little comfort" (285, 294). Hejinian explains that it is precisely the "puzzling, even obscurantist, rather than overdetermining aspect of the allegorical that has the greatest political – and, perhaps, artistic – potential" (285). As both an artistic and a political practice, then, allegory combines temporally divided events or situations "by making use of an occasion" to ultimately become "the epitome of counter-narrative" (282, 295).

Drawing on Hejinian's suggestions for thinking about allegory as an "act, not an exegesis" (296), this study conceptualizes *A Mercy* as a literary experiment that tries to get to the core of white Western modernity's conceptual entanglements between liberal personhood and possession. As an "occasion" in Hejinian's sense, the novel makes use of every instance during which it can interrogate the property paradigm. *A Mercy*'s circular construction is a case in point here. The slave girl Florens' text unfolds over forty-six pages in six different textual fragments, which cut through the other figures' texts. Her last fragment ends only to connect back to the very first one, which opens the novel. In this way,

---

[79] For a general introduction and history of allegory, see Tambling; Haselstein, "Gegenöffentlichkeit," "Vorbemerkungen"; Owens, "Allegorical Impulse 2"; for a discussion of the functions of allegory in postcolonial literatures, see e.g., Sedlmeier.

A Mercy not merely offers a counternarrative to modernity's property paradigm, but it *becomes* allegorical anti-narrative. That is, *A Mercy* becomes the site and practice of an epistemic critique of modernity's calculus of property that is ongoing, a critique that is constantly being revisited, revised, and recalibrated. I claim that *A Mercy*'s allegorical anti-narration emphasizes abandonment—but not in the manner that Best suggests ("Failing" 467). Rather, I think about abandonment in terms of a rejection or unmaking of fictional characters that are or aspire to become the subject of the property paradigm. Accordingly, I discuss *A Mercy*'s allegorical figures in terms of what I call a "refusal of narrativization." Under this rubric, I hope to account for a literary maneuver that needs to be understood as a refusal to restage and thus to partake in hegemonic, dominant discourses about North American beginnings and its liberal, possessing subjectivities.[80] Again, if form plays such a crucial role in the cultural, political, and epistemic work that *A Mercy* does, I suggest that this does not have to do with what Best identifies as the novel casting aside its readers to a "foreclosed position with regard to the slave past" but precisely with making visible the earliest stages of slavery's racial and reproductive calculus and practices of propertization at the New World colonial scene. Instead of "failing to make the past present," I argue that *A Mercy refuses* to make a past present that is bound by notions of ownership.

There is a fundamental tension at work in this project's endeavor to study the ways in which *A Mercy* resorts to strategies of allegorical characterization when presenting its critique of the liberal property paradigm. This tension is caused by what I have identified as *A Mercy*'s allegorical anti-narration, on the one hand, and the fact that the vocabulary available to talk about narrative

---

80 The *OED* offers the following definitions of the word "refusal," among others: the "action or an act of refusing; a denial or rejection of something requested, demanded, or offered"; the "repudiation or renunciation of a contract, allegiance, obligation, etc."; "Something which has been refused or rejected" ("refusal, n."). In the chapter, I use and think about the word "refusal" in all of these ways, and it is from the notion of refusal as "an action or an act" that I came up with the combination of a "practiced refusal" and of "practicing refusal," as the heading of the chapter reads. Importantly, my use of the compound "practicing refusal" also needs to be understood as my acknowledgement of, my bowing to the groundbreaking work done, conducted, and, indeed, practiced, by the "Practicing Refusal Working Group" at Columbia University's Barnard Center for Research on Women. This working group aims at "creat[ing] a new exploratory space for Black thought and to theoriz[ing] different conceptual models for thinking beyond conventional notions of resistance. [...] Practicing refusal names the urgent desire to rethink the time, space, and fundamental vocabulary of what constitutes politics, activism, and theory, as well as what it means to refuse the terms given to us to name these struggles" (Campt, "Introduction"; see also Campt, *Images Matters, Listening to Images*; Moten and Harney).

form does engage with post-slavery thinking's notions of "accounting for" and social death, on the other. How does one address the absence of narrative (social death) in what is, after all, a narrative text? How does one address, in a study that is concerned with literary narrative and/as epistemic critique, the notion that narrative itself is conscripted by the episteme in which it is produced (Wilderson, *Red* 27–28)? What is the relation between allegorical anti-narration and the (un-)making of liberal, possessive subjectivities?

From Fictional Character to [Character]

In light of this tension, what follows is an attempt to lay the foundation, on a methodological level, for my close reading of *A Mercy*'s allegorical figures. While I draw on several narratological concepts, terms, and definitions as tools that will help me examine the ways in which they challenge Western liberal conceptions of private property, (self-) possession, and subjectivity, I can do so only by acknowledging, from a critical perspective that understands antiblackness as a structuring modality of Western modernity, that I use these concepts, terms, and definitions provisionally. With this kind of "explanatory sortie" (a term that I borrow from Best), I seek to account for the notion that the assumptive logics of the narratological terms, concepts, and definitions that I draw on are part of white Western liberal modernity's episteme. In other words, in following *A Mercy*'s lead, I need to work with the existing tools and vocabulary on narrative form and fictional character to ultimately be able to read and write against them. Accordingly, the purpose of this section twofold: First, it enters into dialogue with widely accepted definitions of fictional character. As I will show, the literary criticism archive of what constitutes fictional character is vast, but it needs to be questioned throughout for the ways in which it cannot account for social death.[81] Second, I seek to establish a working definition of fictional character that encompasses the notion of a *refusal to narrativize* characters that represent propertied liberal subjectivities in *A Mercy*. In what follows, I first turn to narrative theory and to different conceptualizations of fictional char-

---

[81] A paradigmatic example of such definitions comes from the work of the Canadian narratologist Uri Margolin, who writes: "In the widest sense, 'character' designates any entity, individual or collective – normally human or human-like – introduced in a work of narrative fiction. Characters thus exist within storyworlds, and play a role, no matter how minor, in one or more of the states of affairs or events told about in the narrative. Character can be succinctly defined as storyworld participant [...] 'Character' in the narrower sense is restricted to participants in the narrated domain, the narrative agents" (Margolin 66).

acter from both the North American and European academies and their debates on how to think about characters and meaning-making before getting to my readings of *A Mercy*'s allegorical figures proper in a second step.

Narrative Theory and the Study of Fictional Character

Narrative theory needs to be understood as a strongly heterogeneous field that straddles a wide range of approaches to the study of narrative texts (e.g., fictional, visual, dramatic).[82] It currently ranges from structuralist-leaning and pragmatically oriented paradigms over psychoanalytic approaches to narrative, feminist narratology (e.g., Fludernik, "Genderization"; Lanser, *Authority*, "Sexing Narrative," "Sexing Narratology"), cultural-studies oriented approaches to narrative and postcolonial narrative theory (e.g., Birk and Neumann; Prince, "Postcolonial") to more recent approaches to the study of narrative, which draw on cognitive linguistics or examine the narrativity of legal discourse, culminating in new research areas such as Law and Literature (e.g., Brooks and Gerwitz; Dimock; Hyde; Thomas, *Law, Cross-Examinations*; Weisberg *Failure, Poetics*).[83] This broader reconfiguration of the discipline of narratology began in the wake of the various cultural turns of the 1980s and 1990s. That is, narrative theory has moved from "a description of textual phenomena to broader cultural questions, various contexts and a growing concern with processes rather than products," pushing

---

[82] In general, I use "narrative theory" and "narratology" interchangeably in this section of the chapter. However, I am also aware that narrative theorists such as Ansgar Nünning have suggested that "narrative theory" is much more suitable as an umbrella term for "theoretical work done on the forms and functions of narrative" than "narratology" ("Taking Stock" 258–259). If we follow Nünning, "narratology" needs to be understood as a "particular kind of narrative theory and the analysis and interpretation of narratives" (259). In this context, Nünning also criticizes the "inflationary use of the term 'narratology'" (241). These terminological debates originate in the proliferation of approaches to the (systematic) study of texts and narrative, burgeoning since the 1990s, that have expanded on as well as critically refined the previously dominant structuralist approaches to narrative commonly known as the discipline of narratology.

[83] This list is by no means comprehensive: for an overview of the emergence and early stages of narratology as a scholarly discipline that constituted itself from different "schools" (French structuralism, Russian formalism, as well as Anglo-American formalist literary theory), see Herman, "Histories." For an overview of the development of the discipline and its key theoretical paradigms, especially the more recent cognitivist ones, see Fludernik, "Histories"; cf. also McHale. For an attempt to map these new approaches to the study of narrative as well as an excellent bibliography of relevant works, see Nünning, "Taking Stock."

context-oriented "postclassical" narratologies rather than a text-centered, structuralist paradigm of "classical" narratology (Nünning, "Taking Stock" 243).[84]

While the study of character and of strategies of characterization[85] tends to remain a "somewhat underresearched" area of narrative theory (Fludernik, "Histories" 43; see also Jannidis 1–3), narratologists largely have approached "character" from a variety of different theoretical perspectives that reflect the above new analytical trajectories of the field.[86] What these approaches have in common is an understanding of character that diverges from a mimetic or realist one, which would regard "characters as imitations of people and tends to treat them [...] as if they were our neighbours [sic] or friends" (Rimmon-Kenan 32).[87] Canadian narratologist Uri Margolin outlines three prominent strands within the study of literary character that developed in conjunction with the previously mentioned broadening of the field of narrative theory: character as a "literary fig-

---

[84] David Herman coined the term "postclassical narratologies" in *Narratologies: New Perspectives on Narrative Analysis*. For a more recent assessment of postclassical narratologies see, e.g., Alber and Fludernik. I borrow the coarse division of text-centered versus context-oriented paradigms from Nünning ("Taking Stock" esp. pp. 243–46), fully aware of the fact that this division can only be a provisional one that helps to broadly map the ever flourishing and diversifying scholarly endeavor of the study of narrative.

[85] In general, the term characterization as used within narrative theory "includes all information associated with a character in a text. [T]his includes information about time, place, actions, and events connected to the character" (Eder, Jannidis, and Schneider 31). This also includes the "ascription of [...] psychological or social traits to a character by a text" (Eder, Jannidis, and Schneider 30). As Eder, Jannidis, and Schneider furthermore elaborate, the term characterization may also refer to "information about [a character's] habitual actions, the circumstances of a person and his or her social relationships" (31). In a broad sense, strategies of characterization thus refer to the "process of connecting information with a figure in a text so as to provide a character in the fictional world with a certain property, or properties, concerning body, mind, behavior, or relations to the (social) environment" (32).

[86] I am referring here to twentieth- and twenty-first-century contributions to the study of character that mainly stem from the Anglo-Saxon academies. For a selective overview of major contributions to the discussion see, e.g., Jannidis 86–98.

[87] In general, E. M. Forster's classic distinction between "flat" and "round" characters continues to remain one the most widely known proposals on how to conceptualize as well as categorize fictional character. It does so despite having been vastly criticized and reconfigured by scholars from the field of narratology (Jannidis 86–87). In the approaches to fictional character discussed in this section, Foster's distinction frequently serves as a point of reference. Forster argued that flat characters "are constructed round a single idea or quality; when there is more than one factor in them, we get the beginning of the curve towards the round" (Forster 67). Round characters, by contrast, change in the course of the narrative and they are constructed around more than one idea or quality. For critical evaluations of Forster's general classification as well as of its limitations, see Bal (*Narratology* 115); Jannidis (86–87); Jahn; and Rimmon-Kennan (40–41).

ure or artifact"; character as an "individual within a possible world"; and character as a "text-based construct or mental image in the reader's mind" (66). In the first instance, a character is a "semiotic [construct] or [creature] of the world, and it is the socially and culturally defined act of fictional storytelling that constitutes and defines them" (67). On this view, the existence of a character is based on texts, and they materialize in the mind of the reader so that "the end result is a relatively stable and enduring inter-subjective entity" (67). In the second instance, character is understood as "an individual existing in some world or set of worlds, both individual and world being very close or very far from the actual world in terms of properties and regularities" (71). In contrast to the previous analytical trajectory, this approach to character as an individual within a "possible world" scenario "touches on the grammar of virtuality" (Fludernik, "Histories" 48). And lastly, approaches to character based on the study of the cognitive-psychological dimensions of narrative view character as "text-based mental models of possible individuals, built up in the mind of the reader in the course of textual processing" (Margolin 76). While the involvement of the reader in the process of characterization certainly also plays an important part in the aforementioned two theoretical frameworks of character analysis, this last one not only explicitly deals with "actual readers and reading" but it is also principally "open to empirical testing" (76; see also Jannidis).

[Character] Orthographies

As two paradigmatic examples of the above discursive field of the study of fictional character, it seems that Gerald Prince's concept of "the disnarrated" (as put forth in an eponymous 1988 essay) as well as Dutch narratologist Mieke Bal's strongly cultural studies-oriented approach to the study of fictional character offer a ready set of analytical tools for my reading of *A Mercy*'s allegorical figures:

Prince's suggestions for thinking about fictional strategies concerning the narratability of events, characters or facts in story worlds in terms of the "disnarrated" resonate with the notion that the fictional beings that people *A Mercy*'s narrative orbit create the effect of being "insubstantial," vaporous, or elusive (Mantel). According to Prince, "terms, phrases, and passages that consider what did not or does not take place [...] whether they pertain to the narrator and his or her narration [...] or to one of the characters or his or her actions [...] constitute the disnarrated" ("Disnarrated" 3). As an analytical tool, the disnarrated "shows that narrative is not only a matter of counting, accounting, and recounting, but also one of *dis*counting" and it "insists upon the ability to conceive and manipulate hypothetical worlds or states of affairs and *the freedom to reject various models of intelligibility,* of *coherence* and significance, various norms, conventions, or codes for world- and fiction-making" ("Disnarrated" 6; emphasis mine). As an "antimodel," Prince's concept of the "disnarrrated" makes explicit how in narrative "choices [are] not made, roads [are] not taken, possibilities [are] not actualized, [and] goals [are] not reached" ("Disnarrated" 5, 6). As such, it generally offers me a way to think about how *A Mercy*'s allegorical figures map the routes that the narrative deliberately chooses *not* to take. A case in point is the early disposal of the allegorical figure of Jacob Vaark from the text's story world after his transformation from a settler looking for greener pastures into a white possessing man, whose property decidedly will include enslaved human beings. As I will show below, the text needs this allegorical figure and representation of the quintessential Lockean liberal subject to disappear to examine the other allegorical figures and how they negotiate the property paradigm.

Bal suggests using the term character "for the anthropomorphic figures provided with specifying features the narrator tells us about. Their distinctive characteristics together create the *effect* of a character" (*Narratology* 112; emphasis mine). More often than not, fictional characters, as "fabricated creatures made up from fantasy, imitation, memory: paper people, without flesh and blood," resemble human beings (113). However, because this resemblance may invite critics and readers to approach fictional characters simply as if they were real human beings (what Bal calls "flat realism") or from the vantage point of psychological criticism, Bal submits that any conception of fictional character should only be based on "those facts that are presented to us in the actual words of the text" and that fictional character should be understood as a text-based "complex semantic unit" (114, 113). For Bal, the dynamic interaction between the reader and the text as part of the creation of fictional character matters in this context, too. She writes: "On the basis of the characteristics they have been allotted, they each function in a different way with respect to the reader. The latter['s] [...] direct or indirect knowledge of the context of certain characters contributes significantly to their meaning" (113, 119). That is, Bal introduces the term "referential character" to account for the ways information about a character is produced based on the reader's previous knowledge about a narrative situation (120–125). As Bal states, "there is information that is 'always-already' involved, that relates to the extra-textual situation, in so far as the reader is acquainted with it" (120). Fictional characters can be understood as referential, in other words, if they relate to information that to some extent may be called "communal," such as general knowledge about a prominent figure from a past historical time period (e.g. Napoleon Bonaparte) or the current U.S. president

4 Narrative Interrogations of the Property Paradigm in *A Mercy* — 113

(the examples are Bal's; *Narratology* 121). In this way, referential characters "fit a pattern of expectation, established on the basis of our frame of reference" (121–122). According to Bal, strategies of characterization in fictional narrative ultimately need to be understood as sites of confrontation, on which fictional characters either fit or clash with the image of the fictional being created in the mind of the reader based on previously acquired, extra-textual knowledge about the narrative situation, and vice versa (122).

We can recognize some limitations inherent to Bal's and Prince's conceptualizations, if we place them into conversation with post-slavery thinking's interrogations of narrative. That is, even though Bal seeks to address formations of power, social hierarchy, and ideology as fundamental to the narratological study of character, as illustrated by her emphasis of a "communal reference frame" underlying the making of fictional character[88]; and even though Prince's concept of "disnarration" aptly describes the "freedom to reject various norms, conventions, or codes for world- and fiction-making," including such things as coherent fictional characters, I want to suggest that both their theorizations ultimately remain grounded in "liberal humanist notions of the universal integrity of the human" (Wilderson, "Aporia" 139). One of the core questions arising in this context is, *What* is that "communal basis for our frame of reference" that Bal speaks of and for *whom* does it functions as such? Again, the prism of post-slavery Black thinking reminds us that narrative as ontological coherence and as transformative promise does not exist for the enslaved. This is because narrative cannot account for the violence that positions the enslaved in the world. Following Wilderson, who writes that "for Blackness, there is no narrative moment prior to slavery" (*Red* 27), the slaves are not part of the community of the Human. By extension, we can also say that they also are not part of assumptions of community that Bal puts forward. In other words, I suggest that Black Studies' questions about the "emplot-ability" of social death also concern the realm of fictional character. I claim that as long as cultural and literary critics like Bal cannot explain "how the Slave is of the world" and thus of (the structure of) narrative, any assumption or conceptualization of fictional character needs to be understood as being fraught with similar explanatory lacunae (Wilderson, *Red* 11).

With this in mind, I use square brackets as a way of connoting the tension created by the notion that most often when fictional character comes into view in

---

88 Bal writes, moreover: "[T]he description of a character is always strongly coloured [sic] by the ideology of critics, who are often unaware of their own ideological hang-ups. Consequently, what is presented as a description is an implicit value judgment" (*Narratology* 119).

literary studies discourse, this discourse's vocabulary cannot speak to post-slavery thinking's arguments about whether accounting for social death in narrative is possible. I also use square brackets as a way of connoting the various demands or claims to New World self-making and private property, or the structurally induced absence thereof, that *A Mercy*'s fictional entities make in their respective textual fragments. I use "[character]" whenever I generally discuss *A Mercy*'s allegorical figures and "[name of a character]," for example [Sorrow], whenever I talk about a specific [character]. Bracketing self-making and the concept of fictional character in this way helps me wrestle with the aporia of not having adequate terms to account for the notion that not all "beings are on the same side of social life," both in narrative and beyond (Wilderson, "Aporia" 141). It helps me locate fictional character in *A Mercy* orthographically as a site of confrontation, destabilization, or break with epistemic, fictional, aesthetic, as well as political formations of the (white) liberal self. I also use square brackets to speak to the status of literary character in antiblack Western modernity that Afropessimist thinking urgently points to. Finally, I use bracketed characters to signal and demarcate *A Mercy*'s active practicing of a refusal of narrativization of fictional character in a literary criticism sense and to emphasize my reading of the novel as an allegorical occasion in Hejinian's sense. In this way, I hope to account for the notion that *A Mercy* does not rely on "fixed" versions of subjectivity but that it considers various im/possibilities of New World self-making over, against, and beyond a grammar of property. What happens in the break of the (assumed) narrative coherence of fictional entities, on the one hand, and the absence of it as created by *A Mercy*'s allegories, on the other? What does it mean for *A Mercy* to linger in the narrative space of im/possibility? What is at stake in the various versions of New World self-making that *A Mercy*'s allegories speak to on the literary level of representation? How might narrative form lead to a breaking open of formations of property, ownership, and subjectivity? How do *A Mercy*'s strategies of characterization become such a powerful tool for confronting, and perhaps also for redefining, the connections between property and personhood—or do they? In what other ways may *A Mercy*'s narrative create subversive spaces over and against liberal practices of propertization and self-making—spaces that allow for what Sylvia Wynter calls "a new frontier to be opened [...] onto the possibility [...] of our fully realized autonomy of feelings, thoughts, behaviors" ("Unsettling" 331)?

Reading *A Mercy*'s Refusal

> For people of color have always theorized—but in forms quite different from the Western form of logic. And I am inclined to say that our theorizing (and I intentionally use the verb rather than the noun) is often in narrative forms, in the stories we create, in riddles and proverbs, in the play with language, since dynamic rather than fixed ideas seem more to our liking. How else have we managed to survive with such spiritedness the assault on our bodies, social institutions, countries, our very humanity?
> — Barbara Christian, "Race for Theory"

> We've got a history of refusing what it is that was refused to us. If we study that history, we can develop some practices that will be useful and it won't be ineffable at all. We actually got something with which to work.
> — Fred Moten and Stefano Harney, *Poetics of the Undercommons*

A Mercy is set in colonial North America (esp. Virginia) in the second half of the seventeenth century. The novel begins *in medias res* and tells the story of the Black slave girl [Florens], who is the property of the Anglo-Dutch farmer, money-lender, and trader [Jacob Vaark] (also called Sir). [Florens] arrives at the [Vaark] farm as part of a partial debt settlement between [Vaark] and the Portuguese slave and tobacco trader [Senhor D'Ortega], who is unable to repay [Vaark] in any other way than with human flesh after one of his slave ships, including the enslaved human cargo on board the ship, has sunk (*AM* 14–15). [Florens] lives on [Jacob Vaark]'s farm with a group of women, all of whom [Vaark] has assembled to live, to birth, and to serve on his estate: [Rebekka] (also called Mistress), whom [Vaark] has shipped to his part of the New World from England to become his wife; an Indigenous woman servant called [Lina], whom [Vaark] buys from a group of Presbytarians; and [Sorrow/Twin], a shipwrecked girl whom [Vaark] does not buy but "accept" from a family of sawyers (49). In addition, two white indentured servants, [Willard Bond] and [Scully] also populate the novel's narrative orbit. [Vaark] regularly makes use of their services even though they serve and belong to the household of a neighboring farm. Moreover, a blacksmith, whom one of the other [characters] describes as a "free African man" (43), also appears on the [Vaark] farm in the novel. [Vaark] commissions [the blacksmith] to forge the gate to a new house that he is building. Finally, the narrative is framed by an account related by the [*minha mãe*], [Florens'] mother, at its close. While [Florens] frequently addresses her mother in her own textual fragments, she is in fact unable to hear what the [*minha mãe*] is trying to tell her in the novel's final chapter.

[Florens'] first-person, autodiegetic text describes her errand to fetch [the blacksmith], who is ordered to help cure her mistress [Rebekka] of the "pox" (*AM* 37). [Rebekka] contracts the disease from her husband [Jacob], who dies of it at the beginning of the novel, leaving the women on his farm to fend for themselves. It seems that only [the blacksmith] will be able to help [Rebekka] survive. The text also suggests that her servants and slaves [Lina], [Sorrow], and [Florens] stand a chance to stay, live, and serve on the [Vaark] farm *only* if [Rebekka] recovers from the disease (57). In opening the novel Florens herself describes her fragment "a confession [...] full of curiosities familiar only in dreams and during those moments when a dog's profile plays in the steam of a kettle" (1). In subsequent chapters, [Florens'] first-person text takes turns with a third-person narrator who "provides the back-stories for Florens [...] and the other characters who live or work on [Jacob Vaark]'s burgeoning Virginia estate" (Jennings 646). Apart from [Florens'] confessions, the [character] of the [*minha mãe*] also takes control of the text in a first-person account. *A Mercy*'s fragmented narrative requires the reader to piece together the information that the text provides on various levels (as, indeed, many of Morrison's later novels require her readers to do, as Jennings reminds us (646)). The novel's narrative time roughly spans a period of eight years from 1682 when [Florens] first arrives at the [Vaark] farm and is "maybe seven or eight" years old to 1690 when [Florens] is sixteen years old (*AM* 3) and it frequently shifts between the fictional present in [Florens'] textual fragments and the fictional past in the narratives of the other [characters].

As for the structure of the close readings, I generally discuss *A Mercy*'s [characters] in the order that they are represented in the novel. The exception here is [Florens], whom I discuss next to last as well as in tandem with the fragment of her mother. To give a brief overview of the coming chapters: I will begin with [Jacob Vaark], whom I discuss as a paradigmatic settler figure in ""The Chagrin of Being Both Misborn and Disowned": [Jacob Vaark], Freedom, and the Pursuit of Property." I also discuss [Vaark] first because the novel appears to position him as the one [character] that keeps the [Vaark] farm and household together. At the same time, however, [Vaark] is disnarrated from the text early on in what I suggest is a narrative maneuver that enables the text to meditate on the other [characters] and their existence at the New World scene in the wake of his death. In "I am Exile Here": [Lina], Self-Inventions, and Dispossession," I turn to [Lina], whom I discuss for the ways she navigates *dispossession* at *A Mercy*'s New World colonial scene. [Lina]'s existence within the novel's seventeenth-century landscapes is fundamentally shaped by the genocide and complete eradication of her tribe by the European colonizers. I argue that within the novel's experimental setup this [character] becomes a representation of *dispossession* as that which describes the capacities or "powers [that subjects] have or lack" (Wilderson, *Red* 8). Next, I analyze the [character] of [Rebekka Vaark]. In "The Promise and Threat of Men": [Rebekka], Subjectivity, and the Ruse of Solidarity," I demonstrate not only how *A Mercy* stages [Rebekka] as an allegory of the space and the place that English women in colonial North America held in the social strata of their nascent environment. I also show how the novel powerfully suggests that [Rebekka]'s struggle for subjectivity is part of the New World's grammar of property, opening for this [character] an avenue towards co-mastery. In the next chapter, ""My Name is Complete": [Sorrow], Anticipating Generations, and the New World Grammar of Property," I turn to the [character] of [Sorrow/Twin]. As probably the most (racially) ambiguous [character] within *A Mercy*'s narrative orbit, I claim that it is [Sorrow]'s very ambiguity which fugitively opens up an utopian moment of the possibility of making generations beyond the property paradigm. This is a moment, which is simultaneously foreclosed by the novel's complicated workings of time and anticipation of a historical future yet to come—a future in which "kinship relations would be subordinated to property relations" (Sharpe, *Monstrous* 34–35). Next, I return to Afropessimism's claim that there is no transformative promise for the slave in narrative in my reading of [Florens]. My argument follows these conceptualizations in the second to last chapter entitled "I am a Thing Apart": [Florens] and the Ruse of Belonging." I here take my second cue from [Florens] herself, who states that she is "a thing a part" and I suggest that [Florens] is void of a transformative narrative promise (*AM* 113). Also following the ways in which the text develops *belonging* as one of its critical themes with the [character] of this enslaved girl child, I suggest that [Florens'] question in *A Mercy* is not a question about subjectivity (as a Human) but that hers is one about being and lasting in/as social death. Then, I closely read the fragment of the [minha mãe] in my final analytical chapter, "There is no protection": The [*Minha Mãe*], Slave Narratives, and the Sexual Economies of Atlantic Slavery." Here, my overall argument is that *A Mercy* brings Atlantic slavery and specifically the (im)possibility for vertical motherhood for enslaved women to its textual orbit with the fragment of the [minha mãe]. By way of engaging with how *A Mercy* here turns to the script of the African American slave narrative, I argue that the function of this script in the [minha mãe]'s fragment needs to be understood as the novel's insisting on the active afterlives of the slave past as its ethical frame of reference.

4.1 "The Chagrin of Being Both Misborn and Disowned": [Jacob Vaark], Freedom, and the Pursuit of Property

[Routing the Argument] In this first close reading of my analysis of *A Mercy*'s [characters], I turn to [Jacob Vaark,] the Anglo-Dutch farmer, trader, money lender, and businessman, who assembles a group of women (his wife, servants, and slaves) on his patroonship and in his household in colonial Virginia. I argue that [Vaark] embodies the quintessential liberal subject as theorized by early Enlightenment thinkers like John Locke—a subject whose claims to freedom are made over and against the systems and practices of New World chattel slavery. In this way, the chapter confronts readings of [Jacob Vaark]'s [character] as a representation of someone, who is morally corrupted by his desire for material wealth at the New World colonial scene, and which stress that it is this greed which pushes him towards investing in rum and, therefore, in slave labor. Finally, I suggest that [Vaark]'s disnarration from the novel's textual orbit is a narrative maneuver that enables the text to meditate on the other [characters] and their existence on the New World scene in the wake of his death.

> Whatever the reasons, the attraction was of the "clean slate" variety, a once-in-a-lifetime opportunity not only to be born again but to be born again in new clothes, as it were. The new setting would provide new raiments of self. This second chance could even benefit from the mistakes of the first. In the New World, there was the vision of a limitless future, made more gleaming by the constraint, dissatisfaction, and turmoil left behind. It was a promise genuinely promising. With luck and endurance one could discover freedom; find a way to make God's law manifest; or end up rich as a prince. The desire for freedom is preceded by oppression; a yearning for God's law is born out of the detestation of human license and corruption; the glamor of riches is in thrall to poverty, hunger, and debt.
> — Toni Morrison, *Playing in the Dark*

> [T]he meaning and the guarantee of (white) equality depended upon the presence of slaves. White men were "equal in not being slaves." The slave is indisputably outside the normative terms of individuality to such a degree that the very exercise of agency is seen as a contravention of another's unlimited rights to the object.
> — Saidiya V. Hartman, *Scenes of Subjection*

Introduction

Within the first few pages of *A Mercy*, one of the novel's [characters] who is introduced as "[t]he man" slowly accesses the novel's narrative orbit as he moves "through the surf, stepping carefully over pebbles and sand to shore" (*AM* 7). His entry is arduous: slowed down by "Atlantic and reeking of plant life [which] blanketed the bay," he carefully moves through a fog that is different from "the English fogs he had known since he could walk, or those way north where he lived now" (*AM* 7). This one, which he needs to penetrate to find his way, was "sun fired, turning the world into thick, hot gold" (7). At first blush, it might seem that the metaphor of the fog marks his crossing from one world to another, from the old world to the New World. However, it also marks another watershed in the man's life. As the narrative progresses, the man makes his way

4.1 "The Chagrin of Being Both Misborn and Disowned" — 119

through the mud and swamp grass "stepping gingerly until he stumbled against wooden planks leading up beach toward the village" (7). Once the fog lifts, his movements become more and more confident. Upon his arrival "in the ramshackle village that sleeps between two huge riverside plantations" (8), the man buys a horse. He will continue his journey, which he makes to meet his business partner, the Portuguese slave and tobacco trader [Senhor D'Ortega], on horseback, "[m]ounted" (8). When the man signs a note to finalize the act of sale, the reader finally learns his name: "Jacob Vaark" (8). At this particular moment of the purchase, [Jacob Vaark] not only signs himself into being but also he signs himself into being as an owner. In the process of acquiring a horse, that is, [Vaark]'s signature gets intricately connected to the transaction and his name comes to signify ownership as well as personhood. [Vaark]'s arduous journey from a Northern to a Southern colonial scene thus comes to signify a radical transformation fueled by the modalities of possession. This becomes even clearer shortly after [Vaark] has arrived at the [D'Ortega] plantation. The purpose of the meeting is to settle a debt that [D'Ortega] owes to [Vaark]. When [D'Ortega] offers to repay [Vaark] with an enslaved girl, [Vaark] initially refuses, stating that "[f]lesh was not his commodity. [...] My trade is goods and gold" (20, 23). [Vaark] ultimately relents, however. Apart from echoing his first journey from England to the American colonies, then, [Jacob Vaark]'s crossing through the Southern fogs symbolizes his transition from a settler looking for greener pastures to a white possessing man, whose property decidedly will include enslaved human beings. Even more so, on his way back to his own farm "a plan was taking shape. Knowing full well his shortcomings as a farmer [...] he [now] fondled the idea of an even more satisfying enterprise. And the plan was as sweet as the sugar on which it was based" (33). As the text suggests, [Vaark]'s plan to invest in rum is fired by the envy he feels for [D'Ortega]'s palatial mansion, which he visits during their business meeting: "He had never seen a house like it" (13). And while [Vaark] seems to harbor concerns about his new business venture he nevertheless reassures himself that "there was a profound difference between the intimacy of slave bodies at Jublio [D'Ortega's plantation] and a remote labor force in Barbados" as he dreams of his third house, a "grand house of many rooms rising on a hill above the fog" (33). [Vaark]'s arduous journey through the Southern fogs, his coming into being through an act of acquiring property, and his dreams about his new house 'on a hill above the fog' fueled by his investments in chattel slavery in Barbados, then, signify his bringing unfreedom to early colonial unsettled hierarchies on the North American mainland.

[Jacob Vaark]'s narrative fragment is the first with which [Florens'] first-person text takes turns. Spanning twenty-seven pages, it is situated in the novel after [Florens'] opening segment and before the other [characters] and [Florens] tell their respective stories. My overarching thesis for this chapter began with the idea that [Vaark] undergoes a significant transformation in his segment of the novel from being "in many ways a good man" into a morally corrupted and greedy version of himself—again, a transformation that precisely is driven by the desire for a more elegant house, sparked during the above encounter with his Portuguese business partner (Gustafson and Hutner 212). [Vaark] wants this house to be similar to, yet not "as ornate as D'Ortega's. None of that pagan excess, of course, but fair" (*AM* 25). Put somewhat differently, I initially was following critical readings that set out to trace [Vaark]'s desire for material wealth and subsequent moral corruption in my own discussion of this [character] and his function in *A Mercy*'s storyworld. As I move through the chapter, it will become obvious how my initial thesis has expanded to be something much more unsettling—unsettling in the sense that it tries to think about [Vaark] as being a part of the New World grammar of property and the property paradigm from the very start and in this way fundamentally questions any "innocent" conceptualization of him as a morally and economically corrupted [character].

Overall, critics have discussed [Jacob Vaark] mainly along the lines of his being corrupted by his greed for economic prosperity and his subsequent reliance on the fast-gained profit from the sugar economies on Barbados.[89] That is, critics mostly read "along the grain" of the narrative in that they first stress how the "impoverishment and outsider status of his youth have inculcated in him a sensitivity to social justice" and then trace [Vaark]'s gradual moral corruption and capitalistic greed for wealth generated by remote enslaved labor forces in the West Indies (Babb 154). As Bellamy has it, for example, [Vaark] "best represents the nascent American spirit of adventure and self-reliance, while his egalitarian ethos enables him to create a household modeling harmonious relations between the races represented in the colonies" before he ultimately acquires "the damning trait of capitalist exploitation" (18, 19). And Strehle in another paradigmatic articulation of this strain writes that "Jacob reflects the best traits and intentions of the American pioneer, particularly the commitment to finding his own way in the new land without falling into the corrupt practices that he associates with Europe. [...] Claiming exceptional status, Morrison's American Adam purchases it with slave labor" (113, 114). Relatedly, some critics in their readings have followed [Vaark]'s narrative self-fashioning as a "ratty orphan become landowner" (*AM* 10), connecting his life as an abandoned and socially dispossessed

---

89 See, e.g., Anolik; Babb; Bartley; Bellamy; Gallego-Durán, "Representations"; Gustafson and Hutner; Karavanta; G. Moore; Strehle; Tally, "Contextualizing"; Tedder; Waegner. In general, it is fair to say that most critical readings of the novel tend to focus on [Florens] and her movement across New World territory and that critics tend not to engage with [Jacob Vaark] as much.

child to his assembling the women on his homestead to argue that [Vaark] is uniquely positioned to create and support a family-like community (Bellamy 18). Indeed, the text appears to suggest that it is [Vaark]'s "egalitarian ethos" and his "self-restraint and gentleness [which] lead people to bring him vulnerable young women who need protection" (Bellamy 18; Gustafson and Hutner 212). [Vaark] functions as an external hold to a community of *A Mercy*'s motley crew of New World outcasts, including his wife, servants, indentured servants, and slaves, all of whom are connected to each other through him (Neary and Morrison). This leads Strehle, for instance, to reason:

> Morrison's representation of American origins in *A Mercy* invokes the community that might have been: in the sense of good fortune, hope, and kindliness brought to the New World by the Vaarks, in the easy harmonies among the multiply ethnic laborers on the farm, in the teasing acceptance of differences among the women in the ship's hold, and in the laughing Indian boys' generosity to Florens. (Strehle 122)

Thus rising from poverty to wealth, [Vaark] and his segment in the novel have often been read as a different and more progressive "version of what will become the quintessential American dream" as well as representing "the possibility of an alternative white maleness that does not take advantage of arbitrarily constructed race and gender privilege" (Strehle 113; Babb 154). In a similar vein, Justine Tally situates [Vaark] in the historical context of American beginnings as she reads him as an allegory of foundational political moments within U.S. history. She continues to argue that "[n]otwithstanding the allusions via Jacob Vaark to the foundations of the future U.S. as a secular, tolerant society, almost everything else in the book 'screams' of Biblical reference on the one hand, and intolerance, on the other" ("Contextualizing" 66). Following Tally, with such [characters] as [Jacob] and [Rebekka], who are "both primary figures of the Old Testament," *A Mercy* offers "a rewriting of the Biblical model" ("Contextualizing" 66), so that *A Mercy*'s allegories ultimately become a "clear disruption of the Biblical myths, aimed at questioning both a divine legitimization of the ownership of property and right of dominion by the descendants of the original patriarchs" ("Contextualizing" 70).[90]

By contrast, my reading of [Jacob Vaark] aims to show that, rather than to think about [Jacob Vaark] as a version of the quintessential self-made man which represents the possibility of an alternative white maleness, this [character]

---

[90] In this critical context of *A Mercy*'s Biblical allusions, see Emerson for a discussion of "*A Mercy* and the Question of a Female Job" (2017); see also Stave's essay ""More Sinned Against than Sinning": Redefining Sin and Redemption in *Beloved* and *A Mercy*" (2013). See Bassard for a discussion of Toni Morrison's *Love* and the Bible (2014).

dons what Toni Morrison elsewhere calls "new raiments of self" (Morrison, *Playing* 34; see also first epigraph to the chapter). Lingering with the above description of the metamorphosis that his [character] undergoes as he moves South in order to conduct his business produces insights that shed a different light on [Vaark]'s existence at *A Mercy*'s New World colonial scene. We come to recognize his [character] as a propertied free man in this way. I argue that [Vaark] represents the quintessential Lockean subject, whose claims to freedom are made over and against the systems and practices of New World chattel slavery. [Vaark]'s freedom to opt for and invest in the West Indian sugar economies needs to be understood as a necessary step in the creation of his liberal self. The idea that I am trying to mobilize in the chapter, in other words, is that there is in fact no dissonance between, on the one hand, [Vaark]'s alleged "social justice ethos" and, on the other, his economic expansion and investments in Barbadian rum and, therefore, in slave labor. For him, ownership of slaves and reliance on their labor capacities becomes a matter of course. I also suggest that [Vaark]'s disnarration from the novel's textual orbit is a narrative maneuver that enables the text to meditate on the other [characters] and their existence on the New World scene in the wake of his death. In order to show how this works, I will first read his [character] against the mythical background of American colonial beginnings to then closely interrogate how *A Mercy* positions him as a quintessential Lockean subject, in a second step.

"A Ratty Orphan Become Landowner": Situating [Jacob Vaark]

[Jacob Vaark] enters *A Mercy*'s storyworld as the son of "a girl of no consequence who died in childbirth" and a "father, who hailed from Amsterdam [and] left him with a name easily punned and a cause of deep suspicion" (*AM* 30–31).[91] Orphaned and poor, the text represents [Vaark] as being lucky enough to have escaped this past when he is "taken on as a runner for a law firm. The job required literary and led to his being signed up by the Company" (31).[92] This job, as the

---

[91] This [character]'s name – [Vaark] – appears to conjure up the Dutch words *varken*, which means *pig* or *swine*, as well as *vaak*, which means *frequent* or *often* in English (my translation).
[92] I would suggest that *A Mercy* quite deliberately offers multiple, ambivalent readings or hints at which historical trading company it refers to. Most likely in my view, the text here alludes to The Virginia Company. After the first and failed attempt to settle at Roanoke, the English founded The Virginia Company, "a joint-stock venture, an early version of today's corporations. Wealthy London gentlemen would buy a share in The Virginia Company, thus giving it the capital monies to start and supply a colony, and they hoped the colony returned a profit to them.

first manifestation of his upwardly mobile existence, leads him to consider a life as a Company man in Barbados (9). Before he is able to act on this consideration, however, an "uncle he had never met from the side of his family that had abandoned him died and left him one hundred and twenty acres of a dormant patroonship" (9–10). And this settles it for [Vaark], who explains that "[i]nheriting land softened the chagrin of being both misborn and disowned" (31). Defying a life that had been "a mix of confrontation, risk and placating," [Vaark] is now "making a place out of no place, a temperate living from raw life" (10). As a trader, businessman, and landowner, [Vaark] relishes his independent life as a constant traveler in and across the North American colonial landscapes, "never knowing what lay in his path, who might approach with what intention. A quick thinker, he flushed with pleasure when a crisis, large or small, needed invention and fast action" (10). The text also suggests that he does so as someone who is not interested in the European's colonial scrambles for land: "Since land claims were always fluid, except for notations on bills of sale, he paid scant attention to old or new names of towns or forts: Fort Orange; Cape Henry; Nieuw Amsterdam; Wiltwyck" (11). Instead, he follows "his own geography [...] moving from Algonquin to Sesquehanna via Chesapeake on through Lenape since turtles had a life span longer than towns" (11). The text furthermore represents [Vaark] – who from "his own childhood he knew there was no good place in the world for waifs and whelps other than the generosity of strangers" (30) – as humble, sensible, and kindhearted, as someone who is readily infuriated by "the brutal handling of domesticated animals" (26) and who would dismount his horse "to free the bloody hindleg of a young raccoon stuck in a tree break" (9).

[Vaark]'s resolve not to let his past as a poor orphan determine his future as well as his compassion handily and easily situate him within a wider context of the hegemonic cultural scripts of the early European settler and of self-made manhood. These scripts or myths, as Paul reminds us, are the "popular and powerful narratives of U.S.-American national beginnings, which have turned out to be anchors and key references in discourses of 'Americanness,' past and pre-

---

King James I granted The Virginia Company a royal charter for the colonial pursuit in 1606. The Company had the power to appoint a Council of leaders in the colony, a Governor, and other officials. It also took the responsibility to continually provide settlers, supplies, and ships for the venture. The Company's plan was to identify profitable raw materials such as gold and silver in Virginia to repay the investors back in England. The first settlers included artisans, craftsmen, and laborers alongside the gentlemen leaders" ("Virginia Company"). By extension, *A Mercy* here may also hint at other European trading companies such as the Dutch West India Company, thus referring the reader to the broader historical frame of European colonial, economic, as well as transoceanic expansion across the globe.

sent" (Paul 11). Emerging from the dense fog covering the Chesapeake shores, [Jacob Vaark]'s story also invokes many of the early settlers' narratives of the alleged "discovery" of the New World.[93] As the prototypical European settler, [Vaark]'s narrative fragment adamantly summons European stories and reports of American discovery. Cathy Covell Waegner notes accordingly that [Vaark]'s "arrival is suggestive of the European explorers' first footsteps in original landfall [...] with [a] self-centered focus on his own 'breath and tread'" (Waegner 94).[94] A paradigmatic articulation of such narratives of original European landfall is John Smith's famous *A True Relation of Such Occurrences and Accidents of Note as Happened in Virginia* (published in 1608 and re-printed many times in various versions and editions, see Paul 95). As the "earliest published work relating to the colony at Jamestown, Virginia (the first permanent English settlement in North America), which is known to bibliographers; having been issued the year after the settlement was made" (Deane ix), Smith tells his readers about his expedition's arrival at the shores of the Chesapeake area and their first encounters with the Indigenous populations:

> within four or five daies after we set sail for Dominica, the 26. Of Aprill: the first land we made, wee fell with Cape Henry, the verie mouth of the Bay of Chissapiacke, which at that present we little expected, having by a cruell storme bene put to the Northward [...] The two and twenty day of Aprill, Captain Newport and my selfe with divers others, to the other number of twenty two persons, set forward to discover the River, some fiftie or sixtie miles, finding it in some places broader, & in some narrower, the Countrie (for the moste part) on each side plaine high ground, with many fresh Springes, the people in all places kindely intreating us, daunsing and feasting us with strawberries, Mulberries, Bread, Fish, and other their Countrie provisions wherof we had plenty: for which Captaine Newport kindely requited their least favours with Bels, Pinnes, Needles, beades, or Glassas, which so contented them that his liberallitie made them follow us from place to place, ever kindely to respect us. (Smith, *True* 1–6)

Here, Smith paints a vivid picture of how his expedition of twenty-two people sets out to discover what we will come to know as the James River and its immediate surroundings, assessing its flora and fauna as well as encountering local

---

93 On the "myth of discovery," see generally Paul, who explains that "the discourse of discovery and [the] powerful European projections [...] envision a new kind of paradise, a utopia somewhere across the Atlantic that alleviates the grievances of the 'old world' and that promises boundless earthly riches" (43; see also Morrison, *Playing in the Dark*).

94 In *A Mercy* we read: "As mud became swamp grass, he turned left, stepping gingerly until he stumbled against wooden planks leading up beach toward the village. Other than his own breath and tread, the world was soundless. It was only after he reached the live oak trees that the fog wavered and split" (7–8).

4.1 "The Chagrin of Being Both Misborn and Disowned" — 125

peoples who 'kindly intreat and feast' Smith and his crew with local produce and fruit, among others. Overall, Smith narrates those initial encounters between the native populations and his people as a mutual, friendly meeting during which all parties involved treat each other with respect and kindness. Like Smith, [Jacob Vaark] "carefully" and "gingerly" navigates the allegedly uncharted regions of the Chesapeake and he is "mindful" of his surroundings and the original inhabitants: "When he sailed the South River into the Chesapeake Bay, he disembarked, found a village and negotiated native trails on horseback, mindful of their fields of maize, careful through their hunting grounds, politely asking permission to enter a small village here, a larger one there" (*AM* 7, 11). And while the reader does not learn how the native inhabitants of the area that he travels respond to his presence on their territory (and in contrast to the passage from John Smith's narrative), *A Mercy*'s representation of the early colonial scene on several occasions suggests that [Vaark] seems to be particularly concerned with the well-being of the animals that he encounters during his travels. As mentioned already, [Vaark] would help free a young raccoon stuck in a tree break and when he sees "a man beating a horse to its knees," he is disgusted by the cruel mishandling of the horse (26). As I continue to move through the chapter, my reading of this [character] will question this narrative fashioning of [Vaark] as a settler who respects the original inhabitants, both human and animal, of the landscapes that he crosses on his way. I suggest that we look at the portrayal of [Vaark]'s [character] as *enacting* or *staging* that kind of representation in order to create the *effect* of him being mindful when, in fact, the text simultaneously moves to position him as someone for whom the colonial landscapes of Maryland and Virginia promised "control of one's own destiny" and offered "freedom and possibility" (Morrison, *Playing* 35, 34). (After all, [Vaark] claims a right to creating a new life and a future of freedom for himself in the New World as he moves through the Chesapeake colonies and thus "from social ostracism to social rank" (*Playing* 35). This will become clearer over the following pages.)

In many ways, [Vaark] also typifies the myth of the quintessential self-made man so prominent in the American imagination of itself, the paradigmatic example of which arguably is Benjamin Franklin (1706–1790) in the eighteenth century (Paul 370–373). In *The Myths that Made America* (2014), Paul notes that the "myth of the self-made man – with a story based on trust in the incentives of the capitalist market, adherence to the Protestant work ethic and luck – may be *the* prototypical modern American fairy tale" (379). As a "prominent figure of empowerment, emancipation, self-reliance and autonomy in the American cultural imagination," the mythical figure of the self-made man both represents and celebrates a certain kind of self-fashioning that is invested in individualism, independence, and free will (391). It also is deeply intertwined with "self-realization based on an ethic of self-interest that aims at the sheer accumulation of property, recognition, prestige, and personal gain without any concern for others" (310). Paul also reminds us how the core building blocks of this myth – self-reliance, individualism, and volition – and the "Pursuit of Happiness" (Greene) that it promises are not only connected to the formation of the modern nation state that the U.S. will become but also how they are inextricably bound by notions of private property. She writes:

> In many ways, the notion that individuals can determine their own future and change their lives for the better is a modern idea and presupposes modern notions of culture, society, and the individual [...] Coupled with the Calvinist work ethic, *the pursuit of happiness constructs the modern individual's path to happiness as the pursuit of property and allows for self-realization in new ways.* (Paul 369; emphasis mine)

In *A Mercy*, a third-person narrator tells the reader about [Vaark]'s journey to his business partner [D'Ortega] at the same time that it relates his life story as a rags-to-riches narrative of upward mobility and liberal, self-made manhood.[95] Refusing to be "sentimental about his own orphan status," [Vaark] begins his new life on English colonial soil on an inherited farm that "was sixty cultivated acres out of one hundred and twenty of woodland that was located some seven miles from a hamlet founded by Separatists" (*AM* 30, 31). As the text suggests, it is [Vaark]'s orphan(ed) status that has inculcated in him a sensitivity to equality. It is this sensitivity, which governs the ways in which he navigates his environment and that will determine how he builds the community that will people his homestead. Indeed, the text states that [Vaark] "found it hard to refuse when called on to rescue an unmoored, unwanted child" (31). Against this backdrop, [Vaark] slowly yet continuously expands his business ventures so that, over time, he becomes a "traveling man [...] Knowing full well his shortcomings as a farmer – in fact his boredom with its confinement and routine – he had found commerce more to his taste" (32, 33). In other words, *A Mercy* positions its readers to bear witness to this [character's] "pursuit of happiness" in the

---

95 Paul explains that as an important dimension of U.S. foundational mythology, the "popular phrase 'rags to riches' describes social mobility in analogy to geographical mobility in the discourse of westward expansion, the difference being that the latter refers to horizontal and the former to vertical mobility. Historically, the notion that upward mobility in US society is unlimited regardless of inherited social and financial status has been used to contrast the US to European societies with rigidly stratified social hierarchies, and to support the claim that the American economic system leads to a higher standard of living in general as well as to a higher degree of individual agency and economic opportunity" (367).

form of his broadening of his mercantile vocations from an orphan and hardworking farmer to a "seasoned" trader and businessman.

From the very beginning of the novel, this kind of broadening situates [Jacob Vaark] not solely within the above mythical context of the self-made man and settler but *strategically* also within a wider historical context of European mercantile expansion to the Americas, the triangular slave trade out of the African continent, and nascent racial capitalism. First, [Vaark]'s continuously expanding trading business both alludes to and represents the mercantile connections and the violent social, political, as well as cultural and capitalist economies of the transatlantic slave trade and colonialism that the European trading nations developed between the African continent, the Americas, and Europe from at least the fifteenth century to the nineteenth century (*Transatlantic Slave Trade Database*; Gilroy; Guasco; Wynter, "Unsettling," "1492"). With [Vaark]'s calculated investments in Barbadian rum, that is, the text strategically references the various phases of European colonization of the West Indies and of Barbados in particular. Historically, English colonizers came to establish permanent settlements on the island of Barbados in the early seventeenth century, taking over the colonization of the island from the Portuguese and the Spanish, who first came to Barbados in the first half of the sixteenth century and left the island "uninhabited, its original inhabitants having been enslaved or driven away [...] in the previous century" (Blackburn 225; see also Beckles; Fuentes; A. Taylor 206–218). Second, [Vaark]'s inheritance of the patroonship from a presumably Dutch relative unknown to him furthermore places him in the broader historical context of Dutch colonialism and slave trade.[96] What is more, it also places him in relation to the then simultaneously emerging legal systems that would regulate life in the New World colonies as well as various arenas of the triangular trade (*AM* 31; see also Bellamy 18, Strehle 113; Harris, "Whiteness"; Morgan, "*Partus*"). In general, while property laws in the English Atlantic have been important from the beginning, "especially since many new citizens did not or could not own property in their countries of origin," there also was "[d]isagreement among the colonies about continuing British legal traditions [which] resulted in differences in colonial laws—some colonies wanted to remain true to British legal tradition, whereas others chose to abandon some or all of the traditions" ("Property Law"). And

---

96 Dutch involvement in the transatlantic slave trade dates to the 1570s (*Transatlantic Slave Trade Database*). In early seventeenth-century North America, the Dutch would settle an area that would span parts of New York, New Jersey, Connecticut, Pennsylvania, and Delaware in today's United States. Among many others, they were active in the fur trade in this region as they built their existence and accumulated wealth and property through their business ventures (*New Netherland Institute*; Shorto; Tally, "Contextualizing").

while property commonly "is divided into two major areas: realty and personalty. Realty is land, whereas personalty is possessions—for instance, jewelry, money, furniture, or (formerly) slaves," state laws would regulate "who may purchase property, who may own it, and how it will be distributed upon the death of the owner or owners" on the colonial scene ("Property Law"). As Black feminist scholars continue to remind us, colonial law would also inaugurate a new symbolic order of property, gender, and race that would regulate descent, heritability, and status through *maternal* instead of paternal lineage (Morgan, "Partus"; Nyong'o; Spillers, "Mama's Baby"). (I will discuss the intricate legal and conceptual connections between private property and inheritance – specifically the inheritance of property/slave status and the regulation of maternal descent through colonial law – in my reading of [Sorrow] below). Third, [Vaark]'s [character] in *A Mercy* furthermore needs to be understood as referencing the discourses and literatures of discovery, exploration, and conquest produced by and in these broader historical and legal contexts (cf. generally Hulme, *Colonial*, "Spontaneous"; Mackenthun; Shields). Those discourses would typically be "not so much about the 'hosts' [...] [but] about their 'visitors,' i.e. those Europeans who arrive and 'discover'" and they would establish and represent colonial formations of power through a variety of metaphors and tropes, including tropes that would render colonial space as gendered territories to be taken, tamed, and dominated by European settlers, among many others (Paul 43, 92).

Against these historical-legal as well as "mythical" discursive backgrounds, therefore, the notion that [Vaark] comes into the possession of land through inheritance is not an innocently narrated fact told by the narrator in passing, as it were. Instead, it clearly is strongly interwoven with European mercantile expansion to the New World as well as with the mechanics that would simultaneously regulate the distribution and inheritance of property on a judicial plane. Joining the ranks of his historical forerunners like John Smith as well as of iconic self-made men from the archive of America's founding myths, [Vaark] represents a version of the quintessential European settler, adventurer, traveler, and businessman, who would make their way to the New World both in search for and driven by the longing for and a "vision of a limitless future, made more gleaming by the constraint, dissatisfaction, and turmoil left behind" (Morrison, *Playing* 34). As part of his vision of a limitless future, [Vaark]'s kindness to orphans furthermore needs to be read as something that he can afford *only* because he is *positioned* by slavery's and colonialism's formations of power as *being one of those who are in power*. While [Vaark] may indeed have some kind of residual "sensitivity of social justice" instilled in him, as many critics have it, the fact that he is able to afford kindness needs to be understood as a colonial act of bending down to some of the weakest—a weakness, however, that has been produced by the very acts of colonial settlement that he is a part of. For someone like [Vaark], affording kindness needs to be read as a kind of self-aggrandizing gesture essential to a position of colonial might. With this in mind, I expand on this situating of [Jacob Vaark] and read his [character] in relation to the New World liberal imagination of self, whose "texture of freedom is laden with [...] slavery," in the following section (Hartman, *Scenes* 116).

"New Raiments of Self": Claiming Freedom

Consider the following passage from the narrative fragment of [Jacob Vaark]'s wife [Rebekka], which follows [Vaark]'s and [Lina]'s fragments. In it, the reader learns that [Vaark]'s path to a new life as a trader and businessman, who "did what was necessary: secured a wife, someone to help her, planted, built, fathered [...] [and who] had simply added the trading life" (*AM* 32), also is a path toward the pursuit of property. As focalized through [Rebekka], we read:

> If on occasion he brought her *young, untrained help*, he also brought home gifts. A better chopping knife, a hobbyhorse for Patrician. It was some time before she noticed how the tales were fewer and the gifts increasing, gifts that were becoming less practical, even whimsical. A silver tea service which was put away immediately; a porcelain chamber pot quickly chipped by indiscriminate use; a heavily worked hairbrush for hair he only saw in bed. A hat here, a lace collar there. Four yards of silk. Rebekka swallowed her questions and smiled. When finally she did ask him where this money was coming from, he said, 'New arrangements,' and handed her a mirror framed in silver. (*AM* 86; emphasis mine)

When [Rebekka] subsequently questions her husband's motivation to build a third house, [Jacob] responds, "What a man leaves behind is what a man is" (87). While these lines lend themselves to a reading in which [Vaark] may indeed be "titillated by [the] luxury of riches" (G. Moore 8), I suggest that they point to the importance of private property as part of the making of liberal subjectivities. What this catalogue of practical as well as frequently more expensive and even "whimsical" gifts, ranging from household items and toys for his still-alive children over more extravagant things like a silver hairbrush and a lace collar, explicitly includes is the property of servants and/as slaves, as the phrase "young, untrained help" hints at. [Vaark] is not simply seduced into becoming a morally less pure version of himself through the prospect of material wealth, as suggested by most readers and critics. Instead, it is the space of the New World, which for this [character] opens up *the possibility to claim himself as a liberal self*. As his Indigenous servant [Lina] observes: "It was not a sudden change, yet *it was a deep one*. The last few years he seemed to be moody, less gentle, but when he decided to kill the trees and replace them with a profane monument to himself, he was cheerful in every waking moment" (*AM* 42; emphasis mine). What [Vaark] gains in the course of his narrative fragment – the *deep* change that he undergoes, if we adopt [Lina]'s perspective – is the freedom to become a master of himself and of his human, plant, and animal environment.

The way in which [Vaark] is able to achieve this, the way in which he is able to peel off his old-world status of being "misborn and disowned" (*AM* 31), is by re-inventing himself through his investment in Barbadian rum. When [Vaark] meets the slave and tobacco trader [D'Ortega] at his estate to settle the debt between them, he senses that his business partner must be unable to reimburse him in the way that they had agreed upon because "for some reason, he had been invited, summoned rather, to the planter's house—a plantation called Jublio. A trader asked to dine with a gentleman? On a Sunday? So there must be trouble, he thought" (12). [Vaark], who is exhausted, dirty, and sweaty from his travels, becomes acutely conscious of his status vis-á-vis his gentleman adversary, as the image of [Senhor D'Ortega]'s palatial plantation-mansion suggests. For even though [Vaark] "had heard how grand it was, [he] could not be prepared for what lay before him. The house, honey-colored stone, was in truth more like a place where one held court" (12). The dinner that follows "was a tedious affair made intolerable by the awkwardness Jacob felt. His rough clothes were in stark contrast to embroidered silk and lace collar. His normally deft fingers turned clumsy with the tableware. There was even a trace of raccoon blood on his hands" (15). When the business negotiations finally begin, [Vaark], as mentioned before, initially rejects slaves as compensation:

> D'Ortega had described with attention to minute detail the accidents beyond his control that made him unable to pay what he owed. But how Jacob would be reimbursed had not been broached. Examining the spotted, bug-ridden leaves of tobacco, it became clear what D'Ortega had left to offer. Slaves. Jacob refused. His farm was modest; his trade needed only himself. (19–20)

In this context, it is important to recall that [Vaark] never refuses slavery or the slave trade on principal grounds but that this [character] does not consider them part of his line of trade. When [D'Ortega] offers slaves as compensation, [Vaark] responds by saying: "'My trade is goods and gold, sir,' said Jacob Vaark, landowner. And he could not resist adding, 'But I understand how hard it is for a Papist to accommodate certain kinds of restraint'" (23). At this point, the reader is still led to believe that [Vaark] has no desire or use for slavery and the profits it generated for the likes of him—yet. Subsequently, however, we learn that [Vaark], like [D'Ortega], will not be able to restrain himself any longer.

4.1 "The Chagrin of Being Both Misborn and Disowned" — 131

When [D'Ortega] insists that he show him which slaves he has to offer [Vaark] and orders "some two dozen or more to assemble in a straight line [...] identifying talents, weaknesses and possibilities, but silent about the scars, the wounds like misplaced veins tracing [his slaves's] skin" (20), [Vaark] not only becomes angry at his adversary at the prospect of this kind of repayment but he also begins the feel "the shame of his weakened position like a soiling of the blood" (21). In an attempt to retaliate, [Vaark] asks that he is given the enslaved female cook, whose "clove-laced sweat" he recognizes from their dinner during which that woman served their food and to whom, [Vaark] suspects, "there was more than cooking D'Ortega stood to lose" (22). As expected, [D'Ortega] refuses. But when the woman suddenly offers her little girl to [Vaark] instead of herself, [D'Ortega] jumps at this chance and settles the deal. It is important to remember in this context that [Vaark]'s asking for the cook as compensation signifies much more than a simple act of retaliation. As suggested by the notion that [D'Ortega] "stood more to lose than cooking" if he gave his cook to [Vaark], *A Mercy* pushes its readers to recognize Atlantic slavery's sexual economies and reproductive calculus in this scene. What also is at stake here, then, is [Vaark]'s going into direct competition with [D'Ortega] for the fulfillment of white male sexual desire and for the reproduction of slave property.

When [Vaark] finally leaves Jublio, [D'Ortega], and their warring trade negotiations behind, he once again, "in spite of himself, envied the house, the gate, the fence" (25). As the text suggests, however, [Vaark] is confident, envisioning for himself a future in which he is equal to the likes of [D'Ortega]. We continue reading:

> For the first time he had not tricked, not flattered, not manipulated, but gone head to head with rich gentry. And realized, not for the first time, that only things, not bloodlines or character, separated them. So mighten it be nice to have such a fence to enclose the headstones in his own meadow? And one day, not too far away, to build a house that size on his own property? (25)

Indeed, [Vaark] resolves that this house, his third mansion, will not be "not *as compromised as* Jublio was. [...] Thin as they were, the dregs of his kind of Protestanism recoiled at whips, chains, and armed overseers. He was determined to prove that his own industry could amass the fortune, the station, D'Ortega claimed without trading his conscience for coin" (25, 26; emphasis mine). *Not as compromised as*. Herein, I suggest, lies the crux of the matter. For while [Vaark]'s own resolve to become [D'Ortega]'s equal (in terms of status and of wealth) *without* "trading his conscience for coin" perfectly lends itself to a reading that holds on to the notion that [Vaark] gets morally corrupted in the course of his section in the novel, I submit that this desire for a morally strong settler-[character,] who "recoils a whips, chains, and armed overseers," is in fact disrupted by the text's contrapuntal conjuring up of processes of liberal subject formation.[97] One such "contrapuntal movement" is the first of two questions in the above passage (Wilderson, "Aporia" 139). It suggestively points at the first of the English so-called enclosure movements, during which in the sixteenth and seventeenth centuries previously commonly owned and used ground would be converted into individually owned land. "With enclosures, fences were replaced by hedges and walls that make clear the owner's right to exclude other members of the community from it at any time. [...] Ownership of enclosed land did not depend on membership in any larger group; it was an exclusive right of access held by a single owner 'against all the world'" (Graeber 35). As such, the enclosure movements constituted one of several developments that led to notions of private property becoming integral to conceptualizations of liberal personhood (Graeber 35–36).

Another such textual contrapuntal conjuring occurs towards the very end of [Vaark]'s text in *A Mercy*. On his way back home, [Vaark] spends the night at an inn. Convinced by a "hawker turned middle man eliminating all hesitations and closing all arguments with promises of profit quickly," it is here that he makes the decision to invest in Barbadian rum (*AM* 29). We read:

> Sand moved under his palms; infant waves died above his wrists, soaking the cuffs of his sleeves. By and by the detritus of the day washed off, including the faint trace of *coon's blood*. As he walked back to the inn, nothing was in his way. There was the heat, of course, but *no fog, gold or gray, impeded him*. Besides, a plan was taking shape. [...] And the plan was as sweet as the sugar on which it was based. And there was a profound difference between the intimacy of slave bodies at Jublio and a remote labor force in Barbados. Right? Right, he thought, looking at a sky vulgar with stars. Clear and right. The silver that glittered there was not at all unreachable. And that wide swath of cream pouring through the stars war his for the tasting. [...] [H]is dreams were of *a grand house of many rooms rising on a hill above the fog*. (33; emphasis mine)

Taking up the previous image of traces of raccoon blood on [Vaark]'s hands, the text here contracts the words "a trace of raccoon blood" it into "coon's blood." It thus evokes the nascent economies and affiliations of liberty and bondage that would, among others, enact systems of stereotypical classification and racist, ab-

---

[97] Again, the slave makes possible the existence of the white liberal subject, whose understanding of itself as a free individual and as an equal among its peers within Western social formations is tied to colonial regimes of law and of claimed rights to objects owned, of claimed rights to the enslaved. As Hartman has it, "The slave is the object or the ground that makes possible the existence of the bourgeois subject and, by negation or contradistinction, defines liberty, citizenship, and the enclosures of the social body" (*Scenes* 62).

jecting social and epistemic formations. These were to become, as Spillers writes in "Mama's Baby, Papa's Maybe," an American grammar or "a meeting ground of investments and privations in the national treasury of rhetorical wealth. My country needs me, and if I were not here, I would have to be invented" (65). While this shift may appear to some critics to represent how [Vaark] washes off his concerns about being involved in the practices of enslavement (a notion that seems to be supported a few lines below by his self-reassuring internal dialogue about the "difference between the intimacy of slave bodies at Jublio and a remote labor force in Barbados"), I suggest that the text here in fact depicts the making of the liberal subject that I traced in my reading of paradigmatic texts from the late seventeenth-century English Atlantic. Contracted into one short phrase, that is, [Vaark] literally washes his "dirty unfreedom" off his sleeves as he dons his new liberal self. What [Vaark]'s [character] effectively embodies in this moment in the narrative is the white liberal subject's "assumption of freedom, i.e. the generative semiosis of an individual subject as the owner of a right to freedom" (Broeck, "Property" par. 3). With this "self-authorizing gesture [...] par excellence," [Vaark] comes to exist as a free man, whose movement across the New Word colonial landscapes no longer is obstructed by fog, neither gold [n]or gray (Broeck, "Property" par. 3). Instead, in washing "coon's blood" off his hands, the text renders [Vaark] as being able to master his own affairs, as a master of himself who is able to pursue "his solitary, unencumbered proficiency" (*AM* 20–21).

Put another way, by figuratively washing off his residual resistance in dealing with "coons" in the slave trade and his profiting from their labor power, [Vaark] wants the reader to believe that he does not have blood on his hands because he does not resort to "slave bodies" in the same way that [D'Ortega] does. What his textual fragment ultimately suggests, however, is that there is in fact not much of a difference between these two [characters] and their respective investments in the slave trade. At stake, then, is not so much the degree to which [Vaark] lets himself be seduced by the prospect of material wealth that his encounter with [D'Ortega] arguably portrays. At stake is the constitutive force that slave ownership generates for the liberal imagination of/and the self. "The silver that glittered there was not at all unreachable. And that wide swath of cream pouring through the stars was his for the tasting." What [Vaark] wants is a piece of the cake of freedom—and freedom, as suggested by these lines, is his for the tasting. For [Vaark], the silver that glitters high above him in the sky represents his vision of what Morrison in the first epigraph to the chapter calls a "limitless future." While materializing in his seemingly innocent aspiring to more material wealth and autonomy, these line show that his existence as a free man in the New World has in fact never been independent from modernity's grammar of private property. [Vaark] comes to take slave property for granted in the same way that Locke's liberal subject assumes slave ownership as a matter of course (see previous chapter). As remote as the labor forces that he will depend on may be, the step from a farmer to a trader involved in the transatlantic sugar economies is represented as both inevitable and logical in his segment of the novel. In this sense, [Vaark]'s gifts of "untrained help" (read: slave property) to his wife logically follow from the "new arrangements" he has made and, as such, need to be understood as part of the liberal subject's claim to freedom.

[Vaark]'s dreams about "a grand house of many rooms rising on a hill above the fog" in many ways resonates with the Pilgrims and Puritans' vision of their exceptional, morally righteous and religiously free New World community (cf. Gura). That is, it resonates with their vision of a "city upon a hill." As put forth John Winthrop's famous lay sermon "A Model of Christian Charity" (1630), this vision has become one of the hegemonic narratives of American beginnings (Paul). Indeed, the notion that the Vaark farm represents an "exceptional community" and a refuge provided by [Vaark] in the strange as well as nascent capitalist environment of colonial Virginia analogously has been picked up on in the novel's critical discourse. As Bellamy has it, for example, [Vaark]'s "embrace of the landscape and the possibilities of the New World, tempered by his general disdain for slavery and sensitivity to the struggles of other abandoned children, positions Vaark *to create and support an idealized New World family*" (18; emphasis mine). We can also recognize in the above lines echoes of Benedict Anderson's thoughts on how modern nations need in fact be conceptualized as *imagined*, as political *communities* formed in connection with a vast range of political, social, and ideological constellations, conceived by their members "as a deep, horizontal comradeship" (7). However, my reading of this [character] and its narrative fragment ventures to suggest something different. At *A Mercy*'s "pre-national" New World colonial scene, the [Vaark] community, at best, gestures towards the Pilgrims and Puritans' communal vision as well as towards the modern nation state that English colonies would become in the next century on allegorical terms. That is, it gestures towards them in allegorical anticipation and with the knowledge that their visions and versions of national, political community will become deeply tethered to a system of racial capitalism and slavery. And while the text arguably points the reader to the notion that [Vaark]'s assembling of the women could be read as [Vaark] offering them refuge on his homestead (promising them a "second chance" as they face a violent restructuring of their existence that was set in motion by settler colonialism and slavery, respectively), *A Mercy* contradicts this impression when [Vaark] contracts the smallpox disease and dies. The reader learns about his death in the narrative fragments of the other [characters]. Towards the end of the novel we continue reading accordingly: "Such were the ravages of Vaark's death. [...] They once thought they were a kind of family because together they had carved companionship out of isolation. But the family they *imagined* they had become was false" (153–154; emphasis mine). I suggest that the text here purposefully situates the community on the [Vaark] farm to represent the liberal subject's coming into being. Put another way, there are at least two ways in which notions of community need to be thought together with the fashioning of [Vaark]'s liberal self. First, if we connect the creation of the community on this homestead to [Vaark]'s embrace of financial investments in the New World's sugar economies at the end of his narrative fragment, what comes to the fore is that his assembling of the/ "his" women essentially both prefigures and configures the making of his new life and self. Again, we read that [Vaark] "did what was necessary: secured a wife, someone to help her, planted, built, fathered" (32).[98] Second, the community forged on the [Vaark] farm and, by extension, the protection it affords those women (if we can call it that) is temporary and frail. It is of no consequence. It is granted from a position of power and not afforded or created in shared lateral community. This reality violently surfaces as soon as [Vaark]'s untimely death removes him from the novel's plotting, leaving the other members of "his community" behind to fend for themselves (e. g., *AM* 87). What, then, are the "ravages" of [Vaark]'s death and subsequent removal from the text? What is at stake in the novel's refusal to keep him in the narrative plotting?

[Coda]: Disnarrating [Jacob Vaark]

So far, I have been trying to make a case that [Vaark]'s narrative fragment in *A Mercy* constitutes a paradigmatic instance of liberal self-making. I have argued that his desire to no longer be "disowned" but to be a self/himself needs to be understood as being fueled by his claim of right to freedom through property. I also suggested that the text deploys concepts of "community" and "family" in service of [Vaark]'s liberal fashioning of self and not as a means to create a refuge for the women that he congregates on his homestead. As I have tried to show in response to dominant critical readings of *A Mercy*, there are no "easy harmonies" (see Strehle, above) on the [Vaark] patroonship. Arriving at the end of the chapter, I want to think about and unpack some more the relationship between

---

[98] Ruth Bienstock Anolik observes in this context, "Morrison plants early clues, including his name, to suggest that Jacob does not partake of the communal ethos" (420).

[Vaark]'s liberal self-fashioning and his untimely death and subsequent removal from the narrative.

Shortly after construction work on his third house begins, [Vaark] falls ill with the smallpox. As he collapses, he orders the women on his farm to bring him to his half-built new mansion. We continue reading:

> All the while he croaked, hurry, hurry. Unable to summon muscle strength, he was deadweight before he was dead. They hauled him through a cold spring rain. Skirts dragging in mud, shawls asunder, the caps on their heads drenched through to the scalp. There was trouble at the gate. They had to lay him in mud while two undid the hinges and then unbolted the door to the house. [...] His eyes shifted to something or someone over [Rebekka's] shoulder and remained so till she closed them. All four – herself, Lina, Sorrow and Florens – sat down on the floor planks. One or all thought the others were crying, or else those were raindrops on their cheeks. (*AM* 87–88)

In line with what I have previously discussed as his assembling the motley crew of women on his farm as part of his liberal self-fashioning, the paragraph shows how [Vaark] continues to make use of these women on the brink of his death when he orders them to take him to die in his unfinished third house, the emblem of his liberal being. As such, he is a burden that the women must carry, quite literally, as illustrated by the phrase "he was deadweight before he was dead." Furthermore, [Vaark]'s elimination from *A Mercy*'s text extends onto his immediate offspring. All the [Vaark] children die prematurely, with their deaths either resulting from disease or, as in the case of [Vaark]'s daughter, an accident (*AM* 53–54, 79, 85, 87, 90). Even though [Vaark] "was confident [Rebekka] would bear more children and at least one, a boy, would live to thrive" (19), he ultimately will not leave behind "acceptable heirs" (Waegner 95). Without heirs, [Vaark]'s [character] is unable to continue his liberal self-fashioning. Spillers reminds us of the intricate connections between liberal self-making and heritability, that is, between liberal self-making and the

> *vertical* transfer of a bloodline, of a patronymic, of titles and entitlements, of real estate and the prerogatives of "cold cash," from *fathers* to *sons* and in the supposedly free exchange of affectional ties between a male and a female of *his* choice [that would] become the mythically revered privilege of a free and freed community. (74)

I will return to the entanglements between white liberal self-making and the formation of recognized kinship structures in my reading of [Sorrow]. For now, I would like to suggest that the novel cancels out any possibility for a continuation of the script of the liberal subject within its own textual orbit by way of disnarrating both [Vaark] and his children.

Finally, [Vaark]'s version of the quintessential liberal subject is not meant to last. His death, and by extension, the deaths of his children constitute a literary maneuver with which the text refuses to inscribe or rewrite fashionings of Western modernity's liberal self. There is, in other words, no future for this representation of liberal selfhood, which has brought unfreedom to *A Mercy*'s still settling racial hierarchies, within the novel's narrative orbit. The phrase "he was deadweight before he was dead" thus takes on yet another meaning in this context. It is the [character] of [Jacob Vaark] itself that needs to be understood as becoming "deadweight" for *A Mercy*'s allegorical project. What I suggest is that [Vaark] needs to disappear from the narrative in order for *A Mercy* to meditate on the ways the property paradigm positions the other [characters] at the New World colonial scene. Without [Vaark], *A Mercy* is able to interrogate how the other [characters] strive towards or embody and deconstruct or reject, respectively, the property paradigm. The following chapters will examine how the novel scrutinizes the New World's grammar of property and its positioning power for [Lina,] [Rebekka Vaark,] [Sorrow,] [Florens,] and, finally, the [minha mãe]. By way of examining what is at stake in the various versions of New World self-making that *A Mercy*'s allegories speak to on the literary level of representation, the close readings that follow will trace *A Mercy*'s insistent questioning of the relationship between private property, the violence of Atlantic slavery, social death, and literary narrative.

4.2 "I am Exile Here": [Lina], Self-Inventions, and Dispossession

[Routing the Argument] The chapter reads the [character] of [Lina] for the ways in which it navigates *dispossession* at *A Mercy*'s New World colonial scene. I argue that this [character] becomes a representation of *dispossession* as that which describes the capacity or "powers [that subjects] have or lack" within the novel's experimental setup (Wilderson, *Red* 8). [Lina]'s existence within the novel's seventeenth-century landscapes is fundamentally shaped by the genocide and complete eradication of her tribe by the European colonizers. In the wake of these violent events, [Lina] completely reinvents her self (*AM* 48), becoming not only a servant to [Jacob Vaark] and his wife [Rebekka]; she also imagines herself to be a part of the "small, tight family" on the [Vaark] homestead. Here, she becomes the one who "ruled everything and decided everything Sir and Mistress did not" (56, 120). I suggest that [Lina]'s [fragment] contrasts with those of, for example, [Florens] and the [minha mãe], who need to be understood as being positioned by the "replaceability and interchangeability endemic to the commodity," i.e. by their *fungibility* (Hartman, *Scenes* 21). I claim that [Lina]'s fragment shows us what it means to be dispossessed *but not the fungible property of someone* within *A Mercy*'s representation of seventeenth-century colonial Virginia. [Lina]'s [character] pushes the reader to ponder questions of *What could have been? What else could have happened?* by way of showing us that [Lina] gains status as someone who effectively runs the [Vaark] farm, becoming [Rebekka]'s loyal, docile servant and "friend." However, I argue, the text ultimately exposes [Lina]'s "exceptional" status on the [Vaark] homestead as a "folly" (*AM* 56).

> Had I not cradled you in my arms,
> oh beloved perfidious one,
> you would have died.
> And how many times did I pluck you
> from certain death in the wilderness—
> my world through which you stumbled
> as though blind?
>
> ....
>
> I spoke little, you said.
> And you listened less.
> But played with your gaudy dreams
> and sent ponderous missives to the throne
> striving thereby to curry favor
> with your king. I saw you well. I
> understood the ploy and still protected you,
> going so far as to die in your keeping—
> a wasting, putrefying death, and you,
> deceiver, my husband, father of my son,
> survived, your spirit bearing crop
> slowly from my teaching, taking
> certain life from the wasting of my bones.
> — Paula Gunn Allen,
> from "Pocahontas to Her English Husband, John Rolfe"

> For though the Indian exists liminally in relation to the Settler, as do the Settler's children and "his" Old World peasants, he or she remains ontologically possible. That is to say, the "Savage," unlike the Slave, is half-alive.
> — Frank B. Wilderson, *Red, White, and Black*

Introduction

The chapter reads the [character] of [Lina] for the ways it navigates *dispossession* at *A Mercy*'s New World colonial scene. [Lina], whose fragment follows that of her master [Jacob Vaark], is the only Indigenous [character] in the novel. In the wake of the eradication of her village and tribe by European colonizers, she first is taken to a group of "kindly Presbytarians," where she becomes their servant, and is subsequently sold to serve [Jacob Vaark] (*AM* 45, 50). It is here, the text suggests, that she finds "a way to be in the world" (46). What follows is my attempt at unpacking how [Lina] needs to be understood as a dispossessed presence within the [Vaark] household but not as the fungible property of [Jacob Vaark] and his wife [Rebekka].

Within the American imagination, the historical figure of Pocahontas – or Matoaka, who was born around 1595 – functions as a "Native American foundational figure" (Paul 90, 89). Historians of American colonial beginnings have shown how at the time of the first English settlements colonial discourse would often resort to exoticizing images and metaphors, which "depicted the Americas as an allegorically feminized space" to be conquered by the European settlers (Paul 91. In this way, the encounter between the European colonizers and the indigenous populations was often represented in highly eroticized terms (Paul; Hulme, *Colonial Encounters*; Mackenthun). Against this backdrop, Pocahontas would become quintessential to representations of Indigenous and European relations, moving to "the core of an American foundational myth that for a long time has been considered the first love story of the 'new world' and thus paradigmatic for casting intercultural relations in the early colonial history of the Americas as harmonious and peaceful" (Paul 89). At the center of this myth essentially are Pocahontas' encounters with two male English settlers: the soldier and writer Captain John Smith, who was fundamentally involved in establishing Virginia as England's first permanent colony on the North American mainland, and the tobacco planter and secretary and recorder general of Virginia, John Rolfe, respectively (McCartney; Salmon). While Pocahontas famously rescues the former from death at the hands of her father Powhatan, allegedly maintaining a live-long friendship with Smith, she would get married to the latter in 1614, give birth to their son in 1615, and travel to England where she died in 1617 (Paul 90; see also Hulme, *Colonial Encounters*; Larkins). At the center of the myth of Pocahontas, in other words, are notions of {seemingly} peaceful colonial encounters and intercultural romance (Mackenthun).

The above epigraph comes from Native American poet Paula Gunn Allen's poem "Pocahontas to Her English Husband, John Rolfe."[99] The poem is Allen's rendering of the mythically famous marriage between Pocahontas and John

---

[99] I first came across this poem when I read Heike Paul's *The Myths That Made America* (2014). My thanks go to Paul for drawing my attention to Paula Gunn Allen's work.

Rolfe. In it, Pocahontas is the one who is takes care of her European husband, as Allen "constructs a stance of superiority on the part of Pocahontas vis-à-vis her husband" (Paul 117). The above verses from the poem show that the speaker "reverses well-known [colonial] stereotypes: it is [Rolfe] who is the 'other'—ignorant, childlike, helpless, and dependent; it is [Pocahontas] who rescues him not once, but many times; and yet, in his world/discourse, she does not have a voice. Ultimately, she holds him responsible for her death, which is intricately connected to his acquisition of fame and fortune" (Paul 117). With the notable exception of John Smith who "actually refer[s] to words she ostensibly addressed to him verbatim" in his works (Paul 89), Pocahontas is voiceless in most reports, treatises, and other texts written and published by the early English settlers in and about colonial Virginia. In giving Pocahontas a voice, Allen's poem fundamentally confronts such discursive erasures.

Allen's poem as well as the mythical rendering of [Pocahontas] as a symbol for successful intercultural relations help me frame my reading of [Lina]'s fragment as a text that interrogates notions of *dispossession* at *A Mercy*'s New World colonial scene. In the novel, we encounter this discursive legacy with the [character] of [Lina] and we do so in the context of the novel's negotiation of what it means to be dispossessed but *not* someone's disposable and usable property. I suggest that, with [Lina], the text urges the reader to consider how early European modernity's grammar of property plays out and determines her status of being and her existence in *A Mercy*'s landscapes of colonial Virginia along the lines of *dispossession* rather than in terms of the "replaceability and interchangeability endemic to the commodity," i.e. *fungibility* (Hartman, *Scenes* 21). In Allen's poem, we read "with your king. I saw you well. I / understood the ploy and still protected you" (Allen 7). Like Pocahontas in Allen's poem, [Lina] sees through her master's inabilities and helps him to get his farm running, passing on her indigenous knowledge to him. Early on in [Lina]'s fragment, that is, we learn that it was [Lina] who "taught [Jacob Vaark] how to dry the fish they caught; to anticipate spawning and how to protect crop from night creatures. [...] Lina didn't know too much herself, but she did know what a poor farmer he was" (*AM* 47). Over the course of the twenty-four pages of her fragment, we also learn that [Lina] not only believes her relationship with [Rebekka Vaark] to be a friendship (51) but that she remains devoted to her mistress even when [Rebekka] suddenly treats her merely as a servant after her recovery from the smallpox (158). Like Pocahontas in Allen's poem, then, [Lina] is the one who 'plucks' her master and mistress from the wilderness, 'protecting' and supporting them when and where she can, nursing her mistress back to health. She also sees through [Jacob Vaark]'s 'gaudy dreams' of building a third house: "Killing trees in that number, without asking their permission, of course his efforts would stir up malfortune. Sure enough, when the house was close to completion he fell sick with nothing else on his mind. He mystified Lina. All Europes did" (42).

My argument in the chapter runs counter to readings of the novel that view the congregated group of female [characters] on the [Vaark] farm as a community, whose members, in their respective attempts to negotiate their existence in the uncharted waters of the New World, are being subjugated, commodified, or dispossessed in similar or equal ways (e. g., Babb; Bartley; Cholant; Gallego-Durán; Waegner). Such readings privilege questions concerning notions of female bonding, cultural hybridity, and solidarity over an ensemble of questions that tries to get to the core of what "being dispossessed" actually means for each of the novel's female [characters]. An example of this is Waegner's article in which we read that "[o]ne of the items on Morrison's agenda in [*A Mercy*] is to document that slavery and forced servitude were not necessarily tied to a particular race or color line in the 17th-century Americas," which then turns into an argument about "self-empowerment and solidarity among the women" in the novel (Waegner 101, 98). In ""I Am a Thing Apart": Toni Morrison, *A Mercy*, and American Exceptionalism" (2013), Strehle takes this argument into another direction when we read that [Jacob Vaark] "buys one slave (Lina) because he needs help on the farm, and he accepts two (Sorrow and Florens) in payment for debts" (114). However, if we think about [Lina]'s positioning in critical relation to post-slavery Black Studies thinkers' arguments about the ways in which slavery's foundational violence singularly positions the enslaved through their *fungibility* (see Chapter 3), we arrive at the notion that [Lina] and her positioning in the novel's representation of colonial Virginia cannot be read in the same way as, for example, [Florens]. Although [Jacob Vaark] buys her from the Presbytarians when he is looking for help for his farm (*AM* 50), [Lina] is not a slave but is positioned by the fact that she survives the genocide of her village. This event leaves her traumatized and makes for the fact that she essentially is "rescued as much as bought" by various European settlers and colonizers (Anolik 419). I argue that in oscillating between, on the one hand, (the survival of) genocide as well as the loss of her tribe, her kin, and her freedom and, on the other, being rescued, being made a servant, and craving to find a place and a space where to exist, her [character] fundamentally speaks to the novel's governing question of *What else could have happened?* That is to say, in urging us as readers to think about what it means for [Lina] to be dispossessed at the New World colonial scene, *A Mercy* leads us down a path that history could have taken and it does so by exposing us to the notion that [Lina], at least temporarily, does in fact become a member of the "small, tight family" that [Jacob Vaark] and [Rebekka] create on their farm (*AM* 56). This kind of imaginative probing into historical paths not taken raises the following questions, among others: What kind of formation of "family" is this, exactly? How can we think about the liminal space between, on the one hand, surviving genocide and the eradication of one's kin as well as experiencing the fundamental loss subtended by these events and, on the other, being able to embrace Human capacity to some degree—in [Lina]'s case, being able to carve out a space for her self on the [Vaark] farm? In what follows, I will engage these questions as I trace the ways in which her [character] opens an avenue for us as readers to think about *dispossession* at *A Mercy*'s New World colonial scene.

Surviving Genocide, Inventions of Self: [Lina] at the New World Colonial Scene

On Being Both Rescued and Bought

In his *Generall Historie of Virginia* (1624), John Smith has famously included a scene in which he, after being captured by Pamunkey chief Opechancanough, is taken to the chief's brother Powhatan and almost gets killed by his captors. Smith writes that after a "long consultation [...] two great stones were brought before Powhatan: then as many as could layd hands on him, dragged him to them, and thereon laid his [John Smith's] head, and being ready with their clubs, to beate out his braines" (*Generall* 49). Legendarily, Pocahontas comes to his rescue when she, as "the Kings dearest daughter, when no intreaty could prevail, got his head in her armes, and laid her owne vpon his to saue him from death" (*Generall* 49). While the incident described in the scene appears to have occurred in 1607, it was not included in Smith's *True Relation of Virginia* published in 1608. Ever since it has been added to the *Generall Historie* seventeen years later (Hulme, *Colonial Encounters* 140), the rescue scene has been "turned into the first great American romance" and thus gained mythically prominent status within hegemonic narratives of American beginnings (Hulme, *Colonial Encounters* 138).[100]

In *A Mercy*, we also encounter a rescue scene of some sort. In this one, however, it is [Lina] who, in line with colonial discourse's gendered representations of the New World (and Virginia, especially) as a "mysterious feminine/feminized space to be penetrated, conquered, and domesticated by the English settlers"

---

[100] For a detailed discussion of the implications of the scene's addition to John Smith's descriptions of his adventures and discoveries in the New World, see Hulme, *Colonial Encounters*; Paul.

4.2 "I am Exile Here": [Lina], Self-Inventions, and Dispossession — 143

(Paul 92), is being "rescued" by different groups of European settlers and colonizers. As the first woman to live and work at the [Vaark] farm, [Lina] enters the novel in the wake of the annihilation of her tribe by European settlers. At the beginning of her fragment in *A Mercy*, when the smallpox hit her village as a result of a gift of infected blankets which her people "could neither abide nor abandon" (*AM* 44), [Lina] together with two young boys first attempts to keep the wildlife from feasting on the bodies before "[n]ews of the deaths that had swept her village had reached out" and a group of French soldiers arrives (44).[101] Relieved at first when the soldiers come to take her with them, [Lina]'s "joy at being rescued collapsed when the soldiers, having taken one look at the crows and vultures feeding on the corpses strewn about, shot the wolves then circled the whole village with fire" (44–45).

Thus orphaned by the genocide committed against her people by European settlers as well as "rescued" by French soldiers, [Lina] consequently is forced to adjust to her new surroundings at the dwelling place of "kindly Presbyterians" to which the soldiers take her (*AM* 45). We read:

> They named her Messalina, just in case, but shortened it to Lina to signal a sliver of hope. Afraid of once more losing shelter, terrified of being alone in the world without family, Lina acknowledged her status as heathen and let herself be purified by these worthies. She learned that bathing naked in the river was a sin; that plucking cherries from a tree burdened with them was theft; that to eat corn mush with one's fingers was perverse. That God hated idleness most of all, so staring off into space to weep for a mother or a playmate was to court damnation. Covering oneself in the skin of beasts offended God, so they burned her deerskin dress and gave her a good duffel cloth one. They clipped the beads from her arms and scissored inches from her hair. Although they would not permit her to accompany them to either of the Sunday services they attended, she was included in the daily prayers before breakfast, midmorning and evening. (*AM* 45–46)

For [Lina], "being rescued" means to be captured and to be domesticated. As the paragraph shows, [Lina] tries everything to adapt to the Presbyterian's Christian worldview, culture, and religious traditions in the refuge that the missionaries appear to provide her with because she is afraid of 'being alone in the world.' This includes the forced negation of her Indigenous customs and beliefs,

---

101 Strehle explains how *A Mercy* refers to the widespread historical narrative that one way in which Indigenous peoples were wiped out was through the gift of blankets by European settlers. Strehle also more generally notes that "[h]istorical evidence is inconclusive about whether the disease was actually transmitted through blankets, since smallpox affected both Europeans and Indians throughout the area before the gift of blankets. It is certain, however, that the British tried to "Extirpate" Indians in order to seize their lands and were largely successful; Indian populations were decimated by disease and warfare" (116).

which is illustrated not least by the fact that the Presbyterians cut off her hair and equip her with European clothes. "Being rescued" not only from the scene of the extermination of her kin (caused by Europeans in the first place), the paragraph also seems to suggest that the Presbyterians believe that they rescue [Lina] from her Native traditions and "heathenish" customs. For not only do they pray for her regularly but they also give her a new name signaling 'a sliver of hope.' Later in the novel, we learn that [Lina] is subjected to physical (and, presumably, sexual) violence while staying at the Presbyterians' and although we do not learn who inflicts this violence on her, we read that this person is a "man of [...] learning and position in the town." Their meetings happen in "secrecy and when he comes to the house I [Lina] look him in the eye. I only look for the straw in his mouth [...] or the stick he places in the gate hinge as the sign of our meeting that night" (102). In [Florens]' fragment, this person is described as [Lina]'s "lover" (103). However, we also encounter the notion that [Lina] also has to serve at his place and is severely beaten by him when she loses money with which she had been entrusted: "The Spanish coin is lost through a worn place in her apron pocket and it never found. He cannot forgive this" (102).[102]

Over the course of her textual fragment, the text tethers the notion that she is "rescued" by the Presbyterians to her being purchased by [Jacob Vaark]. As mentioned earlier, [Lina] is the first woman whom [Vaark] brings to his farm to live and to serve. However, unlike the other women [characters] in *A Mercy*, [Lina] is represented as the only one of those [characters] whom [Vaark] buys "outright and deliberately" when prior to the arrival of his wife-to-be [Rebekka] he purposefully searches for a domestic servant to work and reside on his estate (*AM* 32, 50). We read:

> [Lina] had been a tall fourteen-year-old when Sir bought her from the Presbyterians. He had searched the advertisements posted at the printers in town. "A likely woman who has had small pox and measles. . . . A likely Negro about 9 years. . . . Girl or woman that is handy in the kitchen sensible, speaks good English, complexion between yellow and black. . . . White lad fit to serve. . . . Wanted a servant able to drive a carriage, white or black. . . . Sober and prudent women who. . . . Healthy Deutsch woman for rent . . . stout healthy, healthy strong, strong healthy likely sober sober sober . . . " until he got to "Hardy female, Christianized and capable in all matters domestic available for exchange of goods or specie." (*AM* 50)

---

[102] The above paragraph from *A Mercy* also points the reader to the violent histories and legacies of boarding schools founded by the federal U.S. government, to which Indigenous children were deported from the 1870s. Once at the federal boarding schools, students were forbidden to express their Indigenous cultures in any way and forced to completely transform their existence, forcefully assimilating to white American mainstream society and culture (Bear).

4.2 "I am Exile Here": [Lina], Self-Inventions, and Dispossession

As the paragraph shows, [Vaark] finds [Lina] through an advertisement in which she is described as a 'Christianized' and 'hardy' female domestic servant. With the lines immediately following this paragraph, the text furthermore establishes a connection between [Lina] being put up for sale by Presbyterians and the notion that she has been abused and injured either by them or, more likely, by the man she meets secretly at night. We learn that there "is not rum the second time nor the next […] but those times he uses the flat of his hand when he has anger, when she spills lamp oil on his breeches or he finds a tiny worm in the stew. Then comes a day when he uses first his fist and then a whip" (102) and that the Presbyterians do not ask her what had happened to their servant and instead immediately put her up for sale (50, 102–103). We continue reading that by the time that [Vaark] "acquires" [Lina], her "swollen eye had calmed and the lash cuts on her face, arms and legs had healed and were barely noticeable" (50). With [Lina] being subjected to this kind of violence, then, the text tethers the notion that she is being rescued to an early colonial marketplace, as illustrated by words like "Spanish coin" as well as through the above list of all kinds of (indentured) labor for sale at the village printer.

Self-Inventions
Her experiences of genocide, the loss of her kin and tribe, her suffering from the forced assimilation into a European missionary community, as well as her being subjected to extremely violent abuse "under a Europe's rule" (*AM* 102) set in motion a process of self-invention that will eradicate [Lina]'s existence and identity prior to the arrival of the European settlers. [Lina] undergoes this forced transformation subtended by the violence of genocide and settler colonialism in order to create something that will actually help her survive and cure the obliteration of her previous life. That something, the text suggests, is a new self or identity, which [Lina] forges for herself some time after she has arrived at the [Vaark] farm "while branch-sweeping Sir's dirt floor, being careful to avoid the hen nesting in the corner, lonely, angry and hurting" (46). This is mentioned in her fragment at various instances, which I will need to quote at some length to show how [Lina] in the terrain of the [Vaark] farm discovers a space and a means for herself to survive an old world turned into the New World. Some time after she enters the [Vaark] household, that is, we read how [Lina] decides to

> fortify herself by piecing together scraps of what her mother had taught her before dying in agony. Relying on memory and her own resources, she cobbled together neglected rites, merged Europe medicine with native, scripture with lore, and recalled or invented the hidden meaning of things. Found, in other words, a way to be in the world. […]

> Solitude, regret and fury would have broken her had she not erased those six years preceding the death of the world. The company of other children, industrious mothers in beautiful jewelry, the majestic plan of life: when to vacate, to harvest, to burn, to hunt; ceremonies of death, birth and worship. She sorted and stored what she dared to recall and eliminated the rest, an activity which shaped her inside and out. By the time Mistress came, her self-invention was almost perfected. Soon it was irresistible. (*AM* 46, 48)

As the first paragraph shows, [Lina] turns to her mother's (and by extension her tribe's) native rites and knowledges and merges those with her "new" culture's traditions in her effort to brace herself for her new life. The text suggests that she finds a way to recuperate her native culture from the nothingness that the annihilation of her family and tribe left her with. Put another way, it is on the [Vaark] farm that [Lina] begins to actively map a space for herself and appears to shed her previous experiences with the Presbyterians. In the second paragraph, we can see how this endeavor is inextricably linked to her erasing all memories of her indigenous culture prior to 'the death of the world.' This becomes the only way in which she is able to cope with having survived the mass slaughter of her families. In 'shaping herself inside and out,' [Lina] manages to create a new identity for herself, an almost perfect and irresistible 'self-invention.' To deal with the "shame of having survived the destruction of her families," [Lina] replaces the memories of her "village peopled by the dead" with the "single image [of] [f]ire. How quick. How purposefully it ate what had been built, what had been life. Cleansing somehow and scandalous in beauty" (47).

[Lina]'s inward turn to nature (as illustrated by this image of the fire) comes with her outwardly laboring hand in hand with her master and mistress in order "to bring nature under [their] control" and to build up the Vaark patroonship (*AM* 47). Working, serving, and living on the [Vaark] farm, that is, [Lina] puts her existence in relation to both [Jacob Vaark] and, later, to his wife [Rebekka] and she assumes her role as a docile, loyal servant in their household. It is through the relationships she builds with both her master and her mistress, I contend, that [Lina] "finds a way to be in the world," forging her new self close to the people who bought her. Initially, the text leads us to believe that [Lina] regards both of them as "exceptions to the sachem's revised prophecy. [...] [The sachem] had apologized for his error in prophecy and admitted that however many [European settlers and colonizers] collapsed from ignorance or disease more would always come" (52). That is, [Lina] initially believes that her master and mistress are not entirely driven by a longing for economic success, property, or by an insatiable hunger for native territory and its subsequent incorporation into any given European kingdom. As she muses, "They seemed mindful of a distinction between earth and property, fenced their cattle though their neighbors did not, and although legal to do so, they were hesitant to kill foraging swine. They hoped to live by tillage rather that eat up the land with herds, measures that kept their profit low" (52–53).

In other words, what separates the [Vaark] dwelling from other Europeans for [Lina] is the notion that their endeavor appears to be driven less by a longing for profit than by the desire and belief that "they could have honest free-thinking lives" (*AM* 56). [Lina]'s previous experiences with other European settlers leave her bewildered at the fact that they could rescue her at the same time that they could "calmly cut mothers down, blast old men in the face with muskets louder than moose calls, but were enraged if a not-Europe looked a Europe in the eye. On the one hand they would torch your home; on the other they would feed, nurse and bless you" (44). Based on those experiences, [Lina] ventures to judge the [Vaarks] "one at a time, proof being that one, at least, could become your friend" (44). Indeed, the text suggests that both [Lina] and [Rebekka] consider the relationship developing between them to be a friendship (51, 71). That is, while initially the "hostility between them was instant," [Lina] soon learns that "[t]he fraudulent competition [between them] was worth nothing on land that demanding" (*AM* 51). [Rebekka Vaark] and [Lina] ultimately become a "united front" against the other residents (existing as well as yet-to-come) of the [Vaark] household (51). The text also suggests that [Lina] will show enormous devotion to [Rebekka Vaark]. This plays out, for instance, when [Lina] helps [Rebekka] deliver, and bury, her "short-lived infants" (51) as well as in the wake of [Jacob Vaark]'s death when she nurses her sickened mistress back to health. In [Lina]'s devotion to and friendship with [Rebekka Vaark] I read that she factually becomes a vital element of the [Vaark] turf and territory, on which she is able even to assume some degree of authority. As [Sorrow] reminds us in her fragment, [Lina] would come to "[rule] and [decide] everything that Sir and Mistress did not" (120). In other words, the text configures devotion to her master and mistress as a kind of survival strategy for this [character], enabling her to find a space and a place where to exist after the annihilation of her previous life. [Lina] will help shape life on the [Vaark] farm by inventing herself in relation to as well as working alongside both her master and mistress, putting her in close proximity to their respective claims to liberal subjectivity at the New World colonial scene (see Chapters 4.1 and 4.3). For [Lina], I suggest, this proximity will come to delineate capacity – a certain degree of power that she gains as well as claims she is able to make within the microcosm of the [Vaark] patroonship – and it thus offers a preliminary answer to the novel's core question of, *What else could have happened?* And yet, it also raises the question, to be investigated in the next section of this chapter, as to what, exactly, is at stake in the intricate connection between the notion that [Lina] is being rescued from her previous experiences of genocide, loss, and abuse at the hands of European colonizers and the fact that this "rescue" occurs by matter of her being bought by [Jacob Vaark]?

On Being Dispossessed, But not Property

I turn now to a scene that is positioned, rather strategically, at the beginning of the second half of [Lina]'s fragment. In it, [Lina] directly addresses her ailing mistress [Rebekka Vaark], pleading her not to die. At the same time that the plot follows [Lina]'s fortification of her new self, to which the devotion to her master and mistress is constitutive, I suggest that this scene illuminates her position as a [character] capable of claiming proximity to liberal subjectivity while simultaneously drawing attention to the provisional nature of [Lina]'s new version of self. What we encounter in this scene, in other words, is the notion that [Lina]'s self-invention may not be as "irresistible" after all and that what lurks in the mortar of her "fortified self" is *dispossession* as subtended by genocidal violence. Here is the scene in question, which I quote at some length:

> Don't die, Miss. Don't. Herself, Sorrow, a newborn, and maybe Florens—three unmastered women and an infant out here, alone, belonging to no one, became wild game for anyone. None of them could inherit; none was attached to a church or recorded in its books. Female and illegal, they would be interlopers, squatters, if they stayed on after Mistress died, subject to purchase, hire, assault, abduction, exile. The farm could be claimed by or auctioned off to the Baptists. Lina had relished her place in this small, tight family, but now saw its folly. Sir and Mistress believed they could have honest free-thinking lives, yet without heirs, all their work meant less than a swallow's nest. Their drift away from others produced a selfish privacy and they had lost the refuge and the consolation of a clan. Baptists, Presbyterians, tribe, army, family, some encircling outside thing was needed. Pride, she thought. Pride alone made them think that they needed only themselves, could shape life that way, like Adam and Eve, like gods from nowhere beholden to nothing except their own creations. She should have warned them, but her devotion cautioned against impertinence. As long as Sir was alive it was easy to veil the truth: that they were not a family—not even a like-minded group. They were orphans, each and all. (AM 56–57)

At the beginning of the paragraph, [Lina] directly calls on her mistress not to die. The first phrase of this passage – 'Don't die, Miss. Don't' – is written in the present tense, drawing the reader to the novel's diegetic present only to then continue with [Lina]'s reflections on her existence within the [Vaark] household in the past tense. The scene confronts the reader with [Lina]'s fear for her future should her mistress die. Having long 'relished her place in the small, tight family' of the [Vaark] patroonship, [Lina] tells us how she now comprehends that her relying on this structure had been foolish. That is, the scene exposes how her investment in and devotion to this kind of fantasy ultimately is doomed to fail. Without her mistress, [Lina] will not be afforded the protection that she had hoped for and needs in the wake of her experience of genocide. Instead, she fears that she will become an 'unmastered woman' and thus, 'belonging to no one,' will become 'wild game for anyone.' [Lina] here clearly aligns herself with the other women serving on the [Vaark] farm by explaining how neither [Sorrow,] her infant girl child, [Florens] ('maybe'), nor herself had any kind of legal status within the colonial landscapes of their dwelling. Following [Lina], they are 'female and illegal' and they would be 'subject to purchase, hire, assault, abduction, exile' if [Rebekka Vaark] did not survive.

[Lina]'s interior monologue furthermore draws our attention to the intricate connections between New World family formations, racial capitalism's nascent marketplace, and private property. In the above lines we read that 'Sir and Mistress believed they could have honest free-thinking lives, yet without heirs, all their work meant less than a swallow's nest.' The phrase 'yet without heirs' in combination with the notion that neither [Lina] nor [Florens] or [Sorrow], for that matter, 'could inherit,' gestures towards white Western patriarchal genealogies and formations of family and, by extension, to the heritability and transfer of all forms of private property within the white family from one generation to the next (see e.g., J. Morgan, "*Partus*"; Nyong'o; Sharpe, "Lose"; Spillers, "Mama's Baby"). Put another way, the above scene illustrates how the New World's grammar of property fundamentally structures white family formations such as that of the [Vaark] farm from the very beginning (and it continues to interrogate the nexus of private property, heritability, and kinship that the novel began to examine with the [character] of [Jacob Vaark], see previous chapter). The scene also shows that this is the case despite the fact that [Jacob Vaark] and his wife appear to believe that they are able to build a live for themselves in which property does not appear to play such a fundamental role (*AM* 56). As we have seen in my discussion of [Jacob Vaark], however, the text exposes this as a misconception. To recall, in expanding his business activities as part of his becoming a liberal subject, [Vaark] "did what was necessary: secured a wife, someone to help her [Lina], planted, built, fathered" (32), doing so to the extent that ownership of slaves and relying on their labor capacities will become a matter of course for him. [Lina] thus enters the 'family' that she imagines herself to be a part of as 'someone to help [Jacob Vaark]'s wife' and, if we follow [Lina]'s own words, as someone who does not have any standing in law, who does not have a "surname and no one would take her word against a Europe" (50). (Even though [Lina]'s words in the above passage seem to suggest that all of the women working and serving on the [Vaark] patroonship are 'female and illegal' in similar ways and would become 'squatters and interlopers' on the farm if their mistress died because 'none of them could inherit and none was attached to a church or recorded in its books,' the text constantly moves to disrupt this kind of logic in this fragment as well as in those of [Florens] and [Sorrow], respectively. I will return to this notion in a moment.)

A few lines further on, the scene expands on the notion that [Lina]'s existence is in fact intricately connected to that of her mistress (and her well-being) when it takes up and reconfigures the notion that [Lina] shows complete loyalty for her master and especially her mistress. That is, [Lina] thinks that she has seen through the workings of [Jacob]'s and [Rebekka]'s lives and through the fact that they cannot create a life apart from any kind of external hold or structure, a life that – seemingly, initially – was powered by a different grammar than that of private property. Following [Lina], 'pride alone made them think that they needed only themselves, could shape life that way.' It is in this context that we learn how she decided not to warn them of their 'folly,' but that it was 'her devotion [which] cautioned [her] against impertinence.' In other words, 'devotion' in these lines is inextricably bound by the fact that she is the [Vaarks'] servant. Telling them that they needed 'some encircling outside thing' would in fact mean that [Lina] overstepped the bounds of behavior acceptable for her. In that 'impertinence' are the social hierarchies of the [Vaark] household, hierarchies which [Lina] does not dare challenge because she believes that they afford her protection; hierarchies that lead her to become, in the eyes of [Rebekka Vaark], "steady, unmoved by any catastrophe as though she has seen and survived everything" (*AM* 98).[103]

Finally, the above passage also positions [Lina] as a [character] that offers a glimpse of the future when she tells us that her master and mistress's alleged attempt at creating 'honest free-thinking lives' was meaningless in the face of the absence of 'some encircling outside thing' like a religious congregation, a family, or a different kind of structure of community. 'As long as Sir was alive it was easy to veil the truth: that they were not a family—not even a like-minded group.' The complicated workings of time within the novel's diegesis – a text that is set in the past, told in both the present and the past tense, and foreshadows a historical future yet to come and that the reader knows about from the study of history books – here open up an avenue for the reader to scrutinize the fantasy of the [characters] peopling the [Vaark] farm as a 'like-minded' community.

By way of neat narrative maneuvering, then, the scene stages this [character]'s oscillation between being "rescued" and being bought, which really is

---

103 Indeed, we also encounter this before a few pages before when [Lina] tries to tell her mistress that she sees trouble, evil, and danger in [Sorrow] and believes her presence to be responsible for the untimely deaths of [Rebekka]'s infant sons. In response to this, [Rebekka] tells her to "stay" and keep quiet (*AM* 54).

an oscillation between capacity and loss, between claims of right to being close to, and perhaps to even gain access to the realm of, liberal subjectivity (represented by the fact that she becomes a trusted, loyal servant to and friend of the [Vaarks]) and genocidal, colonial violence. It is this hydraulics, I suggest, through which we have to read her [character] and which positions us to recognize (the workings of) *dispossession* at the New World colonial scene. That is, *dispossession* here demarcates a liminal position (to paraphrase Wilderson's words in the second paragraph to the chapter) that "shuttles" (to borrow from Wilderson again, *Red* 50) between some sort of self-determination and making of a self, on the one hand, and complete annihilation, on the other. We are exposed to this again a few pages later when [Lina] remembers how [Florens] had arrived at the [Vaark] farm and how she "had fallen in love with her right away" (*AM* 58). That is, as soon as [Jacob Vaark] brings [Florens] with him when he returns from one of his business trips, [Lina] decides to take care of this little girl. Thus extending her devotion onto [Florens], as well, we read that [Lina] is determined that "this one [...] *could be, would be, her own.* [...] Some how, some way, the child assuaged the tiny yet eternal yearning for the home Lina once knew where everyone had anything and no one had everything" (59, 58; emphasis mine). These words imply that [Florens] incites a desire in [Lina], which she would not be able to stifle despite her otherwise irresistible self-invention. At night, when [Lina] tells [Florens] stories of "wicked men who chopped off the heads of devoted wives" as well as "stories of mothers fighting to save their children from wolves and natural disasters" (59), both [Lina] and [Florens] suffer from "[m]other hunger—to be one or have one [...] [they] were reeling from that longing, which, Lina knew, remained alive, traveling the bone" (61). To me, these last words suggest [Lina] is aware that her self-invention may appear to be irresistible but in fact will never be fully complete, because she will continue to feel that insatiable mother hunger, infinitely. In this context, I also need to point out, first, that many critics have in their readings positioned [Lina] as the "novel's central mother figure" (Montgomery, "Traveling Shoes" 628). What I want to stress in response, and second, is that the text also suggests that [Lina] may in fact be responsible for the death of [Sorrow's] first child. In [Sorrow]'s textual fragment, we read, for instance: "Although Sorrow thought she saw her own newborn yawn, Lina wrapped it in a piece of sacking and set it a-sail in the widest part of the stream and far below the beavers' dam" (*AM* 121). This goes to show that the text positions [Lina] in complex ways as someone who cares for and takes care of others at the same time that she has the capacity to literally make life-and-death decisions vis-à-vis the enslaved women on the [Vaark] farm. It is in her relationship with [Florens], finally, that we can find another clue as to the workings of the New World's grammar of property. That grammar is in her resolve that [Florens] 'could be, would be, her own.' In the 'could be, would be, her own' is the notion that [Lina] will do everything in her power to make this enslaved girl child her own, is the notion that she claims to have the right to become the person in charge of her. In the 'could be, would be, her own,' in other words, is the notion that [Lina] invents herself in dangerous proximity to the formations of and claims to liberal self-making that both [Jacob Vaark] and [Rebekka Vaark] represent.

[Coda]: "A Kind of Death for Herself"

Throughout the chapter, I have in my reading of this [character] so far not only suggested that [Lina] becomes a member of the [Vaark] household after being "rescued" by "Europes" (French soldiers and Presbyterian missionaries alike) (*AM* 42) but that she is able to create a space for herself in which she gains some sort of capacity. That is, through her devotion to and by taking care of her master and mistress, [Lina] imagines herself to be a member of the [Vaark] household after having been purposefully purchased by [Jacob Vaark] to work on his farm. (Another rescue, if you will, albeit one that is intimately tied to a business transaction which makes [Lina] [Vaark]'s servant). I have also suggested that [Lina]'s invention of her new self on the [Vaark] farm is not as complete as it seems and that her [character] in fact exposes the reader to the workings of *dispossession* at *A Mercy*'s New World colonial scene. That is also to say that *A Mercy* suggestively probes into the possibility of claims to access to (liberal) subjectivity for this [character] at the same time that it positions [Lina] through the colonial violence of genocide.

In closing, I want to draw attention to another passage in [Lina]'s fragment, which follows shortly after the above cited long passage. We continue reading:

> Lina's mistress is mumbling now, telling Lina or herself some tale, some matter of grave importance as the dart of her eyes showed. [...] Helpless to disobey, Lina brought it to the lady. She placed it between the mittened hands, certain now that her mistress will die. And the certainty was a kind of death for herself as well, since her own life, everything, depended on Mistress' survival, which depended on Florens' success. (*AM* 57–58)

In moving between the present tense and the past tense, these lines once again allow the reader to witness [Lina]'s positioning as [Rebekka Vaark]'s loyal servant, whose existence entirely depends on the well-being of her mistress. The phrases 'Lina's mistress is mumbling now' and 'certain now that her mistress will die' bring this positioning to the reader's immediate present, where it will remain. (The reader will encounter a similar sense of immediacy in [Florens'] fragment, which is the only fragment in *A Mercy* that is told in the present tense throughout.) The last few lines of the paragraph reiterate the notion that [Lina]'s existence at the New World colonial scene is fundamentally intertwined with that of her mistress (and, by extension, with that of [Florens]), for the certainty that her mistress will die from the smallpox becomes for [Lina] a 'kind of death for herself.' As [Lina] herself observes shortly before her mistress orders [Lina] to give the mirror to her, "gaz[ing] through the way pane of the tiny window" and talking to a "forest of beech trees": ""You [the trees] and I, this land is our home," she whispered, "but unlike you I am exile here"" (*AM* 57).

Finally, this textual fragment thus both configures and positions [Lina] along the lines of genocide, survival, and exile (in her homeland) and thus contrasts this [character] with the representations of the enslaved Black girls and women within *A Mercy*'s narrative frame. That is, the novel positions her as being dispossessed by her master and mistress but decidedly not as their fungible property. As we learn at the very end of the novel, [Rebekka Vaark] puts both [Sorrow] and [Florens] up for sale but decides to keep [Lina] in her household and on her farm. In turn, she "requires [Lina]'s company on the way to church but sits her by the road in all weather because she cannot enter" (*AM* 158). Within the arrangements of power (to paraphrase Frank Wilderson, *Red* 48) on the [Vaark] patroonship, [Lina] ultimately will remain [Rebekka]'s steady servant. Staying with her mistress, [Lina] clings to the sliver of capacity she has forged for her self in this environment, if only provisionally. In this way, her [character] specifically both defies and deconstructs readings arguing that the congregated women on the [Vaark] estate are subjugated, commodified, or dispossessed in similar or equal ways, readings that ignore how the New World's grammar of property positions her in fundamentally different ways than the enslaved female [characters] in *A Mercy*.

4.3 "The Promise and Threat of Men": [Rebekka], Liberal Self-Making, and the Ruse of Solidarity

[Routing the Argument] The critical reception of Toni Morrison's *A Mercy* has largely read [Rebekka], [Jacob Vaark]'s wife, into a paradigm of universal female affiliation or solidarity across racial, religious, and cultural boundaries, among others, which appears to develop among the women on the [Vaark] dwelling in the wilderness of colonial Virginia. I argue that while playing with the possibilities of solidarity among the community of women on the [Vaark] patroonship as well as among other groups of women that [Rebekka] encounters in the novel's fictional orbit, *A Mercy* confronts any notion of solidarity between those women and, in fact, urges its readers to take a more critical stance in this respect. That is to say, I read her [character] as ultimately foreclosing any possibility of such notions of solidarity on the level of narrative. I seek to demonstrate not only how *A Mercy* stages [Rebekka] as an allegory of the space and the place that English women in colonial North America held in the social strata of their nascent environment but also how the novel powerfully suggests that [Rebekka]'s struggle for subjectivity is part of the New World's grammar of property, ultimately leading her to claim white female co-mastery at the New World colonial scene.

> Lina and I look at each other. What is she fearing, I ask. Nothing, says Lina. Why then does she run to Sir? Because she can, Lina answers. – *A Mercy*

Introduction

In general, critics have amply celebrated the publication of *A Mercy* and its narrative politics of bringing back prominent concerns from Morrison's previous novels, such as mother-daughter relationships or Black female (self-)empowerment, by creating a sophisticated discourse on the female community on the Vaark dwelling. As Waegner writes in this context, for instance: "Morrison carefully presents moments of self-empowerment and solidarity among the women in her novel" (98). Zooming in on the ways in which the women [characters] that [Vaark] congregates around himself share a history of having been both subjugated and commodified because of their gender,[104] this discourse more often than not takes the friendship between [Rebekka] and [Lina] that develops on the [Vaark] homestead as a perfect example of cross-racial, cross-cultural, as well as cross-religious affiliations (Karavanta) as it promotes notions of solidarity between those female [characters]. In other words, by stressing notions of solidarity, community, and companionship between these differently positioned women, critics have so far discussed [Rebekka] mainly in terms of "coalitions" forming between her and the women on board the merchant ship that brings the women to the New World (Cox; Strehle; Waegner) as well as between [Rebekka] and the other female figures on the [Vaark] patroonship, especially [Lina]

---

[104] As Babb writes, for example: "*A Mercy* makes another departure from origins narratives in its record of women's voices that illuminates female commodification. The women characters who come together at the Vaark farm all arrive there via transactions: Lina is bought by Vaark; Rebekka becomes his wife through his funding an arranged marriage; Florens is acquired in the settlement of a debt; and Sorrow is given to Vaark free of charge to remove her from the sons of a local sawyer" (156).

(Gallego, "Nobody"; Karavanta). Moreover, another dominant strand in the critical discourse on the [character] of [Rebekka] in *A Mercy* reads her as inverting dominant self-narrativizations of North American beginnings and American exceptionalism by allegorizing, demystifying, and parodying, respectively, predominant notions of spirituality and Christian doctrine in such mythicized fictions of "original beginnings" (G. Moore; Strehle; Tally, "Contextualizing").

What is being asserted in this discourse through the focus on female companionship and solidarity in more or less explicit ways is that all of the female [characters] on the [Vaark] farm are *dispossessed* in at least a similar (if not the same) way, precisely because of the fact that they have all come to the New World by way of having been both subjugated and commodified by men generally and by [Jacob Vaark] in particular because of their gender. Indeed, the novel is taken to suggest that while [Rebekka] has successfully transcended an earlier, more docile version of her self (having been dependent on her father, for example), critics and readers seem to agree that her status of being in the New World is far from being more independent from the wills of (male) others, which is something that this [character] appears to share with *A Mercy*'s other women [characters]. By contrast, I am interested in the functions of the notion of solidarity in the novel or, rather, in how the text uses [Rebekka] to meditate on and play with the *possibility* of solidarity between the women on the [Vaark] dwelling.[105] [Rebekka]'s narrative is cushioned between [Florens'] partitioned first-person text that structures the novel's plot and takes turns with the other [characters'] more self-contained, third-person texts. It is both framed and structured by [Rebekka]'s hallucinations that are induced by the smallpox, a disease and a fever that she has contracted from her late husband. Embedded in these hallucinations, which are narrated interchangeably in both the present and the past tenses over a total of twenty-nine pages, are flashback episodes in which she re-

---

105 The *OED* defines "solidarity" as the "fact or quality, on the part of communities, etc., of being perfectly united or at one in some respect, esp. in interests, sympathies, or aspirations," and as a "perfect coincidence *of* (or *between*) interests ("solidarity, n.")." Sara Ahmed's famous formulation of "solidarity" also comes to mind here: "Solidarity does not assume that our struggles are the same struggles, or that our pain is the same pain, or that our hope is for the same future. Solidarity involves commitment, and work, as well as the recognition that even if we do not have the same feelings, or the same lives, or the same bodies, we do live on common ground" (*Cultural Politics* 189). While these definitions/conceptions resonate in my reading of [Rebekka], I am interested in what happens when "solidarity" meets the grammar of property within *A Mercy*'s representation of the New World, what happens when structural positionality and claims of right to property undermine such conceptions. This shall become clearer over the coming pages.

members her past life in England as well as episodes from her current life in colonial Virginia.

In what follows, I want to think about how these narrative negotiations over solidarity are strongly linked to the narrative's engagement of questions concerning [Rebekka]'s self-making as an emerging liberal subject both in and of the New World. I take my cue from the following passage in *A Mercy*, which is situated in [Florens'] text. In it, [Florens], who is on an errand to fetch the [blacksmith] to help cure her mistress [Rebekka] from the smallpox, meditates on the notion of choice and she remembers a scene in which she, [Lina], and [Rebekka] perform a bathing ritual. We read:

> Now I am thinking of another thing. Another animal that shapes choice. Sir bathes every May. We pour buckets of hot water in the washtub and gather wintergreen to sprinkle in. He sits awhile. [...] She [Rebekka] wraps a cloth around to dry him. Later she steps in and splashes herself. He does not scrub her. He is in the house to dress himself. A moose moves through the trees at the edge of the clearing. We all, Mistress, Lina and me, see him. [...] Mistress crosses her wrists over her breasts. Her eyes are big and stare. Her face loses its blood. [...] I am thinking how small she looks. It is only a moose who has no interest in her. Or anyone. Mistress does not shout or keep to her splashing. *She will not risk to choose.* Sir steps out. Mistress stands up and rushes to him. Her naked skin is aslide with wintergreen. Lina and I look at each other. What is she fearing, I ask. Nothing, says Lina. *Why then does she run to Sir? Because she can*, Lina answers. (*AM* 68–69; emphasis mine)

As focalized through [Florens], this passage underlines how [Rebekka]'s [character] in *A Mercy* fundamentally brings notions of choice to the fore—choice in the sense of having the freedom to make a choice or to opt for a certain way of existence. [Rebekka] 'will not risk to choose' to stay in the bathtub and, therefore, close to [Lina] and [Florens]. Instead, she decides to 'run to Sir,' choosing proximity to her husband [Jacob Vaark] and, thus, to him as a representation of the quintessential liberal subject.

I argue that [Rebekka]'s struggle for liberal subjectivity – from her position as an underclass, poor, white, immigrant, married woman – is significantly different from those of the other congregated women on the [Vaark] farm, even though their lives and status of being in colonial Virginia are also determined by [Jacob Vaark] and his economic and other choices and success. [Rebekka]'s struggle is different from those of the other female [characters] living in the [Vaark] household because [Rebekka], despite the fact that she is poor (at least when she flees England) and appears to be legally disenfranchised through her marriage and, therefore, fully dependent on her husband, does have access to liberal subjectivity. That is, she would take the master's place if she could. In fact, the text suggests that [Rebekka] ultimately does become a (widowed) plantation mistress, or at least something like an allegorical blueprint for what will become the quintessential Southern belle/plantation mistress, as I hope to show as I develop my argument over the following pages. My reading of [Rebekka] both traces and focuses on the intricate connections between the mostly "ad-hoc" and short-lived relationships she builds with various (groups of) women, on the one hand, and her struggle for liberal subject status in the New World, which is intimately tied to her existence as [Jacob Vaark]'s wife, on the other (*AM* 11).

[Rebekka], History, Allegory

In order to understand how [Rebekka]'s narrative navigates a particular, white woman's struggle for full liberal subjectivity while also meditating on and ultimately dismissing the possibilities of solidarity between the differently positioned [Vaark] women, it is important to remind ourselves of the social, cultural, political, as well as legal position women like her occupied historically in seventeenth-century England and, after their emigration to the New World, in colonial Virginia. In general, white English women in [Rebekka]'s historical time and place "constituted a much smaller proportion of the population in Virginia than in Europe" (E. Morgan 163) and most of them came to the New World as indentured servants, servants, or as "mail-order brides" (Zug).[106] Most of the settlers who came to the New World to build England's first colony on the North American continent were men and "because these men were unable to find wives, they were deserting the colony like droves" in the first decades of the seventeenth century (Zug). The colony desperately needed women in order to flourish, which led to colonial administrators and the Virginia Company, respectively, "putting out an advertisement targeting wives" (Zug). Despite these measures, white men would continue to outnumber white women in the colony "by a ratio of four to one" in at least the first half of the century (Scott and Lebsock). As Edmund Morgan has it, moreover, white women continued to be "scarce in Virginia in 1691 and doubtless continued to be for another twenty or thirty years" (336). This also meant that white women in colonial Virginia would not remain on their own or widowed long after their husbands passed away (E. Morgan 164).

---

[106] Law professor Marcia Zug uses the term "mail-order brides" to describe those English white women who came to colonial Virginia to become the wives of the local planters and businessmen. These marriages usually were pre-arranged.

English women who responded to the marital scheme advertised by Virginia's colonial administration or who came to the colony in the years after it had been implemented seem to have gained some economic advantage over their English counterparts. Not only were they able to leave a country that, ridden by a civil war, poverty, and unemployment (induced by both the war and the enclosure movements), would no longer offer prospects for employment for these women, who usually belonged to the lower classes and, in general, sought positions as domestic servants (Scott and Lebsock). They were also promised "a dowry of clothing, linens, and other furnishings, free transportation to the colony and even a plot of land [in addition to] their pick of wealthy husbands[; they also were] provided with food and shelter while they made their decision" (Zug). For those women who came to Virginia as indentured servants and who "survived their term of service, husbands [would also] be easy to come by" and together "they might move rapidly to become what they never could have aspired to be in England—landowners" (Scott and Lebsock). With such a perspective of reasonable economic prosperity at hand, many women made their way to Virginia to start a new life in the New World full of hope.

On a legal plane, moreover, white (servant) women, whether married or single and despite the fact that there were "no legal restrictions on voting in Virginia until 1670," generally were not allowed to vote while a "man who had finished his term of service, whether he had set up his own household or not, could cast his vote" (E. Morgan 145). Most important for the present argument is the notion that all white women in colonial Virginia were legally subjected to the common law system of coverture, which designated a married woman's legal existence as inseparable from that of her husband (see generally e. g., Stretton and Kesselring; Broeck, *Gender* 50–67; Zug).[107] As both a legal system and a legal fiction at work roughly from the thirteenth to the nineteenth centuries in both England and, later, in its New World colonies, coverture at common law needs to be understood as a system that regulated a woman's access to and control of (her) property. As Stretton and Kesselring explain in this respect, which I am quoting here at some length:

> Upon marriage a wife lost the ability to own or control property, enter into contracts, make a will, or bring or defend a lawsuit without her husband. A married woman's real property – her lands – fell under her husband's control. He did not own them and he could not sell them without her consent, to be given freely before a judge in her husband's absence, but

---

[107] It is important to note that the laws of coverture affected not only married women but single and widowed women as well because it was generally assumed that women would (re-)marry and, thus, would "fall under a husband's control" (Stretton and Kesselring 5–6).

during his lifetime he could do with them what he wished, planting them or leaving them fallow, renting them out, and taking to himself any profits they produced. He could even lease them to another for an extended period of time, so that if his wife outlived him she could claim the revenues but would never enjoy possession. A woman's movable property – her money, livestock, and personal possessions – became her husband's outright. He had total control over any cash she brought to marriage or inherited or earned thereafter. He could sell her possessions, including her clothes and personal effects, or make bequests of them in his will without her permission. [...] The only limits on this ownership came in the custom of paraphernalia, which allowed a wife to keep clothes and personal items after her husband died (but not while he lived), and the practice of returning to a widow any 'chattels real,' interests such as debts or bonds that she brought to marriage, if they had not vested in her husband before she died. A husband's control over his wife's real property and ownership of her personal property helped explain her inability to make a valid contract, sue or be sued in her own name, or make a will without his permission. Jurists later cited it as a reason to deny a wife custody over her children, too. [...] A husband could not legally be denied the right of sexual access to his wife's body. (Stretton and Kesselring 8, 10)

What is being asserted in the paragraph above, then, is that white women in England and in colonial Virginia were subjected to a legal system that would often dispossess them of their (personal) property as well as of their identities as fully recognized subjects before the law. What Stretton and Kesselring fail to mention in this context is that slaves were often part of a married woman's private property (in colonial Virginia or elsewhere) and that she would often find ways to take control of her property—even though she may not legally have been allowed to decide what to do with them (Jones-Rogers).

Historians of American beginnings and of slavery additionally remind us of the following two points: First, white women in colonial Virginia would often not only inherit a large portion of their deceased husbands' estates but, functioning as administrators, they would also be in charge of them legally. In some cases this could mean that "claimants against the estate had to make their claims to the [widow], and she, by delaying payment, might continue to enjoy the whole for some time" before re-marrying eventually (E. Morgan 166). Second, white English women not only became administrators of their late husbands' fortunes, be they land, livestock, money, or, indeed, enslaved human beings; but they would also appear before the courts to claim their individual rights to property, both in seventeenth-century England and in colonial North America. In her essay on "Women and Property Litigation in Seventeenth-Century England and North America," Lindsay Moore explains that while "women in the seventeenth-century English world remained subordinates at the level of both household and the state, they nevertheless appeared as litigants before the courts to protect their rights to property" (113). Moore points out that because married women in England and North America could not appear before a common law court to pursue litigation over their property as they were bound by the system of coverture, English women would alternatively turn to "ecclesiastical courts, which routinely allowed even married women to appear independently from their husbands" (115).[108] Seventeenth-century colonial English women, however, had less legal opportunity to pursue litigation over property because of a strong "focus on common law in the colonies" (Moore 126), which meant that they made claims of right to their property much less frequently than their counterparts in the mother country.[109]

In her recently published *They Were Her Property: White Women as Slave Owners in the American South* (2019), historian Stephanie E. Jones-Rogers fundamentally revises existing scholarship on the role of white slave-owning women in American South. Her study thus critically supplements our thinking about the places those women held in eighteenth and nineteenth-century Southern society as well as about the ways they were involved in, depended on, and profited from chattel slavery. Jones-Rogers contends that previous historical scholarship has usually focused on "the wealthiest single or widowed women" and often assumed that "the authority women held over their slaves [was] obligatory, rather than voluntary or self-initiated, management and discipline of enslaved people" (xi, xii). By contrast, Jones-Rogers' groundbreaking study focuses on and demonstrates that Southern married white women from different social classes took part in "the most brutal features of slavery, they [...] profited from them, and defended them" (ix). It shows, in other words, how those slave-owning women

> contended with husbands, male employees, community members, and officials about their ownership of slaves, as well as about how much control such men could exercise over their property and who else would be afforded the privilege of doing so. [...] They fully embraced the institution of slavery and all the economic benefits that came along with it. (Jones-Rogers 204)

---

108 Moore discusses equity courts and ecclesiastical courts as providing different avenues (mostly) for married women to seek legal redress in matters concerning their property. While equity courts "upheld the doctrine of 'separate estates,' a legal instrument that allowed a married woman to retain independent control of her property for her separate use and prevented it from falling under the control of her husband[,] [...] ecclesiastical law virtually ignored the assumption that married women should be legally represented by their husbands" (118).

109 On a slightly different view, historians have also argued in this context that in the economic arena colonial women in Virginia had a significant advantage over English women, precisely because they would inherit a larger share of their late husbands' estate (E. Morgan 164–165; Zug). Marcia Zug writes accordingly: "Because malaria, dysentery, and influenza were widespread in colonial Virginia, early death was also common. This meant that most marriages were short, but the morbid upside was that colonial law and practice ensured widowed women were uncommonly well provided for."

Jones-Rogers also shows that white married and slave-owning women fought for their right and access to slave property within the realm of the law: "Slave-owning women brought legal suits against individuals, both male and female, who jeopardized their claims to human property, and others sued them in kind" (xvi). Put another way, white married slave-owning women would unflinchingly invest in and claim their right to their slave property.

Against this larger historical backdrop, I propose to read [Rebekka]'s [character] as an allegory of white English colonial married women. In the fictional arena of colonial Virginia in *A Mercy*, [Rebekka] represents aspects of gender, dispossession, and class as part of the struggle that white colonial women began to fight on the social, political, cultural, and legal frontlines as they tried to survive and exist in the New World. As we have seen in the preceding paragraphs, for many white married women colonial Virginia promised them unheard of economic advantages at a time when they were legally disenfranchised as well as subjugated in and by marriage. As Toni Morrison elsewhere reminds us, the New World opened up a space where "habit of genuflection would be replaced by the thrill of command. Power – control of one's own destiny – would replace the powerlessness felt before the gates of class, caste, and cunning prosecution" (*Playing* 35). [Rebekka] allegorizes this struggle for a new kind of subjectivity, personhood, and/or being that so many settlers sought in and on new territory from the distinct perspective of colonial English women. As I hope to show over the coming pages and following Jones-Rogers's arguments, [Rebekka Vaark] ultimately emerges as co-master and as "co-conspirator" at *A Mercy*'s seventeenth-century colonial scene and, in this way, offers the reader an allegorical look into the future and onto white married women in nineteenth-century America and their profound involvement in "American slavery and the marketplace" (Jones-Rogers 205, xvii).

[Rebekka] grows up in a lower-class household as the daughter of a "waterman" in an urban English landscape (*AM* 72). The text suggests that her childhood is strongly shaped not only by her parents' "fire for religious matters" but also by endemic violence kindled by civil war and religious animosities as part of the gradual transition from feudal rule to early capitalist socio-political formations in England from at least the seventeenth century. [Rebekka] has witnessed multiple hangings, executions, and "a drawing and quartering" by the time that she is sixteen years old; "execution was a festivity as exciting as a king's parade" in the poor household that she is raised in (73). In this civil war and poverty-ridden environment, [Rebekka]'s prospects in life come to be even more limited than they would have been at the beginning of the century and she is desperately looking for a way out, for without money or the inclination to peddle goods, open a stall or be apprenticed in exchange for food and shelter, with even nunneries for the upper class banned, her prospects were servant, prostitute, wife, and although horrible stories were told about each of those careers, the last one seemed safest. (*AM* 75–76)

And so, when news reaches her parents that a Virginian settler and landowner is looking for a young woman willing to become his wife in the New World, [Rebekka] is thankful for this avenue of escape despite frequent news and reports of violence between the settlers and the native populations (*AM* 72–75). [Rebekka] considers those New World "squabbles" to be trivial because, for her, "brawls, knifings and kidnaps were so common in the city of her birth that the warnings of slaughter in a new, unseen world were like threats of bad weather" (74, 73). The text thus positions [Rebekka]'s desire to escape her current situation by way of becoming the wife of a New World settler from the outset against any kind of future she might have lived in her home country.

While the plot follows [Rebekka]'s past life in England, the novel also suggests that her "escape route" is motivated by economic factors, too. That is, the text explicitly maps her future marriage as some kind of a business transaction. It reads:

> Already sixteen she knew her father would have *shipped her off* to anyone who would book her passage and *relieve him of feeding her*. A waterman, he was privy to all sorts of news from colleagues, and when a crewman passed along an inquiry from a first mate – a search for a healthy, chaste wife willing to travel abroad – *he was quick to offer his eldest girl*. The stubborn one, the one with too many questions and a rebellious mouth. Rebekka's mother objected to the *sale* – she called it that because the prospective groom had stressed '*reimbursement*' for clothing, expenses, and a few supplies – not for love or need of her daughter, but because the husband-to-be was a heathen living among the savages. (*AM* 72; emphasis mine)

Again, as the "object" of a transaction that is geared towards both "relieving" her family of providing for her and towards creating some kind of a living for her, [Rebekka] reconstructs and allegorizes the historical position of many English women who came to the colonial Virginia as "mail-order brides" (Zug) seeking a more prosperous life in the New World. Moreover, I read the fact that [Rebekka]'s father appears to seize the first opportunity to marry his daughter off as speaking to the notion that her future is severely limited by her class status as well as by her gender. It is a future in which she treads well-trodden paths as 'prostitute, servant, or wife' (see above previous paragraph) and in which she is dispossessed by the choices that men like her father make for her for their own economic benefit or because of ulterior motives. We continue reading that "[a]s with any future available to her, it depended on the character of the man in charge" (*AM* 76).

A Mercy leads the reader to consider how the [Vaark] dwelling in colonial Virginia, initially at least, becomes an arena in which [Rebekka] creates for herself a new identity and subject position as [Jacob Vaark]'s wife that appears to enable her to leave her past behind while also being independent of other settler communities around them. I say initially because, as I will argue in the next section of the chapter, it turns out that the inhabitants of the [Vaark] dwelling are not as independent of other settlements in their colonial surroundings as they thought they would be. That is, after her husband's death, [Rebekka] will ultimately turn to a group of Anabaptists in her struggle to become a subject in this New World environment (*AM* 96–98). As soon as [Rebekka] lands at the shores of colonial Virginia, stepping off the ship that took her there, she and [Jacob Vaark] get married: "It was seal and deal. He would offer her no pampering. She would not accept it if he did. A perfect equation for the work that lay ahead" (84). With [Rebekka] settling into her new life and into "the long learning of another: preferences, habits altered, others acquired; disagreement without bile; trust and that workless conversation that years of companionship rest on" (85), the text projects her existence as the wife of the landowner and farmer [Vaark] as fulfilling if lonely in the face of her children's subsequent deaths, common but not whimsical, sustainable but not profitable (95, 86, 85). Together, [Vaark] and [Rebekka] create their lives independent of the (spiritual) companionship of others, "lean[ing] on each other root and crown. Needing no one outside their sufficiency. Or so they believed" (85). For [Rebekka], then, her past in English society – where, when finally offered a place to be apprenticed as a domestic servant, she was "running from the master and hiding behind doors" (75) – is one that she gladly trades for her present life as [Vaark]'s wife. In this new and bountiful life, "the cost of a solitary, unchurched life was not high" (91) and promises of a different kind of subjectivity that is shielded by a husband from the "threatening world" and wilderness beyond the confines of her dwelling abound (87). Put somewhat differently, *A Mercy* suggests that [Rebekka] feels free, whole, and happy once she has established a new life for herself in colonial Virginia.

"Women of and for Men": The Ruse of Solidarity, or Struggling for Liberal Subjectivity

[Rebekka] travels to the New World on board a merchant ship called the *Angelus*. On her long transoceanic passage on the way to meeting her husband, she is accompanied by a group of seven other women who also travel across the Atlantic to America, albeit for different reasons than her. As [Jacob Vaark]'s wife-to-be, [Rebekka] is the only one of these lower-class and poor white women whose passage has been "prepaid [...] The rest were being met by relatives or craftsmen who would pay their passage—except the cutpurse and the whore whose costs and keep were to be borne by years and years of unpaid labor" (*AM* 80). On board the ship, then, those prostitutes and thieves and otherwise disgraced females as well as indentured servants and mail-order-brides become part of a temporary community of "exiled, throw-away" English women (80). And even though they will never meet again after disembarking "their" oceangoing vessel, this impermanent group helps [Rebekka] navigate "her own female vulnerability, traveling alone to a foreign country to wed a stranger, these women corrected her misgivings. [...] [T]he company of these exiled, thrown-away women eliminated" her fears (80–81). Here, the reader is lead to believe that what keeps the community of women on board the *Angelus* together is an increased awareness of their existence as "[w]omen of and for men" (83).

I take my cue from this last quotation, which also supplies the title to this section of the chapter, to argue that *A Mercy* also allows us to think through the ways in which the text links [Rebekka]'s struggle to find a place in the world that lies ahead of her to the (im)possibility of female solidarity or companionship across religious, cultural, and racial boundaries at the Virginian colonial scene. I am thinking here in particular about the relationship that develops between [Rebekka] and the Native American woman [Lina] in the [Vaark] household, on which, as I suggested earlier, critics have amply commented on in their readings of *A Mercy* as a paradigmatic example of female bonding in the "mixed-race community of have-nots" on the farm (Grewal 192). Despite the fact that the novel renders this congregated group as 'women of and for men,' too, it stands in stark contrast to [Rebekka]'s shipmates. In fact, it represents them as women of and for *one man:* [Jacob Vaark,] who legally owns them. In other words, I suggest that the difference between those women lies in the ways that they are positioned in relation to the New World's grammar of property, to the ways in which property orders Human existence in the Western hemisphere, and that it is this kind of structural positioning that *A Mercy* elaborates on.

I turn now to two scenes situated towards the beginning and the end of [Rebekka]'s fragment, respectively. Those scenes illustrate how [Rebekka] first forges her friendship with [Lina] before ultimately breaking the strong emotional bond that has developed between them. In the first, [Rebekka] comes to her new household after [Lina], who is the first woman [Vaark] acquires to help him run his farm, and [Rebekka] initially feels deep hostility against her. Only gradually, a closer relationship develops between the two:

> [Rebekka] bolted the door at night and would not let the raven-haired girl with impossible skin sleep anywhere near. Fourteen or so, stone-faced she was, and it took a while for trust between them. Perhaps because both were alone without family, or because both had to please one man, or because both were hopelessly ignorant of how to run a farm, they became what was for each a companion. A pair, anyway, the result of the mute alliance that comes of sharing tasks. Then, when the first infant was born, Lina handled it so tenderly, with such knowing, Rebekka was ashamed of her early fears and pretended she'd never had them. (*AM* 72–73)

[Lina] becomes for [Rebekka] a loyal and trusted companion, on whose opinion, skill, and knowledge she comes to rely (*AM* 70–71). The passage shows that, from [Rebekka]'s point of view, this relationship is borne out of the pressing necessity for companionship in a potentially hostile and challenging environment in which the two of them "had to please one man." While both [Lina] and [Rebekka] will ultimately consider their companionship a firm friendship (51, 71), the text also suggests that "hostility," "fraudulent competition," and "fear" lurk between the mortar of this friendship—in spite of [Rebekka]'s disavowal of these notions when she "pretends she'd never had them" (51, 71). If what haunts the friendship between those two women are those early moments in which fear and mistrust determined their encounters, what does this tell us about the notion of solidarity so plentifully celebrated by many readers and critics of the novel? Is the companionship between [Rebekka] and [Lina] really evidence of what Waegner might call a moment of self-empowered solidarity (98)? How are [Rebekka]'s struggle for full liberal subjectivity and the possibility of female solidarity, epitomized by her close friendship with [Lina], connected? In the New World, can such moments of solidary friendship – which really are "ephemeral" in the narrative as Lynn Neary has it in a televised *NPR* interview with Toni Morrison on the novel shortly after its publication – be at the foundation of the different struggles for liberal subjectivity as the "earliest version[s] of American individuality, American self-sufficiency" in this period (Neary and Morrison)?

While the text wallows in the possibility of those moments, it also complicates them as it adds new layers of meaning to them with every page on which [Rebekka]'s narrative unfolds. This shows in the second scene under scrutiny here or, rather, in a series of meditations that occur throughout [Rebekka]'s narrative. They represent how [Rebekka] "between fever and memory" ruminates on her status in the world, past and present, as well as on the future ahead of her in the wake of [Vaark]'s death (*AM* 71). The narrative frequently switches narrative time and perspective from third-person to first-person narration and vice versa as her thoughts

> bled into one another, confusing events and time but not people. [...] How could she not know the *single friend* [Lina] she had? [...] The best husband gone and buried by the women he left behind; children rose-tinted clouds in the sky. Sorrow frightened for her own future if I die, as she should be, a slow-witted girl warped from living on a ghost ship. Only Lina was steady, unmoved by any catastrophe as though she has seen and survived everything. [...] And though she understood that her thoughts were disorganized, she was also convinced of their clarity. That she and Jacob could once talk and argue about these things made his loss intolerable. Whatever his mood or disposition, he had been the *true meaning of mate*. Now, she thought, there is *no one except servants*. (AM 70, 71, 98, 97; emphasis mine)

In the passage [Rebekka]'s emphasis on her husband being a 'true mate' overrides all other considerations of what [Rebekka] might consider a close connection with the women surrounding her. The text underlines this by suggesting to the reader that [Rebekka]'s thoughts are lucid despite the fever. It also once again plays with the possibility for female solidarity on the Vaark dwelling by juxtaposing [Rebekka]'s friendship with [Lina] – the 'single' and 'steady' friend she has in the world – with her intimate and 'true' relationship with [Vaark]. However, with a dead husband and dead children, there seems to be no future for 'real companionship' on this land—not only not for [Rebekka] herself but also not, e.g., for [Sorrow], whose future [Rebekka] knows will now depend on her completely. I read the novel's pitting of one version of affinity that [Rebekka] feels for the different people around her against another as an instance in which the text rules out the possibility of any kind of cross-cultural or cross-racial solidarity between the women in the [Vaark] household. This is presented in the text via [Rebekka]'s musings that now, after her husband's death, there is 'no one except servants.' She thus effectively cancels out the friendship with [Lina], turning it into a relationship between mistress and servant and elevating her own social and moral status.

In the above interview with Lynn Neary, Toni Morrison also comments on the importance of community for the people that inhabit the [Vaark] patroonship. For Morrison, notions of community, society, and belonging are crucial for the women at the center of [Jacob Vaark]'s household. However, Morrison also points out how vulnerable this "little society" is because what actually holds them together externally is just one "peg" (Neary and Morrison). This peg is [Jacob Vaark] and after his death, as already suggested in the previous chapter, the fragile structure of his household falls apart immediately. So in addition to negating any previous emotional ties with [Lina] by way of reducing her to the status of servant, I suggest that the sense of solidarity between those women, assembled as they are by [Jacob Vaark], can only develop against the backdrop of an external, patriarchal hold that [Jacob Vaark] represents. Put somewhat differently, that which makes any version of solidarity possible between the [Vaark women] is [Jacob Vaark] himself, for it is him who assembles and accumulates them on his homestead. In light of this, it seems to me that readings that mainly focus on notions of solidarity, companionship, and cross-cultural alliances among the group of women on the [Vaark] dwelling necessarily fall short: they fall short because by engaging with these concepts as the novel's assumed primary concern they do not reckon with the ways that the New World grammar of property already is at work fundamentally where those women are situated. They also fall short because they fail to recognize how the novel, by way of neat narrative ploy, exposes that grammar at work in [Rebekka]'s struggle to become a liberal self in colonial Virginia.

Mistress in the Making

Let me quote the following excerpt from *A Mercy* in order to further connect [Rebekka]'s struggle for liberal subjectivity to emerging New World regimes of property and ownership. It is situated towards the very end of her narrative and in it [Rebekka] puts herself in relation to all the women in her immediate as well as her more distant social environment, past and present, as she assesses her position as a newly widowed woman. This takes place during her fever-induced hallucinations during which she frequently imagines the presence of her female shipmates on board the *Angelus*:

> Well, her shipmates, it seemed, had got on with it. As she knew from their visits, whatever life threw up, whatever obstacles they faced, they manipulated the circumstances to their advantage and trusted their own imagination. The Baptist women trusted elsewhere. Unlike her shipmates, they neither dare nor stood up to the fickleness of life. On the contrary, they dared death. Dared it to erase them, to pretend this earthly life was all; that beyond it was nothing; that there was no acknowledgement of suffering and certainly no reward; they refused meaninglessness and the random. What excited and challenged her shipmates horrified the churched women and each set believed the other deeply, dangerously flawed. Although they had nothing in common with the views of each other, they had everything in common with one thing: the promise and threat of men. Here, they agreed, was where security and risk lay. And both had come to terms. Some, like Lina, who had experienced both deliverance and destruction at their hands, withdrew. Some, like Sorrow, who apparently was never coached by other females, became their play. Some like her shipmates fought them. Others, the pious, obeyed them. And a few, like herself, after a mutually loving relationship, became like children when the man was gone. *Without the status or shoulder of a*

man, without the support of family or well-wishers, a widow was in practice illegal. (AM 95–96; emphasis mine)

It is in these moments of [Rebekka]'s feverish ruminations that the text once again brings together notions of female solidarity, female-on-male dependency, and questions concerning those women's subjectivities/capacities in the New World. [Rebekka] locates herself in this prism of differently positioned women. Indeed, she envisions men to be the nexus between them, envisions them to be the one and only thing that they share in the world, when she comments that they "had *everything* in common with *one thing:* the promise and threat of men" (*AM* 96; emphasis mine). While her shipmates know how to use men to their own advantage, for example, the Baptist women chose to believe in the biblical story of genesis/history of creation with its focus on Adam. As the last few lines of the paragraph show, moreover, [Rebekka] compares her widowed existence to that of children. She believes that without a husband, without his 'status or shoulder,' she is reduced to the social status of a child, for as a widow she in fact becomes 'illegal.' This brings me back to my earlier discussion of the historical-legal space and place of married white colonial women in Virginia and it reiterates the notion that [Rebekka], like those historical women, does not believe to have an independent standing or subject status before the law. The passage, then, conveys a strong sense that [Rebekka] knows that her status in the New World is inextricably bound to her husband's. It also shows that [Rebekka], who feels "unowned" by her husband's death, begins to search for a new social or legal structure that would contain her and give her some sense of coherence or "role" in the world as she knows it. The narrative quickly provides a resolution for her, which comes in the form of a nearby Anabaptist village congregation. In order to belong she "had only to stop thinking and believe" (*AM* 97), only be "repossessed" by or relate to some kind of external hold.

It is in these moments of [Rebekka]'s feverish ruminations also, I suggest, that the text exposes [Rebekka] as a "mistress in the making".[110] That is, the

---

[110] I borrow the term from Jones-Rogers, who uses it in her study to describe and examine the ways in which white girls developed and learned how to be plantation mistresses and slave owners in the nineteenth-century American South (1–24). Jones-Rogers tells us that white southern girls learned how to become and to be slave mistresses "through an institutional process that spanned their childhood and adolescence. Over the course of these formative years, white girls practiced techniques of slave discipline and management, made mistakes and learned from them, modified their behavior to meet various conditions, and ultimately decided what kind of slave owners they wanted to become. It should come as no surprise that many of them wanted to be profitable ones. […] Ownership and control went hand in hand, and for

text here reveals [Rebekka]'s wish and ability to belong – or what Frank Wilderson would call [Rebekka]'s "aspirations to Human capacity" (*Red* 42) – by way of rendering the assumption of solidarity between those women necessarily ambiguous and, as I suggest, by ultimately dismissing it. That is to say, if given the chance, [Rebekka] follows the rules of the property paradigm. In light of this, perhaps it would seem fair to consider the assertion of a sense of solidarity between [Rebekka] and the women on the Vaark farm a *ruse* on the level of narrative.[111] If given the chance and as part of her own ceaseless struggle for liberal subjectivity in the New World, that is, [Rebekka] ultimately will claim the place of the master in the wake of her husband's death. What becomes clear if one cares to see through the narrative's ploy with female solidarity – looking, in other words, for what it at stake for [Rebekka] at this particular moment in the narrative – is that [Rebekka] could have chosen to continue her friendship with [Lina] but does not; she could have chosen to live and work on the farm together with the other women but decides not to. Instead of choosing to belong with these women, she decides to join the [Anabaptists] and to become a part of their spiritual community. Even more so, [Rebekka] actually is highly aware that, as a widow, she also has the power (legal, absolute) to do with her property what she wants to do, which the text illustrates in the previous passage when [Rebekka] notes that [Sorrow] should be afraid of her own future with her mistress's impending death. Put another way, [Rebekka]'s [character] allegorically represents and proleptically points the reader to the ways white southern married women in nineteenth-century America would come to manage their financial affairs and human property and, therefore, "their direct economic investment in slavery and their pecuniary interested in perpetuating it" (Jones-Rogers 202). After all, the text not only suggests that [Rebekka] is more than capable and willing to handle her economic affairs (as illustrated by her wanting to sell [Florens] and [Sorrow]) but that she also and likely will marry again soon. As we continue reading in an exchange between the indentured servants [Willard] and [Scully]:

> Mistress had changed as well. [...] Rising from her sickbed, she had taken control, in a manner of speaking, but avoided as too tiring tasks she used to undertake with gusto. She laun-

---

white girls who had slaves, developing techniques of management and discipline was an important aspect of their early training. For those who were newly inducted into slave-owning communities, 'the plantation was a school' where they learned how to be propertied women" (1–2, 4).

[111] I borrow the term *ruse* from the first chapter of the first part of Wilderson's *Red, White, and Black: Cinema and the Structure of U.S. Antagonisms* entitled "The Ruse of Analogy." In it, he develops his arguments on the structural incommensurability between Blackness as Slaveness and US civil society (see esp. pp. 35–53).

dered nothing, planted nothing, weeded never. She cooked and mended. Otherwise her time was spent reading a Bible or entertaining one or two people from the village.
"She'll marry again, I reckon," said Willard. "Soon."
"Why soon?"
"She's a woman. How else keep the farm?"
"Who to?"
Willard closed one eye. "The village will provide." He coughed up a laugh recalling the friendliness of the deacon. (*AM* 143–144)

[Coda]: Claiming White Woman Mastery

Some might think that I have gone too far in suggesting that [Rebekka] would in fact take the master's place on the [Vaark] homestead, that [Rebekka] needs to be understood as someone who actually aspires to the status of a coherent liberal subject. As I have tried to show, the property paradigm organizes her lived experience in the New World in fundamentally different ways than the lived experiences of the other women that she meets on her long journey to becoming and belonging in the New World. *A Mercy* throws into relief this long struggle for liberal subjecthood as it maps [Rebekka]'s place and space in colonial Virginia as one in which she initially is both female and illegal yet ultimately full of capacity or possibility to become a liberal subject. Like her husband, that is, [Rebekka] moves within the realm of the white liberal Human. Her existence will remain intricately connected to white men, as gestured towards by the notion that she will soon marry again. *A Mercy*, by this route, critically challenges readings that locate [Rebekka] as being subjected to as well as subjugated by the grammar of property in the same way that, say, [Sorrow] would be.

En route to some kind of a conclusion to this chapter, let me offer two more related points: One) Historians Firor and Scott note that "[e]very newcomer [in colonial Virginia] had to withstand the ordeal of 'seasoning'—catching, then surviving the diseases prevalent in the new environment." If we chose to read [Rebekka] in the way that I have proposed throughout the chapter so far, we find how for [Rebekka] her sickness and subsequent recovery from the smallpox puts in motion a process of self-fashioning. That is, her sickness, fever, and hallucinations might be best understood as part of a process in which [Rebekka] adjusts to the New World colonial scene, as part of her being "seasoned" both literally and metaphorically in this environment. In their own way, [Rebekka]'s feverish hallucinations – while fogging her mind and disorienting her temporarily – nevertheless enable her to clearly see which steps to take to carve out a space for herself as a newly widowed woman in the midst of the [Anabaptist congregation]. Like the fog that envelops her husband-to-be when he makes his first steps on the shores of the colonial Chesapeake before signing himself into liberal being during a business transaction (see Chapter 4.1), I read those feverish hallucinations as well as [Rebekka]'s recovery from them as a moment in which the novel dismisses other options for her existence on [Vaark] territory—options that are not necessarily bound by the New World grammar of property such as carrying on with her life and friendship with [Lina] in the way that she did before [Vaark]'s death. For as soon as the hallucinations stop, [Rebekka] claims her position in the world among the [Anabaptist] spiritual community and village and, by extension, structurally as a member of the Human fold. In other words, [Rebekka] is able to claim relationality within the realm of Human sociability.

Two) As my final hand in this section of the chapter, I want to offer a brief reading of her [character] in which I think about how [Rebekka] both relates to and prefigures dominant configurations of white (Southern) womanhood written and created by a historical future yet to come: chattel slavery in colonial Virginia and, later, in the United States from the early eighteenth century through the nineteenth and beyond. In other words, I also consider [Rebekka]'s moving towards the realm of the Human to be part of a narrative gesture in which *A Mercy* hints at the sexualized and racialized ideologies/economies of womanhood that were subtended by chattel slavery and renewed and refined within the plantation household (e.g., Broeck, "Property"; Fox-Genovese; Jones-Rogers; Painter). I am thinking here, primarily, about dominant and dominating constructions of white femininity as well as white female domesticity and sexuality, which Broeck summarizes as follows:

> The status of white women within the plantation complex [...] was aggressively marked by an almost schizoid antagonism between not having civil rights on the one hand and being extremely privileged, socially and culturally, on the other. [...] [This also included] the parasitical configuration of dominance and oppression which enabled white women's position in the plantation system vis-à-vis Black women, and black men, for that matter. [...] For a white lady, domesticity meant a kind of compulsive but luxurious construction of the white female body, which required extreme efforts at staging this body. [...] Domesticity meant being trained to expect and to accept black labor for one's own sustenance as a matter of course [...] white ladies had the power of representing their oftentimes absent husbands in matters concerning the "big house." (Broeck, "Property")

The white plantation mistress's domesticity, her status within the household, and her sexuality were inextricably bound to her slaves and specifically to Black slave women, whose (reproductive) labor and bodies were perpetually open to and readily available for the master's – and the mistress's – needs, desire, and will, be they sexual, economic or other (Adrienne Davis "Don't Let"; Angela Davis).

Throughout the narrative, *A Mercy* deliberately signposts these racialized as well as sexualized histories of Black and white womanhood within the plantation household, with, for example, the presence of the white plantation mistress of the Portuguese slave trader [Senhor D'Ortega], on the one hand, and the [*minha mãe*], [Florens'] enslaved mother, on the [D'Ortega] plantation, on the other.[112] What concerns me here arguably takes me beyond the narrative story-world proper of the novel, for *A Mercy* itself yields no obvious clues as to whether [Rebekka] will ultimately become a plantation mistress, like [Senhor D'Ortega's wife], with her next marriage or, perhaps, the one after that. Nevertheless, I want to suggest that [Rebekka] also needs to be read as a kind of prototype for the antebellum Southern belle. To repeat one of my earlier arguments, [Rebekka] needs to be read as a "mistress in the making."

Putting *A Mercy* in conversation with Valerie Martin's 2003 Orange Prize-winning novel *Property* might open up an additional conceptual window on these issues on the literary level of representation. Set in the first half of the nineteenth century on both a Louisiana sugar plantation that is threatened by slave rebellion and in the city of New Orleans in the midst of yellow fever and cholera outbreaks, *Property* is narrated by the white plantation mistress Manon Gaudet. As a text by "a contemporary white female writer which tries to come to terms with the legacy of an inextricable connection of white femininity to slavery," the plot follows Manon, her unhappy marriage, and it centers on the complicated, coercive relationship between Manon and "her" slave-servant Sarah (Broeck, "Property"). In their own ways, Christina Sharpe and Sabine Broeck have provided two excellent readings of the novel (Broeck, "Property"; Sharpe, "The Lie"). They critically expose how novels like *Property* need to be understood "to be engaged in *constructing a useable past out of which a post racial present and future might be understood to have been always already coming into existence—even under the most brutal of systems*" (Sharpe, "The Lie" 194); and how *Property*'s narrative politics may be viewed as an attempt at representing "the social, cultural and psychic implications the material fact of property

---

[112] Indeed, *A Mercy* carefully leaves tracks for its readers that would support the present argument but are easily overlooked. For example, in the last textual fragment of the novel, the [*minha mãe*] tells the reader and [Florens] how she came to the English colonies on the North American continent, which is a story littered with perpetual sexual violence and rape (*AM* 161). The text juxtaposes those experiences with the horrible presence of the [D'Ortegas] and it shows that both husband and wife take part in coercing her: "And it would have been alright. It [and it being the sexual violence] would have been good both times, because the results were you and your brother. But then there was Senhor *and* his wife. I began to tell Reverend Father but shame made my words nonsense" (*AM* 164; emphasis mine).

has for the positioning of the white mistress, and her black slave on an axis of gender [and] the problematics of the *splitting* of gender, into white female human beings who have gendered subjectivity, and black slaves who do not" (Broeck, "Property"). Both Sharpe and Broeck caution their readers against the novel's reception which has often positioned the novel within a paradigm of so-called post-racial woman/feminist identity formations while in fact ignoring the white female protagonist's abusive, parasitic, violent behavior towards her slave Sarah. Their arguments are carefully constructed around various scenes in the novel in which Manon either witnesses/participates or actively creates/participates in the sexual, physical or other subjugation or use of her enslaved female property.

My aim here is not to provide another reading of Martin's text (which would not be able to add much to Sharpe's and Broeck's respective powerful arguments) but to gesture at the following: Like the white plantation mistress Manon Gaudet in Martin's text, who becomes a wealthy widow after her husband gets killed in a slave uprising and who is subsequently able to "master her own affairs" (Broeck, "Property"), [Rebekka] begins to manage and to master her own affairs in the wake of her husband's death. A relatively wealthy, landowning widow herself, [Rebekka] carves out a space for herself as part of the [Anabaptist] village congregation which eliminates any previous gender-based connections with the other women on the [Vaark] farm. Like Manon Gaudet, [Rebekka] in *A Mercy* assumes authority and control over her servants *even before* she opts for the spiritual group (*AM* 51). An example of this is a letter she hands to [Florens] when she sends the enslaved girl to fetch the blacksmith for help. The letter clearly states that [Florens] *"is owned by me"* and that *"[o]ur life, my life, on this earthe depends on her speedy return"* (*AM* 110; see also my reading of [Florens]). That is to say, the narrative brings to the fore [Rebekka]'s investments in the property paradigm because her life, literally, depends on her slave property when she falls ill. [Rebekka]'s life depends on [Florens], who needs to find [the blacksmith] to help cure her mistress. There are, of course, many differences between those two literary representations of white Southern womanhood—with Manon staying by herself in nineteenth-century New Orleans and [Rebekka] looking for a seventeenth-century community in which to belong as only one example. What I hope to convey by briefly juxtaposing Manon Gaudet and [Rebekka] in this way is that it seems to be fair to also think about [Rebekka] as a kind of preliminary version of white Southern womanhood that will continue to emerge as white patriarchal capitalist antiblack supremacy materializes in colonial North America. To think about her along those lines once again shows that she can create relationality structurally. Again, Afropessimism tells us that there is no such thing as a true relationship between the subject and the abject because the gratuitous violence of slavery structurally and politically "forecloses upon reciprocity" (Wilderson, "Aporia" 140). Reciprocal relationships, in turn, can only exist between members of the Human fold. At stake in [Rebekka]'s narrative fragment are the ways dominant conceptions of white female subjectivity and liberal self-making, the historical roots of which lie at least partly in colonial Virginia, will develop and travel through time, space, and epistemes.

For [Rebekka Vaark], finally, the New World offers possibility. For her, just like for her husband, the colonial scene of Virginia and Maryland provides the ground for her claims to freedom from feudal rule and religious doctrine as well as for her claiming mastery of herself and others. It is here, in other words, that property or, rather, co-ownership of land, servants, and slaves, as well as ownership of herself open up an avenue towards the fold of the liberal Human for this [character]. Struggling for liberal subjectivity in this way, the relationships that [Rebekka] forges with her female companions/servants and slaves in the [Vaark] household need to be situated not within a paradigm of universal female affiliation or solidarity across racial, religious, and cultural boundaries but in/as a representation of white female claims to co-mastery at the New World colonial scene (*AM* 143). By way of neat narrative maneuvering and contrary to mainstream readings of this [character], I claim that *A Mercy* plays with the possibility of solidarity between the women on the [Vaark] farm only to emphasize how [Rebekka] opts for liberal Humanism and thus ultimately rejects such possibilities. That is to say, [Rebekka]'s narrative segment suggests that notions of choice are fundamental to the making of the liberal *woman* subject at the New World colonial scene. After her husband passes away, [Rebekka] chooses affiliation with a religious congregation and, like many widowed women who lived in colonial Virginia, takes control of her husband's estate and human property (143), which she ventures to sell (157). In this way, [Rebekka] becomes "co-conspirator" in the formation of racial slavery on the North American mainland (Jones-Rogers 205). With the [character] of [Rebekka Vaark], then, *A Mercy* allegorically throws into relief how in colonial North America white female liberal self-making was bound by private property and how ownership and control of their human property establishes for these women capacity for choice in the first place.

4.4 "My Name is Complete": [Sorrow], Anticipating Generations, and the New World Grammar of Property

[Routing the Argument] In the present chapter, I turn to the [character] of [Sorrow/Twin], which probably is the most (racially) ambiguous [character] in *A*

Mercy. In contrast to the novel's critics and reviewers, who most often insist on reading [Sorrow] as either Black or white, I will in my close reading stay with the ambiguity of this [character]. I claim that it is [Sorrow]'s very ambiguity which fugitively opens up an utopian narrative moment on the possibility of making generations beyond the property paradigm—a moment, that is simultaneously foreclosed by the novel's complicated workings of time and anticipation of a historical future yet to come. This future is one in which "kinship relations [would be subordinated] to property relations" (Sharpe, *Monstrous* 34–35). In other words, I also suggest that *A Mercy* utilizes this fragment to speculate about kinship formations at the New World colonial scene. I claim that this [character]'s textual fragment is one which the novel envisions and anticipates the making of Black generations that, while originating from slavery's formations of predatory sexuality and reproduction of human property, persist and flourish; and that they do so while the scripts of white kinship and family formations are discontinued and disnarrated. However, I also suggest that this moment will not last within *A Mercy*'s representation of colonial and New World landscapes. With [Sorrow], that is, *A Mercy* ultimately shows us what was lost. Indeed, her very name refers us to what could not happen.

> By way of history, which is really allegory after all, June's father had told her, "Only read the business section of the newspaper. *That* is the news. [...] The other parts are the casualties and the fantasies."
> — Dionne Brand, *Love Enough*

> In this case, what is at issue is the difference between the deployment of sexuality in the contexts of white kinship – the proprietorial relation of the patriarch to his wife and children, the making of legitimate heirs, the transmission of property – and black captivity —the reproduction of property, the relations of mastery and subjection, and the regularity of sexual violence[.]
> — Saidiya V. Hartman, *Scenes of Subjection*

Introduction

In what follows I turn to the [character] of [Sorrow/Twin] and thus to what critics of *A Mercy* have described as the novel's perhaps most (racially) ambiguous [character]. Indeed, in dealing with [Sorrow], who in [Lina]'s fragment is described as "[v]ixen-eyed [...] with black teeth and a head of never groomed woolly hair the color of a setting sun" and as having "[r]ed hair, black teeth, recurring neck boils and a look in those over-lashed silver-gray eyes that raised Lina's nape hair" (*AM* 49, 51–52), readers and reviewers have devoted considerable critical energy in trying to make sense of (and define?) [Sorrow]'s racial identity. In her survey of the novel's critical reception, Jessica Wells Cantiello, as one of the very few critics who focuses explicitly on [Sorrow] in their respective reading of *A Mercy*, states accordingly that most scholars and reviewers tend to either read [Sorrow] as white or Black ("From Pre-Racial" 173). In this context, she goes on to tell us that critics' attempts to "racially identify" this [character] fail to consider that the text "gestures toward a multiracial America" and "illustrate[s] the reliance on and assumption of a black/white paradigm in what many critical commentators call a 'post-racial' United States" ("From Pre-Racial" 173). Pointing to the fact that *A Mercy* was published within a week of Barack Obama's election as President of the United States, Cantiello brings multiracialism to the novel's discursive field. With it, she sets out to question approaches to reading the novel that are wedded to the ideology and rhetoric of colorblindness and post-racialism, which accompanied Obama's ascent to the White House.[113] Cantiello goes on to tell us:

> While some aspects of the book work well for this type of reading, the tendency to emphasize certain comparisons, particularly the semantic relationship between Morrison's use of *pre-racial* to describe the novel's late seventeenth-century racial landscape and the media's use of *post-racial* to describe Obama's America, simplifies and at times misreads the complexity of the racial relationships Morrison explores in the text. The reviewers' confusion crystallizes around the character of Sorrow, a confusion that Morrison insists upon but reviewers try to explain away. ("From Pre-Racial" 165)

If we follow Cantiello, then, [Sorrow]'s [character] highlights racial mixing as an important historical context of North American colonial beginnings. My intention in bringing this up is not to delve deeply into and/or push a critique of multiracialist discourse. Black Studies scholars and thinkers like Jared Sexton have amply alerted us to multiracialism's proximity to colorblind ideology and the ways in which it

> solicits alliance with other political and intellectual efforts to go "beyond the black-white binary" [...] efforts which, in many cases, have been shot through with an air of antiblackness[.] In the register of contemporary racial politics, black identity appears as an antiquated state of confinement from which the "multiracial imagined community" [...] must be delivered; the negative ideal against which "the browning of America" [...] measures its tenuous success. (Sexton, *Amalgamation* 6)

Rather, what is at issue in the chapter are the ways in which an approach like Cantiello's raises important questions as to how property, race, reproduction,

---

113 For a critique of post-racialism in the law and beyond see, e.g., Cho.

and sexuality act in concert and how their intersections are navigated in [Sorrow]'s textual fragment in *A Mercy*.

With one, if not the most pronounced critical approach to reading [Sorrow] thus briefly sketched[114], what follows needs to be understood as an attempt to deal with the very ambiguity of her [character]. [Sorrow]'s textual fragment in *A Mercy* is situated after that of [Rebekka Vaark] and spans a total of nineteen pages. She enters *A Mercy*'s colonial scene and the bourgeoning [Vaark] estate after being "[a]ccepted, not bought, by Sir, she joined the household after Lina but before Florens and still had no memory of her past life except being dragged ashore by whales" (*AM* 49). [Sorrow]'s ambiguous (racial) status at *A Mercy*'s New World colonial scene installs in the novel a kind of double movement by which it opens a window of possibility that it closes almost instantaneously. In other words, [Sorrow]'s ambiguity pushes the reader to consider and navigate at least two things: First, [Sorrow] returns us to Weinbaum's conceptualization of the "race/reproduction bind" as that which organizes transatlantic modernity's "knowledge about nations, modern subjects, and the flow of capital, bodies, babies, and ideas within and across national borders" as well as to white Western patriarchal genealogies and formations of family and, thus, to the transmission of property, including the white name and slave property, from one generation to the next (Weinbaum, *Wayward* 2; Spillers, "Mama's Baby"; see Chapter 3). With [Sorrow], as I argue, *A Mercy* once again enters into conversation with Black feminist thought on the racialized and sexualized nexus of property and slavery; with [Sorrow], *A Mercy* both suggestively and fugitively probes into the possibility of making generations beyond the white patronymic of transgenerational transmission of private property. (And it is against the backdrop of this theoretical exchange about this nexus on the literary level of representation that we might perhaps think of [Sorrow] as Black). Put another way, this [character] brings to *A Mercy*'s narrative orbit a utopian moment, which invites us as readers to consider the possibility of making Black generations beyond the liberal property paradigm. Second, [Sorrow]'s textual fragment and specifically her two pregnancies shed light on what Hartman in the above epigraph describes as the "different deployments of sexuality in the contexts of white kinship and black captivity" (*Scenes* 84), thus anticipating a historical future yet to come (and doing so in imaginary hindsight, as it were, from our twenty-

---

114 In addition, Otten and Roye have read [Sorrow] within frameworks of "motherlessness" and orphanhood in this context.

first century present).[115] That is, *A Mercy* gestures not only towards racial slavery becoming systematic in North America but also towards the sexualized, racialized ideologies of womanhood in the Deep South, which would render white women's children kin and black women's children property to be "passed between and on among those [white] kin" (Sharpe, "Lose"; Jones-Rogers). (And in this sense, [Sorrow] does indeed speak to the historicity of race and sexuality, as well as interracial sexuality, that multiculturalism's assumptive logics so often obscure, if we follow Sexton's arguments (*Amalgamation* 4)).

Lingering with [Sorrow]'s ambiguities, the following interconnected questions arise: How does [Sorrow]'s fragment navigate kinship formations bound by a calculus of property that negates kinship for some while granting it to others? How does [Sorrow] confront the text of a white Western patronymic? (How) Does she become the locus of a rigorous critique of such kinship formations? Is it possible at all to imagine her [character] as paradigmatic for a different conception of kinship? In grappling with these questions as well as others that will surface as I move through the chapter, I examine this textual fragment's utopian moment of making/anticipating generations, which I here reconstruct and trace through [Sorrow]'s two pregnancies. However, in contrasting the making or anticipating of generations in [Sorrow]'s fragment with the pregnancies of [Rebekka Vaark], I also suggest that the novel ultimately will disnarrate such an utopian vision. As for the chapter's structure, I first situate her [character] within *A Mercy*'s representation of the New World colonial scene. Then, I will examine the ways her pregnancies contrast with those of her mistress [Rebekka Vaark]. I end each of its main parts with additional sets of questions. This reflects my attempt to deal with this highly ambiguous [character], which insistently pushes us as readers to continuously interrogate the intricate connections between private property, race, and kinship.

Situating [Sorrow]

[Sorrow] is the third girl, after [Lina] and [Rebekka] and before [Florens], to enter [Jacob Vaark]'s household. Like [Florens], [Vaark] "acquires" her in a business

---

115 There is another [character] with which *A Mercy* engages such notions: the [*minha mãe*]. It is the [*minha mãe*] who embodies this future. As I will show in my last close reading of the novel, the [*minha mãe*]'s text brings Atlantic slavery and specifically the (im)possibility for individual motherhood for enslaved women to *A Mercy*'s textual orbit (see Chapter 4.6).

transaction (*AM* 32, 31).[116] Apart from the above few very brief descriptions of her physical appearance, which suggest that she may be of mixed racial descent,[117] here's what we know about [Sorrow]: Born on board a ship, [Sorrow] literally comes from the waters of the Atlantic (114–117). It remains unclear whether we are dealing here with a pirate ship or a slave ship, but we learn that it carries stolen "cargo: bales of cloth, chests of opium, crates of ammunition, horses and barrels of molasses" (115). While we do not learn anything about her mother, we know that [Sorrow]'s father is the ship's captain, who "keep[s] her aboard. He reared her not as a daughter but as a sort of crewman-to-be. Dirty, trousered, both wild and obedient with one important skill, patching and sewing sailcloth" (124–125). [Sorrow] arrives at the shores of the New World after their ship has foundered. It is then and there, "beneath the surgeon's hammock in the looted ship. All people were gone or drowned and she might have been too had she not been deep in an opium sleep in the ship's surgery," that she first meets [Twin], her "identical self" who "couldn't be seen by anybody else" (114). Once on land, [Sorrow] is taken in by a sawyer family, who nurse her back to health, equip her with a new name, and make her their servant/slave. It is here that the reader receives another clue about [Sorrow]'s physical appearance, namely that the sawyers had mistaken her for "a lad" (116). In the very first sentence of her textual fragment, we also learn that "Sorrow" is not her real name: "She did not mind when they called her Sorrow so long as Twin kept using her real name. [...] So if she were scrubbing clothes or herding geese and heard the name Captain used, she knew it was Twin. But if any voice called 'Sorrow,' she knew what to expect" (114). For some, like [Lina], [Sorrow] with her "unbelievable and slightly threatening hair" (117) furthermore signifies a kind of mythical danger. [Lina] seems to think that [Sorrow] is not only far from reliable but also somehow responsible for the death of the [Vaark children]. In [Lina]'s textual fragment the reader is told accordingly that in "Sorrow's presence eggs would not allow themselves to be beaten into foam, nor did butter lighten cake batter. Lina was sure the early deaths of Mistress's sons could be placed at the feet of the natural curse that was Sorrow" (53). In this context, reviewers in their attempts to make sense of this [character]'s mythical quality have often made a point about [Sorrow] being "separate and alone" (Cox 115), about her being "psychologically broken"

---

[116] To recall, we learn in [Vaark]'s fragment in the novel, "He believed it now with this ill-shod child that the mother was throwing away [i.e. [Florens]], just as he believed it a decade earlier with the curly-haired goose girl, the one they called Sorrow. And the acquisition of both could be seen as a rescue" (*AM* 32).

[117] The word that the [sawyer] who gives [Sorrow] to [Vaark] in said business transaction uses is "mongrelized" (*AM* 118).

(Roye 223), or about the fact that "Sorrow's interior space is fragmented" (Wardi 95). John Updike, moreover, describes [Sorrow] as "long addled in the head by her shipboard traumas and her illusion of an advisory companion called Twin" ("Dreamy Wilderness"). Finally, we also learn that the sawyers put this eleven-year-old girl (*AM* 118) to multiple tasks, none of which she seems to be able to complete in an adequate way (if we adopt the [housewife]'s perspective). We read:

> Sorrow's bare feet[118] fought with the distressing gravity of land. She stumbled and tripped so much on that first day [minding geese] at the pond that when two goslings were attacked by a dog and chaos followed, it took forever to regroup the flock. She kept at it a few more days, until the housewife threw up her hands and put her to simple cleaning tasks—none of which proved satisfactory. But the pleasure of upbraiding an incompetent servant outweighed any satisfaction of a chore well done[.] (*AM* 117)

Over the following paragraphs, the text quickly exposes the reader to the sexual violence that [Sorrow] is subjected to while staying with the [sawyer family] and being made their "incompetent servant" (117). We continue reading:

> The housewife told her it was monthly blood; that all females suffered it and Sorrow believed her until the next month and the next and the next when it did not return. Twin and she talked about it, about whether it was instead the result of the goings that took place behind the stack of clapboard, both brothers attending, instead of what the housewife said. Because the pain was outside between her legs, not inside where the housewife said was natural. (*AM* 117–118)

These lines reveal not only that [Sorrow] is repeatedly raped by the [sawyers' sons] but also that she becomes pregnant because of these frequent predatory attacks on her, as her 'monthly blood does not return.' By way of neat narrative ploy, these two paragraphs establish a connection between [Sorrow]'s arrival on colonial shores, sexual coercion, and the New World's grammar of property. For when [Jacob Vaark] "accepts" [Sorrow], he does so as part of a business transaction between him and the [sawyer]. In [Vaark]'s fragment we read: "A decade ago now, a sawyer had asked him to take off his hands a sullen, curly-headed girl he

---

118 Like [Florens], who craves for "anybody's shoes" throughout her text and for whom the absence of shoes signifies her being positioned outside of the realm of the liberal Human (see Chapter 4.6), [Sorrow] navigates Virginia's colonial landscapes barefoot. After she has arrived at the [sawyers'] dwelling, the [housewife], while handing her some ill fitting clothes, tells her, ""I'll have to make you something more fitting for there is nothing to borrow in the village. *And there won't be any shoes for a while*"" (*AM* 117; emphasis mine). However, the text does not offer any more hints as to whether [Sorrow]'s feet will remain bare.

had found half dead on a riverbank. Jacob agreed to do it, provided the sawyer forgive the cost of the lumber he was buying" (31). At stake in her text, in other words, are the intricate connections between sexual subjection and kinship as structured by the New World's grammar of property. For it is here that we learn that, in the wake of 'the goings that took place behind the stack of clapboard,' [Sorrow]'s "hurt was still there when the sawyer asked Sir to take her away, saying his wife could not keep her" (118). In fact, then, [Sorrow] arrives at the [Vaark] farm already pregnant (118) and it is here also that the text establishes a continuity between her shipwrecked arrival in the New World, her time at the [sawyers'] and the sexual violence she is subjected to, as well as her being made the currency in the business transaction between [Jacob Vaark] and the [sawyer].

I take my cue from these paragraphs in arguing that *A Mercy* examines the possibility of making Black generations on the New World colonial scene with this [character]. The text places [Sorrow]'s pregnancy (which is the first of two pregnancies that we learn of) and subsequent birth of her child next to that of her new mistress [Rebekka Vaark]. It is [Lina] who tells [Sorrow] that she is pregnant in response to which [Sorrow] "flushed with pleasure at the thought of a real person, a person of her own, growing inside her" (*AM* 121). The other women on the [Vaark] farm largely ignore [Sorrow]'s pregnancy. That is, after breaking the news to [Sorrow], "Lina simply stared at her and, hoisting the basked on her hip, walked away. If Mistress knew, she never said, perhaps because she was pregnant herself" (121). While [Sorrow]'s mistress gives birth to "a fat boy who cheered everybody up—for six months anyway. They put him with his brother at the bottom of the rise behind the house and said prayers," it is [Lina] who tells [Sorrow] that her child did not survive. We read:

> Although Sorrow thought she saw her own newborn yawn, Lina wrapped it in a piece of sacking and set it a-sail in the widest part of the stream and far below the beavers' dam. It had no name. Sorrow wept, but Twin told her not to. "I am always with you," she said. That was some consolation, but it took years for Sorrow's steady thoughts of her baby breathing water under Lina's palm to recede. (*AM* 121)

We can recognize in these lines and in their juxtaposition of both [Sorrow]'s and [Rebekka]'s pregnancies, of their giving birth to, and of the deaths of their respective children the sexualized, racialized ideologies of womanhood in the Deep South and in the New World more generally, which would render white women's children kin and black women's children property (e.g., Jones-Rogers; J. Morgan, "*Partus*"; Painter, *Southern History*; Sharpe, "Lose"). That is, in setting side by side these two women's reproductive abilities the text here anticipates a historical future yet to come—a future in which concepts of kinship, reproduction, and family would become (and have historically and epistemically been) enmeshed with enslavement, the market, and racial capitalism in the United States; a future "in which blood *becomes* property (with all of the rights inherent in the use and enjoyment of property) in one direction and kin in another" (Sharpe, *Monstrous* 29). Again, this would also be a future, as Black feminists continue to remind us in this context, in which the formation of

> 'Family,' as we practice and understand it 'in the West' – the vertical transfer of a bloodline, of a patronymic, of titles and entitlements, of real estate and the prerogatives of 'cold cash,' from *fathers* to *sons* and in the supposedly free exchange of affectional ties between a male and a female of *his* choice – [would become] the mythically revered privilege of a free and freed community. (Spillers, "Mama's Baby" 74)

In the above paragraph from [Sorrow]'s textual fragment in *A Mercy*, we can recognize such conceptions of "family" in the birth of [Rebekka Vaark]'s son. To recall, the novel positions the making of more [Vaark] generations explicitly as part of the creation of [Jacob Vaark]'s liberal subjectivity as well as of [Rebekka Vaark]'s claim to white female mastery (*AM* 32, 76–78; see Chapters 4.1 and 4.3). Letting [Rebekka]'s son perish from a fever (*AM* 54), these lines allow for the phantasmatic possibility that the Western patrilineal order of kinship that this child represents will in fact not manifest itself within the narrative frame of the novel. It is a speculative representation because it opens a path that does not lead down the same route that history has taken. This is reiterated multiple times in the text through the untimely deaths of [Rebekka Vaark]'s daughter and an unspecified number of sons. Despite the fact that [Jacob Vaark] "was confident that [Rebekka] would bear more children and at least one, a boy, would live to thrive" (19), then, the novel ultimately seems to dismiss and refuse future [Vaark] generations, as well as white Western family formations more generally, by disnarrating all of the [Vaark] children from the novel's plotting.

We can also recognize in these lines and in this powerful comment on white patrilineal kinship formations on racial capitalism's emerging landscapes in colonial Virginia something much more unsettling. That is, these lines draw attention to the notion that the alliances forged on the [Vaark] farm are subtended by the liberal property paradigm and that its grammar also determines the ways in which the (non-)relations between the [Vaark] women are structured. While *A Mercy* exposes assumptions of solidarity between the [Vaark] as a ruse in [Rebekka Vaark]'s fragment (see Chapter 4.3), the text here also shows how [Sorrow] and [Lina] (and [Florens], for that matter) are not only positioned differently by the property paradigm in relation to their mistress [Rebekka] but also in relation to each other. Put another way, the text exposes the reader once again to the difference between *dispossession* and *fungibility* and to the ways [Sorrow] and

[Lina] are in fact positioned by different regimes of violence, which structurally work against one another and open up relations of antagonism or conflict, respectively (see Chapter 3). What these lines alert us to, then, is that [Lina] is the one who "ruled and decided everything Sir and Mistress did not" (*AM* 120), despite being profoundly dispossessed by the genocide of her tribe and by her subsequently being made a servant. [Lina] is the one who effectively and silently runs the [Vaark] farm and it is [Lina] who appears to be responsible for the death of [Sorrow]'s nameless newborn. Let me draw your attention again to the following lines, extracted from the above paragraph: "Although Sorrow thought she saw her own newborn yawn, Lina wrapped it in a piece of sacking and set it a-sail in the widest part of the stream and far below the beavers' dam" (*AM* 121). In this configuration, it is [Lina] who wields authority over which children born in the [Vaark] household are given a chance to thrive. With [Lina] doing everything in her power to keep in place the fragile system of the [Vaark] household (including her power to decide which children in the [Vaark] household will live and which will die) we arrive at the notion that some (read: white) family formations will be considered more valuable than others. We arrive at the notion, in other words, that white genealogies of recognized kinship formations and the production of the white family are dependent on the destruction and disruption of Black family formations (see e.g., Nyong'o).

In the text's juxtaposition of these two women's pregnancies and of the deaths of their respective children, we are also exposed to the complicated workings of time in the novel's diegesis more generally. That is, while every reading of *A Mercy* happens in the now, the novel's plotting takes its readers back to a historical past, the outcome of which – racial slavery and the emergence of white Western modernity and its liberal subject along the lines of the property paradigm – the text anticipates. As readers, we are able to recognize this from our contemporary, twenty-first century perspective. And yet, the novel's very plotting once again invites us to consider "what could have been" or "what could have happened" (see Chapter 4.2). With [Sorrow]'s fragment, its pushes us to ask, what would have happened if [Lina] had not interfered? What could have been her reasons to do so? What kind of family structures could have developed on the [Vaark] farm had [Rebekka]'s child not died prematurely and had [Sorrow]'s baby not "breath[ed] water every day, every night, down all the streams of the world" (*AM* 122)? What if [Sorrow]'s infant would have had a name? Would they have been sold or would they have stayed at the [Vaark] farm? Would they have become members of some kind of a "free and freed community," eventually?

[Sorrow]'s Utopia, or Anticipating Generations

I now turn to [Sorrow]'s next pregnancy and to the birth of her second child. This occurs "[y]ears later" around the time that [Jacob Vaark] dies while building his third house and roughly when [the blacksmith] is summoned to help cure [Rebekka] (*AM* 122). At issue here are the ways in which the text mobilizes an utopian moment with the birth of this child.[119] This moment responds to the above questions by showing us "what could have been" had history taken a turn away from the liberal property paradigm. However, I suggest that this moment is fleeting and fugitive despite the fact that is produces a counter-history of care and kinship (see, e.g., Hartman, *Lose*, "Venus") that confronts the liberal subject formations represented in the text through [Jacob Vaark] and [Rebekka Vaark]'s respective claims to freedom and (co-)mastery.[120] It is fleeting and fugitive not merely in the sense that it does not last within the novel's plotting. Echoing what Dionne Brand in the chapter's very first epigraph describes as the "casualties and the fantasies" of history, this moment also is fleeting in the sense that it anticipates what was lost in the making of the New World.

We are told about the birth of [Sorrow]'s second child on the last two and a half pages of her fragment (in my edition of the novel). We do not learn who the father of this child is, but the text suggests that it might be the deacon from the nearby village. As we read: "There were cherries, too, and walnuts from the deacon. But she had to be quiet. Once he brought her a neckerchief which she filled with stones and threw in the stream, knowing such finery would raise Lina's anger as well as alert Mistress" (*AM* 121). That the deacon may well be the father

---

[119] In general, I follow Ruth Levitas in my use of "utopia," who defines it as a concept that expresses "the desire for better way of being" (*Concept* 9). I also use "utopia" and "utopian" here broadly in reference to Fátima Vieira's definition of utopia as a literary genre that "relies on a more or less rigid narrative structure: it normally pictures the journey (by sea, land or air) or a man or woman to an unknown place (an island, a country or a continent); once there, the utopian traveller is usually offered a guided tour of the society, and given an explanation of its social political, economic and religious organization; this journey typically implies the return of the utopian traveller to his or her own country, in order to be able to take back the message that there are alternative and better ways of organizing society" (7). Jaap Verheul expands on this when he explains that "the written utopia remains an idealized and unrealized blueprint" (2). Arguably, [Sorrow]'s [character] can perhaps also be read as being on some kind of journey in the sense that she arrives on the unknown shores of the New World and literally comes from a ship—although she will never be able to return.

[120] For conceptualizations of "fugitivity" in relation to Blackness see generally, e.g., Hartman, *Lose*; Campt, *Listening*, *Image Matters*; Moten and Harney; Kawash.

of this child is reiterated a few pages later when [Sorrow] watches [Florens] and [the blacksmith]:

> The blacksmith and Florens were rocking and, unlike female farm animals in heat, she was not standing quietly under the weight and thrust of the male. What Sorrow saw yonder in the grass under a hickory tree was *not the silent submission to the slow goings behind a pile of wood or a hurried one in a church pew that Sorrow knew.* [...] *In all of the goings she knew, no one had ever kissed her mouth. Ever.* (*AM* 126; emphasis mine)

There is yet another possibility as to who the father of [Sorrow]'s second child may be and that is [Jacob Vaark]. In [Lina]'s fragment, we read that their mistress "said nothing when, to stop [Sorrow] roaming, he said, Sir made the girl sleep by the fireplace all seasons" (*AM* 52). As the above passage suggests, however, whether the deacon or, for that matter, [Jacob Vaark] actually fathered this child does not matter because like the previous "goings" which led to her first pregnancy, this one also appears to be part of the sexual violence that she is subjected to ever since her arrival at the shores of the New World. [Lina] accordingly addresses [Sorrow]'s pregnancy as "another virgin birth" in her fragment (54). (Of course, this phrase yet again references Atlantic slavery's reproductive economies, the production of slave property through systematic rape, as well as the notion that the children resulting from this often grew up side-by-side their white brothers and sisters on the plantation before they were either used/forced to serve their "master-fathers," "master-brothers," or "mistress-sisters," or sold for profit (e.g., Fox-Genovese; Jones-Rogers).

As mentioned already, [Sorrow]'s pregnancy occurs after the death of her master and during her mistress' illness, which leaves the [Vaark] farm "in disarray" (*AM* 129). When finally "in the afternoon silence of a cool day in May, on an untended farm recently swathed in smallpox [...] Sorrow's water broke," she sets out to the near riverbank alone, with her mistress sick, [Lina] untrustworthy because [Sorrow] takes her to be responsible for the death of her first child, and [Florens] on her errand to fetch [the blacksmith] (130). With some help given by [Will] and [Scully], who "heard her moans and poled their raft to the river's edge," [Sorrow] here gives birth to a little girl (130). For [Sorrow], the birth of her daughter requires her full attention and care and it also means that [Twin] disappears, which the text describes an "absence [...] hardly noticed" (131). With [Twin] gone, the text furthermore suggests that [Sorrow]'s daughter is the only child to survive within as well as beyond the novel's immediate narrative frame. As the next few lines show, [Sorrow] believes that giving birth to her baby girl releases her from the power men had had over her, constituting a change in her existence at the New World colonial scene. We read: "All her life she had been saved by men – Captain, the sawyers' sons, Sir and now

Will and Scully – she was convinced that this time she had done something, something important, by herself" (131). The text reiterates this a few lines later by gesturing at the notion that her "new status as a mother" equips her with a sense of previously non-existent "legitimacy." What is now at the center of [Sorrow]'s "new" existence at the New World colonial scene is what she knows and allegorizes as notions of care, of life, and of future generations. This is emphasized, for example, by way of [Sorrow]'s "attending routine duties" at the [Vaark] farm and her "organizing them around her infant's needs, impervious to the complaints of others" (132). Finally, the text pushes this by way of [Sorrow] naming her daughter as well as by the fact that she renames herself "Complete" at the very end of her fragment (131, 132).[121]

Of course, [Sorrow]'s active re/naming of both her daughter and herself references and comments on naming practices during Atlantic slavery. Scholars of slavery have shown that slave masters and mistresses would often give their slaves "Christian names or classical names of the Greco-Roman civilization" ("Naming Practices"), thus erasing the enslaved's previous names. Most slave narratives also speak to naming practices during slavery, as for example in the narrative of Olaudah Equiano (1789), in which we learn that he is renamed Gustavus Vassa after being bought by his master (Carretta 44). In the archive of slavery, moreover, the enslaved (continue to) appear as nameless "cargo, inert masses, and things" (Hartman, "Venus" 10). In "Venus in Two Acts," which is Hartman's struggle with the question of whether the anonymity of the archive can be remedied through the writing of stories, we encounter the enslaved, nameless female/girl/woman of the archive,

> [v]ariously named Harriot, Phibba, Sara, Joanna, Rachel, Linda, and Sally, [who] is found everywhere in the Atlantic world. The barracoon, the hollow of the slave ship, the pesthouse, the brothel, the cage, the surgeon's laboratory, the prison, the cane-field, the kitchen, the master's bedroom—turn out to be exactly the same place and in all of them she is called Venus. ("Venus" 1)

In taking up this history of naming practices during slavery, what is at stake in this moment in the narrative? What does it mean that [Sorrow]'s fragment ends with her new name? What does it mean that [Sorrow] appears to gain "legitimacy" through her new status as a mother and what kind of legitimacy is this? And what does it mean that [Sorrow], in turn, thinks about her child as a "person of her own" (*AM* 121)—a phrase that seems to suggest that, just like [Florens], her

---

[121] In this context, literary scholars have amply commented on the significance of names and naming practices in Morrison's novels (see, e.g., Kirby; Lyles-Scott).

vocabulary and conception of the world around her is saturated with private property's structuring grammar (see Chapter 4.5)? What does it mean that the novel imagines the potential beginning of a future that strives to be separate from the "sanctity of property" (Hartman, *Scenes*) with this [character] and its newborn but does not see such a future through within its plotting? In other words, how does one get from [Sorrow] to "Complete" (*AM* 132)?

[Coda]: "Complete" Fantasy, "Complete" Sorrow

So far, I have in tracing [Sorrow]'s two pregnancies tried to show how *A Mercy* uses her textual fragment to speculate about and conjure up an utopian moment within its plotting. In this moment, the possibility of a future that is not bound by the property paradigm is embodied by [Sorrow]'s second child and appears to also manifest itself in the fact that [Sorrow] ultimately renames herself "Complete." In many ways, then, this [character] can be read as confronting *A Mercy*'s readers with an utopian moment or a version of history, in which motherhood and notions of care become a kind of antidote to her shipwrecked existence as well as to the sexual violence she is subjected to, bestowing on [Sorrow] something like a sense of legitimacy on the North American colonial mainland.

Again, this textual moment is fleeting. Returning to [Lina] and the notion that she is responsible for the death of [Sorrow]'s first child as well as to the fact that [Sorrow] does prevent her from getting too close to and thus, perhaps, from killing this second child, I arrive at the notion that the text's conjuring and idea of a "legitimate future" for [Sorrow] is immediately undermined by the property paradigm. In other words, the novel shows us that there is another way in which [Sorrow]'s motherhood and the phantasma of a future that is not bound by the property paradigm are profoundly eroded. In one of [Florens'] later textual fragments, that is, the reader learns that in the wake of her recovery from the smallpox, [Rebekka Vaark] plans to sell both [Florens] and [Sorrow]. We continue reading:

> Sorrow she [Rebekka Vaark] wants to give away but no one offers to take her. *Sorrow is a mother. Nothing more nothing less.* I like her devotion to her baby girl. She will not be called Sorrow. She has changed her name and is planning escape. She wants me to go with her but I have a thing to finish here. (*AM* 157; emphasis mine)

These lines echo the novel's previous language of ownership and business transactions, of [Sorrow] being "acquired," that we encountered in [Jacob Vaark]'s text. They point us to the notion, brought to the scene of critical inquiry by

Black feminist thinkers and historians of slavery, that "kinship relations [would be subordinated] to property relations" when racial slavery would become systematic (Sharpe, *Monstrous* 34–35). In this configuration, to recall, kinship loses meaning for the enslaved (Spillers, "Mama's Baby"). In the above paragraph we also read that [Sorrow] is 'nothing more nothing less' than a mother. That is, even though she 'has changed her name and is planning escape,' and even though she has given birth to a child, these words suggest that this does not mean that her status of being and existence within *A Mercy*'s colonial landscapes has changed in any way. She is and will continue to be 'nothing more and nothing less' than what she was before, which is [Rebekka Vaark]'s, or perhaps someone else's, property. The same will likely hold true for her child. By extension, then, the above lines and the language and grammar of property they evoke once again refer us the ways in which reproduction would be tethered to questions of race, status, heredity, and descent, as for example by colonial legislation such as Virginia's paradigmatic seventeenth-century *Partus Sequitur Ventrem* act. Put another way, [Sorrow]'s becoming a mother and the birth of her daughter will give her legitimacy in another, even more unsettling sense, namely that she will be subjected to the social, political order that such laws were designed to create and uphold.

[Sorrow]'s textual fragment ends with her new name "Complete." It is this new name which equips her with some sense of empowerment/power, as the text appears to suggest. This shows when she speaks up to her mistress, who "said nothing about the baby, but sent for a Bible and forbade anyone to enter the new house[,] Sorrow […] was bold enough to remark to her Mistress, 'It was good that the blacksmith came to help when you were dying'" (*AM* 131). In response to this, [Rebekka Vaark] stares at her and says, "'Ninny […] God alone cures. No man has such power'" (131). If we connect [Rebekka Vaark]'s words the issue of naming practices under slavery, what emerges is the notion that [Rebekka Vaark], by addressing her as "Ninny," in fact calls [Sorrow] a fool. [122] In light of this, I want to suggest that [Sorrow]'s very desire to change or transcend her status in the [Vaark] household, reflected by her renaming herself "Complete," is disrupted by her mistress calling her a "fool." In [Rebekka Vaark]'s view, [Sorrow] is a fool for believing that she is able to transcend her status as [Rebekka]'s property.[123]

---

[122] The most prominent definitions offered by the *OED* for this word are "a fool" and "a simpleton," respectively ("ninny, n.").

[123] Indeed, [Sorrow] is the only [character] in *A Mercy* who calls herself by her new name. The only other [character] who remarks upon the fact that she has renamed herself is [Florens] in the above paragraph.

With [Sorrow], *A Mercy* attempts to take us down a path that leads us away from the devastating historical choices made in/by/during the formations of racial slavery. This yearning in [Sorrow]'s fragment to create a moment of possibility, of anticipating generations beyond the confines of the New World's grammar of property, is immediately confronted by and disnarrated within the novel's plotting. What this [character] suggests, in other words, is that one cannot create utopia out of the past because utopia imagines the future. Finally, considering this double movement of creating a fleeting utopian moment, of anticipating generations that are not allowed to last, it would perhaps seem fair to ultimately consider this [character] a "Complete Sorrow," a complete fantasy. Or, given that *A Mercy* carries Atlantic slavery's reproductive calculus and its histories and legacies into the future (which is the reader's immediate present) with this [character], we might also think of her as a [Sorrow] "Complete(d)."

4.5 "I Am a Thing Apart": [Florens] and the Ruse of Belonging

[Routing the Argument] In this chapter I draw on Afropessimism's claim that there is no transformative promise for the slave in narrative. My argument follows this argument as I turn to the textual fragments of [Florens] in *A Mercy*. Taking my second cue in the chapter from [Florens] herself, who states that she is a "thing apart" (*AM* 113), I suggest accordingly that [Florens] is void of a transformative narrative promise. In thus taking up post-slavery thinkers' concerns about the connections between narrative and social death, the chapter focuses on how the text develops belonging and unbelonging as critical themes with the [character] of this enslaved girl child. I suggest that [Florens'] question in *A Mercy* is not a question about subjectivity (as a Human) but that hers is one about being and "lasting" in/as social death. In this way, I situate my reading of [Florens] as a critique of the Human and their claim to freedom as (self-)ownership and I contend that *A Mercy* navigates this nexus between Human self-making and ownership in [Florens'] textual fragments through belonging in a proprietorial sense rather than through notions such as identity, female agency, or self-emancipation (from patriarchal formations of power). In other words, I think about notions of belonging in critical relation to Afropessimism's critique of narrative's embeddedness within the fold of the Human and in relation to formations of ownership and the property paradigm.

> Once again, trying to fit into the other's shoes becomes the very possibility of narration.
> — Saidiya V. Hartman and Frank B. Wilderson, "Position of the Unthought"

> It's an old confusion, people turning into things. When folks is gone (sold, dead, run-off), you got a corn husk doll, a walnut-shell ring, fingertips of dirt on the hem of a dress. It happened so much, maybe now things turn into people. The house, Tata—Garlic could hear it speak. All it contained of the brown lives it had eaten; it was a living thing.
> — Alice Randall, *The Wind Done Gone*

Introduction

The present chapter turns to the textual fragments of [Florens] and, thus, to the ways in which *A Mercy* here both navigates and interrogates notions of belonging at the New World colonial scene. In doing so, I turn to another ruse that *A Mercy* exposes: the ruse of belonging (with the first one being the ruse of solidarity that I discussed in my close reading of [Rebekka Vaark]). In "The Position of the Unthought" – an interview conducted by Frank Wilderson and published in the journal *Qui Parle* – Saidiya Hartman draws attention to the impossibility or "problem of crafting a narrative for the slave as subject" (184). In discussing *Scenes of Subjection: Terror, Slavery, and Self-Making in Nineteenth-Century America* and the ways her seminal book both addresses and undermines (national) narratives of individual freedom before, during, and after the Reconstruction Era; how it addresses not only the subjects that those narratives both assumed and fashioned but also the reenactments of subjection and subjugation facilitated by those narratives, the conversation between Hartman and Wilderson fundamentally revolves around the question of "Who does that narrative enable?" Hartman specifically raises this question in thinking about "issues of consent, will, and agency" (*Scenes* 80) in the context of the legal, conceptual, and social entanglements of property and personhood in nineteenth-century America (*Scenes*, esp. pp. 79–124). As Hartman goes on to explain:

> That's where the whole issue of empathic identification is central for me. Because it just seems that *every attempt to emplot the slave in a narrative ultimately resulted in his or her obliteration*, regardless of whether it was a leftist narrative of political agency – *the slave stepping into someone else's shoes and then becoming a political agent* – or whether it was about being able to unveil the slave's humanity by actually finding oneself in that position. In many ways, what I was trying to do as a cultural historian was to narrate a certain impossibility, to illuminate those practices that speak to *the limits of most available narratives* to explain the position of the enslaved. (Hartman and Wilderson 184; emphasis mine)

As previously discussed, post-slavery theoretical trajectories have made explicit how the violent histories and the legacies of slavery continue to shape not only the material realities of Black life in the United States and beyond but also how

4.5 "I Am a Thing Apart": [Florens] and the Ruse of Belonging — 191

knowledge production and transfer in the modern Western world was and continues to be predicated precisely on these histories; how, in other words, (the structure of) narrative about/of Human life is subtended by antiblackness (see also Wilderson, "Aporia"). That is, rather than being a means or a structure that can account for the slave, narrative needs to be understood as being within the purview of the Human, as being part of the Human subject's repertoire of being/becoming. Thinkers like Hartman and Wilderson thus throw into relief how the status of narrative in the liberal imagination of freedom and personhood itself was and continues to be bound by the modalities of slavery.[124]

What follows needs to be understood as a kind of struggle, on the one hand, to deal with narrative and its adjacent conceptual archive of such things as capacity, transformation, movement, character development, change or resolution (Wilderson, "Aporia") and, on the other, to reckon with how the "the world-making and the world-breaking capacities of racial slavery" subtend this archive (Hartman, "Belly" 166). If narrative holds out a transformative promise only to Human subjects, then what does this mean in the context of *A Mercy*'s seventeenth-century plotting and of its staging of [Florens] and her trek? How does [Florens'] telling disrupt or break with the grammar and the narrative of the Human? Might there be a vocabulary with which to actually account for [Florens] that is not bound by this grammar and which does not reinscribe it? In thus taking up post-slavery thinkers' concerns, the chapter focuses on the enslaved girl [Florens] and on how the text develops (un)belonging as critical theme with her [character].[125] I think about belonging not in terms of belonging to a group of people, as in, for instance, being part of the nation or kinship formations (as most critics would, see next paragraph). By contrast, I place belonging in rela-

---

[124] Hartman's concern with how narrative, history, the making of subjects, and power are intricately connected continues to be part of her thinking. In "Venus in Two Acts" (2008) Hartman raises those issues in relation to the archive of Atlantic slavery and the impossibility of writing history from the perspective of the enslaved. In thus grappling with the violence of the archive, Hartman turns to issues of methodology when she suggests "critical fabulation" – "playing with and rearranging the basic elements of the story […] re-presenting the sequence of events in divergent stories and from contested points of view" ("Venus" 11) – as a way or practice of jeopardizing the narrative building blocks conventionally used in hegemonic writing of history (see also Chapter 3).

[125] The most widely circulating understanding of belonging is defined as "[s]omething which belongs to or is connected with another" or which "constitutes part of another"; other definitions include that of an "item of (esp. movable) personal property, a possession, an effect" as well as a "member of one's family, a relative" and the "fact of appertaining or being part; relationship, affiliation" ("belonging, n.").

tion to the property paradigm and accordingly arrive at belonging as belonging *to*, as "being the property or possession of" someone else.

In general, readers and reviewers of *A Mercy* have largely neglected to account for notions of ownership in their discussions of [Florens]. Regardless of their respective readings' specific investment, most critics of *A Mercy* read [Florens'] trek through the wilderness of colonial Virginia as a "journey" or a "quest" for identity, subjectivity, and self-love geared towards remedying [Florens'] "fragmentation and hopelessness" (Nehl 15; see also Goad; Schreiber, "Personal"). More often than not, critics have framed this as her "journey toward a unified self" as well as in terms of a recovery from the traumatic experiences of colonialism and enslavement or as her resisting those formations (Carlacio 130; see also Cholant; Michlin; Müller, "Standing"; Putnam; Wyatt). This is also to say that in their critical articulations readers and scholars have located [Florens] and the movement that she makes in her texts almost exclusively within the fold of individual female (sexual) agency and empowerment. In this context, they think about her [character] and, by extension, the figure of the Black enslaved girl/woman primarily in terms of the acquisition of her own voice (see Eaton; Gallego-Durán, "Female Identity"; Nehl). An example of this is Markus Nehl's study *Transnational Black Dialogues: Re-Imagining Slavery in the Twenty-First Century*. Nehl here locates *A Mercy* among the genre of the neo-slave narrative[126] and argues that Morrison's novel expands those narratives that, "in their original form, primarily deal with the African American experience of slavery in the nineteenth century" (Nehl 55).[127] For Nehl, *A Mercy* is a "multi-perspective, highly fragmented, self-reflexive, non-linear and poetic text full of unresolved tensions and inner ambiguities" that fundamentally explores "the paradigm shift from human bondage to racial slavery that took place in the early North

---

126 Paradigmatically, Melton, Müller (*Presence*, "Standing"), and Peterson also read *A Mercy* as a neo-slave narrative. By contrast, Michlin writes that "*A Mercy* is a tragic fictional *herstory* of irreparable harm inflicted by slavery, but it is not a 'slave narrative,' for, as Toni Cade Bambara rightly insists: 'we've been trained to call [them] slave narratives for reasons too obscene to mention, as if the 'slave' were an identity and not a status interrupted by the very act of fleeing, speaking, writing'" (119).

127 In his study, Nehl references Afropessimism as that with which *A Mercy* "participates in a constructive discussion [...] about the meaning of (anti-)blackness" (57). While Nehl's study offers the only other reading of *A Mercy* in relation to Afropessimism and other post-slavery theorizing (to my knowledge at this point), I would argue that his arguments remain wedded to and focus mainly on conceptualizations of such things as agency and self-empowerment. His arguments, in other words, continue to search for resolution despite the fact they make recourse to the "impossibility of giving a coherent account of Florens's life and of working through and closing the wounds of slavery" (57).

American colonies" (57, 55). As a "powerful black feminist reflection" on this period, Nehl argues, *A Mercy*'s complex narrative form "reflects the black slave characters' experiences of uprootedness, sexual abuse and fragmentation in late seventeenth-century North America. Without denying the possibility of black agency and resistance, *A Mercy* highlights the crushing power of chattel slavery" (56). Starting from the novel's critical theme of abandonment developed in the book's core scene in which [Jacob Vaark] accepts [Florens] as a partial debt settlement and which is restaged multiple times in the text (see also Best, "On Failing"), others have pushed readings that use [Florens'] trek to discuss interrelated topics such as orphanhood (Goad; Montgomery, "Traveling Shoes"; Otten; Stave, "Across Distances"; Vega-González) and motherloss (Cox; Jimenez) and the trauma induced by these experiences that are subtended by slavery in seventeenth-century colonial Virginia. In ""I Am a Thing Apart": Toni Morrison, *A Mercy*, and American Exceptionalism," Susan Strehle shows how Morrison's novel confronts US American cultural, national narratives such as American Exceptionalism and its corresponding founding myth of a "chosen people" and their "errand into the wilderness"[128] within a project of settler colonialism. She discusses *A Mercy* as a text that "emphasizes divisions, distinctions, and distances, as it portrays in the colonies a potential community stifled at its inception by the assumption of an exceptionalist destiny" (109). While all of the novel's [characters] seek to belong to and find their place in this nascent community, it is [Florens] who, believing that she "deserves to be chosen" by the community (117), ultimately remains isolated, as Strehle claims. According to Babb, *A Mercy* navigates the [characters'] attempts at belonging in the wilderness of colonial Virginia as the "realization of the necessity of a group," only to conclude that "even messy community is better than selfish individualism" (Babb 158, 159). In their own ways, Cox and Gallego-Durán, moreover, evoke notions of belonging and unbelonging in their discussions of the assemblage of the [Vaark] women in *Mercy* as an "intra-feminine" (Cox) community of "female outlaws" (Gallego-Durán, "Female Identity") that functions as a site of healing for these women. Put another way, what surfaces across the spectrum of these differently nuanced readings, on the one hand, is an assumption of character development, of individual progress, and of narrative closure and resolution for [Florens], which shows not least in the fact that these readings follow [Florens'] "telling" in the order in which it is represented in the text (*AM* 1). On the other hand, critics conceptualize belonging in this context mainly as belonging to a specific community formation.

---

128 Cf. Perry Miller's eponymous study.

My overall argument in the chapter is that a critique of the Human subject and its assumption and grammar of property and (self-)ownership animates [Florens'] text. At issue are the ways in which [Florens'] text in *A Mercy* defies this assumptive Human grammar. In contrast to the existing body of literature on the novel, my reading of [Florens] does not follow *A Mercy*'s sequencing but, rather, centers on two scenes. The first scene that I examine in the chapter is [Florens'] encounter with a group of [Puritan village folk]. While on her way to [the blacksmith]'s dwelling, [Florens] seeks shelter at [Widow Ealing]'s house, who together with her [Daughter Jane] is expecting a visit from the [village congregation], hoping to prove that her daughter is not the demon that the [village people] believe [Jane] to be. In the scene, [Florens] produces a letter to those people to show them that she is on an errand to save her mistress's life and not, as assumed, the "Black Man's minion" (*AM* 109). In the second scene under scrutiny here, [Florens] has finally arrived at [the blacksmith]'s dwelling. Here she is confronted with [Malaik], a young boy and foundling whom [the blacksmith] has taken in his care. Later in the scene, a massive fight ensues between [Florens] and [the blacksmith] at the end of which it remains unclear whether [Florens] has seriously injured or perhaps even killed [the blacksmith]. By situating my reading of [Florens] as a critique of the Human and their claim to freedom and/as (self-)ownership, I contend that *A Mercy* navigates this nexus between Human self-making and ownership in [Florens'] textual fragments through a representation of belonging in a proprietorial sense rather than through notions such as identity, female agency, or self-emancipation (from patriarchal formations of power). I argue that the early modern "fashioning of the self-possessed individual" is allegorized, deconstructed, and exploded most explicitly with her [character] (Hartman, *Scenes* 4). I also suggest that the text continuously juggles with the possibility (and to some extent the materiality) of [Florens'] opposition to male power over herself and that it ultimately insists upon the impossibility of her escaping or transcending the workings of private property at the New World colonial scene. In other words, such things as agency, empowerment, or resistance, which figure prominently in the critical discourse on the novel, ultimately are inconsequential for this [character]. For [Florens], there is no such thing as the "transformative promise, which narrative holds out to human subjects" (Wilderson, "Aporia" 139). In a first step, I turn to the two scenes in question as I examine how *A Mercy* elaborates on notions of belonging with this [character] and connects those to the property paradigm. Second, I analyze the trope of the shoes that is so prominent in [Florens'] texts and I think about what the absence of shoes signifies with respect to her [character]'s existence. Lastly, I examine how her unbelonging – what she herself terms "last-

4.5 "I Am a Thing Apart": [Florens] and the Ruse of Belonging — 195

ing" (see below) – interrogates the intricate connection between narrative, property, and liberal fashioning of self.

[Florens], Telling, Belonging

Within the first few lines of *A Mercy*, there appears in the text [Florens'] voice and her attempt to address and to reassure a 'you' that she will not hurt them. The reader will most likely come to identify the 'you' as [the blacksmith] but it could at this point also address the reader (there is a third possibility to which I will return later in the chapter). To briefly recapitulate, [Florens'] first-person narration unfolds over forty-six pages in six different textual fragments and in these fragments, she describes her trek through the wilderness of colonial Virginia to the dwelling of [the blacksmith] as well as her return to the [Vaark] farm. Her trek constitutes an errand on which her mistress [Rebekka Vaark] sends her after she has fallen ill with the smallpox in the hope that [the blacksmith] will cure [Rebekka Vaark] from the illness. [Florens'] first meets [the blacksmith] when [Jacob Vaark] hires him to forge the gate to a new mansion he intends to build and it is with him that she is, as one of the other women [characters] on the [Vaark] farm has it, "completely smitten" (*AM* 94). From the very beginning of her text/trek, that is, [Florens] decides that she will stay with [the blacksmith] and not return to her mistress. In those first few lines of her fragment and the novel we read:

> Don't be afraid. My telling can't hurt you in spite of what I have done and I promise to lie quietly in the dark – weeping perhaps or occasionally seeing the blood once more – but I will never again unfold my limbs to rise up and bare teeth. I explain. You can think what I tell you a confession, if you like, but one full of curiosities familiar only in dreams and during those moments when a dog's profile plays in the steam of a kettle. (*AM* 1)

Tensions and ambiguities permeate these lines. The phrases 'You can think what I tell you a confession' and 'if you like' anticipate the text's complex internal energies and tensions. That is, while the former phrase invokes the genre of the confessional narrative, the latter ('if you like') both challenges and disarms such categorization immediately, suggesting that such categorization does not matter to [Florens].[129] Best makes a similar point when he writes: "'You can think what I tell you a confession, if you like' [...] [Florens] invites, only to under-

---

[129] For discussions of confession, cultures of confession, and confessional narratives see generally e.g., Bok; Brooks; C. Taylor; Foster; Renov.

cut that solicitation with the observation that 'confession we tell not write as I am doing now'[...] Confession or not, this chapter certainly anticipates the irresolution of those to follow" ("On Failing" 469). The above lines also set the stage for *A Mercy*'s overall critique and interrogation of the liberal property paradigm. [Florens] calls the act of relating what happened to her, how she thinks about herself, and how she is positioned in the world her "telling" (*AM* 1). Letting [Florens] literally "tell" her story fundamentally speaks to post-slavery thinking's critique of the status of narrative within Western modernity. In her "telling," as I hope to show, we can recognize the critique of the conceptual conflation of (the structure of) narrative and liberal self-making.

I take my next cue from the following passage, which comes from the first scene that I will closely examine in the chapter. As mentioned already, this is [Florens'] encounter with a group of [Puritan village people] at the house of a woman called [Widow Ealing], who offers her shelter and whose [Daughter Jane] has been accused of witchcraft by the [village congregation]. They arrive at [Widow Ealing]'s home with the intention of determining whether [Daughter Jane] is a demon. However, the text quickly shows how, shocked by [Florens'] presence among them, they focus their attention on the enslaved girl child in [Jane]'s stead. Attempting to protect herself from the [village folk] and their "[e]yes that do not recognize me, eyes that examine me for a tail, an extra teat, a man's whip between my legs" (*AM* 112–113), [Florens] produces a letter that [Rebekka] equipped her with in order that she may run her errand as quickly as possible. I will return to the letter in a moment. Suffice it to say for now that the [village people] take the letter with them when they leave the [widow]'s house in order to deliberate what they want to do with [Florens], leaving her devastated (111). The paragraph reads:

> Something precious is leaving me. *I am a thing apart.* With the letter I belong and am lawful. Without it I am a weak calf abandon by the herd, a turtle without shell, a minion with no tell tale signs but a darkness I am born with, outside, yes, but inside as well and the inside dark is small, feathered and toothy. (*AM* 113; emphasis mine)

These lines illustrate that [Florens] knows that she only exists as a "thing" in the racist, objectifying gaze of the village people and, by extension, in the social environment of the [Vaark] household and of colonial Virginia. She knows that the letter is "precious" because it both signifies and establishes belonging in relation to formations of ownership. That is, the phrase 'with the letter I belong and am lawful' draws attention to the ways in which [Florens'] status of being is shaped by possession on both a social and a legal plane. [Florens'] designation of herself as a "thing apart" also fundamentally refers us to notion that she, unlike the liberal subject, cannot claim individual liberty through possession (of self and others). It also refers us to the notion that her sentient existence (to paraphrase Wilderson, *Red*) is elaborated instead by *accumulation* and *fungibility*. These lines raise the subsequent set of interrelated questions: What does it mean that [Florens] considers herself to be a "thing"? What does it mean that Morrison allows a "thing" to tell its story and not a Human subject in a liberal individualist sense? Is it at all possible to imagine [Florens] as a narrator in a literary criticism sense of the term or to conceptualize her textual fragments as narrative? Can narrative/the novel form account for [Florens] as a "thing"?

"Belonging": Part One

Let me turn to the first scene under scrutiny here. Seeking shelter for the night while on her way to [the blacksmith], [Florens] arrives at a "proto-Salem village" emptied of its inhabitants, who "are at evening prayer" (Bross 185; *AM* 104). She knocks at the door of the "single lit house in the village" and the woman who opens the door for her, though hesitant at first, affords [Florens] shelter (*AM* 105). At the house, [Florens] learns that the woman's name is [Widow Ealing] and that she has a daughter called [Jane][130]. While she is provided with food and shelter, [Florens] witnesses how the two women attend to a fresh wound on [Daughter Jane]'s leg. The wound is meant to prove that [Daughter Jane], who has a dark voice and a "wayward eye" (112), is a human being and not the demon that the village people accuse her of being: "I see dark blood beetling down her legs. In the light pouring over her pale skin her wounds look like live jewels. [...] Those lashes may save her life. [...] look at her wounds. God's son bleeds. We bleed. Demons never. [...] They will not come until morning" (106, 108). These words suggest that the two women are put in danger by the [townspeople] who question [Daughter Jane]'s humanity, not least because of her "outspokenness [and] a physical abnormality" (Bross 185). As Kristina Bross notes, *A Mercy* here both takes up and foreshadows "the invasive and humiliating treatment of women accused of witchcraft" during the Salem witch trials in Massachusetts in 1692 (186).[131] However, this also needs to be understood as a moment

---

[130] [Florens] calls her "Daughter Jane" in the remainder of the text (*AM* 105–113).
[131] Bross makes this observation in an article in which she discusses the uses of *A Mercy* in an undergraduate class on seventeenth-century magical beliefs that she taught at the College of Liberal Arts at Purdue University. The full quote, in which she discusses [Florens] as a representation of women accused of witchcraft, reads: "The students found the parallels between Florens's examination as a demon and the search for physical evidence of witchcraft in Salem es-

where to engage with, on the one hand, antiblack formations and practices of racialization and subjugation and, on the other, the making or "the guarantee of (white) equality" in colonial Virginia (Hartman, *Scenes* 62). Historian Edmund S. Morgan reminds us that "white men [in colonial Virginia] were equal in not being slaves" (*American Slavery* 381). The scene under scrutiny here clearly speaks to this historical predicament or what we today might call this sense of white solidarity. When the [townspeople] (one man, three women, and a little girl) arrive at [Widow Ealing]'s dwelling, [Florens] initially is absent from the encounter. As soon as she steps into the room, however,

> each visitor turns to look at me. The women gasp. The man's walking stick clatters to the floor causing the remaining hen to squawk and flutter. [...] The little girl wails and rocks back and forth. [...] One woman speaks saying I have never seen any human being this black. I have says another, this one is as black as others I have seen. She is Afric. Afric and much more, says another. [...] It is true then says another. The Black Man is among us. This is his minion. (*AM* 109)

The paragraph shows how the objectives of the village congregation's visit (i.e. establishing whether [Daughter Jane] is indeed a demon as suspected) are abandoned as soon as [Florens] enters the scene. The visitors turn to [Florens] instead and her appearance causes them distress, which is amplified most in the little girl's "shaking and moaning" upon seeing [Florens] (*AM* 109). In the passage we can thus recognize an early instance (historically speaking) of colonial Virginia's path to a racialized as well as racist social hierarchy, which systematically set apart people of African descent "for treatment different than that accorded to others" and which also made white European settlers and colonists into members of the emerging Virginian social body who would be considered free and equal by virtue of their self-possession (Fields 119). The phrases 'She is Afric. Afric and much more' clearly speak to this genealogy of systematically and structurally abjecting human beings. That this genealogy was also related to discourses embedded in religious doctrine – mobilized in the text by way of its recourse to the Salem witch trials – is illustrated by the phrase 'the Black Man.' As focalized through the [townspeople] and their endeavor to find out whether [Daughter

---

pecially striking, but the moments in the novel in which the parallels were most clear can lead to the realization that the invasive and humiliating treatment of women accused of witchcraft in 1692 is *not* evidence of an isolated, insane mob mentality, but draws on violent practices across the mid-Atlantic culture. Like Rebecca Nurse and Sarah Bishop, Florens is strip-searched for a devil's mark" (185).

4.5 "I Am a Thing Apart": [Florens] and the Ruse of Belonging — 199

Jane] is a demon[132], the text positions [Florens] as a heathen. In this context, the phrase 'the Black Man' appears to signify Satan or the devil. The word 'minion' at the very end of the paragraph accentuates this as it points to the [village people]'s conviction that [Florens] worships the devil.[133]

[Florens] quickly understands that the presence of this group of villagers in [Widow Ealing]'s house no longer signifies danger or violence against [Daughter Jane] but that it instead means danger for herself. We continue reading: "I am not understanding anything except that I am in danger as the dog's head shows and Mistress is my only defense. I shout, wait. I shout, please sir. I think they have shock that I can talk" (*AM* 109). In her despair and in her attempt to prove that she is not the 'Black Man's minion,' [Florens] produces a letter that [Rebekka] equipped her with when she sent her servant on this errand to save her own life. The letter reads:

> *The signatory of this letter, Mistress Rebekka Vaark of Milton vouches for the female person into whose hands it has been placed. She is owned by me and can be knowne by a burn mark in the palm of her left hand. Allow her the courtesie of safe passage and witherall she may need to complete her errand. Our life, my life, on this earthe depends on her speedy return. Signed Rebekka Vaark, Mistress, Milton*
> 18 May 1690. (AM 110)

The letter both states and produces in writing [Rebekka]'s legal claim to ownership of [Florens] as it simultaneously establishes the time and place of this claim to be in the dwelling of Milton on May 18, 1690. It reiterates the spatiotemporal frame of this claim through a repetition of the words 'Mistress' and 'Milton,' which are situated both at the beginning and the end of this short piece of writing and which frame, [Rebekka]'s name. While the letter states that it vouches for 'the female person into whose hands it has been placed,' the first part of the sentence immediately following this phrase further determines the status of this 'female person': "She is owned by me." Historians of slavery remind us in this con-

---

132 Indeed, the text describes [Widow Ealing] as having "red hair" and "green eyes" (*AM* 104), and thus evokes stereotypical renderings of the witch as an emblematic figure of female independence and agency (Federici). Of course, the text here also alludes to the Salem witch trials that occurred in colonial Massachusetts between 1692 and 1693. During those trials, more 200 people were accused of witchcraft, with 20 of them being executed. Later, the colony admitted that the Salem witch trials had been a mistake "and compensated the families of those convicted. Since then, the story of the trials has become synonymous with paranoia and injustice, and it continues to beguile the popular imagination more than 300 years later" (Blumberg).

133 The *OED*'s first entry on the word "minion" defines it as "a person who is dependent on a patron's favour [sic]; a hanger-on" and as "a follower or underling, esp. one who is servile or unimportant" ("minion, n.").

text not only that literacy (reading and writing) increasingly became a means to control the enslaved populations from the colonial period through the antebellum era. They also go on to tell us that

> [n]owhere is the relationship between power and writing demonstrated more clearly than in the restrictions passed by every southern colony – and, later, every slaveholding territory or state – against the movement of the enslaved. In every southern slave code, one of the first sections is devoted to a discussion of the written pass that all slaves had to have in hand in order to leave the premises of their owners. The pass was to include the name of the carrier of the pass, identify the home plantation, and specify the date and time of absence covered by the pass. These passes (also called tickets, letters, or certificates of leave) had to be signed by the owner or a responsible employee, and served as the equivalent, on the local level, of a passport today at the international. (Monaghan 321–322)

While generally enabling the enslaved to trade goods, slave passes also served as a means to "monitor what went into the plantation[.]" Monaghan explains that in Florida, for example, "the only purchases [slaves] could make without a [written pass or] ticket were brooms and baskets, which were considered to be items of slave manufacture" (322). In *A Mercy*, [Rebekka]'s letter, as an allegorical representation of the slave pass and the technologies of surveillance and control that it signifies (see also "Slavery, Institutional Racism"), denotes [Rebekka]'s claim to ownership and control of [Florens] and her movement at the New World colonial scene. It also shows that it is through this claim that [Rebekka] both maintains and renews her capacity as a subject, her corporeal integrity (Wilderson, *Red*), for it states that it is [Rebekka]'s 'life on this earthe [which] depends on her speedy return.'

I suggest that the letter, in turn, signifies the absence of such capacity for [Florens]. It illustrates that [Florens] also needs to be read in terms of what Spillers describes as "the captive body reduced to a thing, becoming *being* for the captor" and not, as many critics have done, in terms of agency ("Mama's Baby" 67). Her ability to move through *A Mercy*'s colonial landscape and to "travel" to [the blacksmith] (granted by her mistress's letter) is bound by her *fungibility*. This is illustrated by her words, uttered as she produces the letter, "Let me show you my letter I say quieter. It proves I am nobody's minion but my Mistress" (*AM* 109). *A Mercy* here both appropriates and riffs on the trope of the slave pass ubiquitous in the literary tradition of the African American slave narrative. In his *Narrative of the Life of Frederick Douglass* (1845), to give a prominent example, Frederick Douglass amply demonstrates his view that "literacy was the high road to freedom" (Monaghan 339). At several points in his narrative, Douglass tells the reader not only how he taught himself how to write but also suggests that he might "have occasion to write my own pass" (F. Douglass 414). Later

4.5 "I Am a Thing Apart": [Florens] and the Ruse of Belonging

in his narrative, the reader witnesses how Douglass's plans to run away from his current master together with a few other slaves are thwarted when they are chased down, attacked, and tied by a group of white "constables" (435). Douglass manages to throw the slave pass that he had written for himself into the fire unnoticed and he shortly after tells one of his comrades to dispose of his slave pass by way of eating it (436).[134] In this way, they manage to conceal their carefully crafted plans to run away in which those self-written passes played an essential part. Later, Douglass will escape to the North with the help of the papers of a free black man and a train ticket (another written document) (F. Douglass; Monaghan 341). In *A Mercy*, by contrast, [Florens] has never learned to write properly[135] and the letter that she is equipped with does not hold out to her a promise of capacity or any kind of phantasmatic claim to freedom. Instead, it configures her existence through [Rebekka]'s proprietorial claim of her belonging to her. (Again, for [Rebekka] to say that [Florens] "belongs to her" means that she owns this enslaved child.)

The text navigates proprietorial claims to belonging not only through the letter but also by referring the reader to slavery's paradigmatic scene of the violent display, the multiple examinations, and the sale of slave-property on the auction block. Bills of sale, which turned into letters of ownership (and even the travel passes which seem to grant mobility on a legal plane when in fact they re-inscribed the slaves' *fungibility*) were part and parcel of the spectacle that is the auction block, which paradigmatically staged the "event" of the sale as well as the "tricks of the trade and, by extension, the related practices that secure[d] and reproduce[d] the relations of mastery and servitude" (Hartman, *Scenes* 40, 41). *A Mercy* invokes this spectacle in [Florens'] telling only a few lines further on in the text when she tells us how the [townspeople]

> point me to a door that opens onto a storeroom and [...] tell me to take off my clothes. Without touching they tell me what to do. To show them my teeth, my tongue. They frown at the

---

134 We read: "When we got about half way to St. Michael's, while the constables having us in charge were looking ahead, Henry inquired of me what he should do with his pass. I told him to eat it with his biscuit, and own nothing; and we passed the word around, "*Own nothing;*" and "*Own nothing!*" said we all" (435–436). Calvin Warren in *Ontological Terror: Blackness, Nihilism, and Emancipation* (2018) takes this quotation from Douglass's narrative as a starting point for his interrogations of Blackness and nothingness within philosophical discourse and in critical relation to Afropessimist thinking.

135 As I will discuss in more detail towards the end of the chapter, [Florens], despite the fact that [Reverend Father] teaches her to read and write (*AM* 4), "tells" rather than writes; moreover, her telling is in disarray: "Sometimes the tip the of nail skates away and the forming of words is disorderly" (*AM* 156).

> candle burn on my palm[.] They look under my arms, between my legs. They circle me, lean down to inspect my feet. Naked under their examination I watch for what is in their eyes. Not hate is there or scare of disgust but they are looking at me my body across distances without recognition. Swine look at me with more connection when they raise their heads from the trough. (*AM* 110–111)

Naked under their probing and their penetrating gaze, the text suggests that [Florens] is firmly positioned outside of the world of the [village people].[136] The phrase 'Swine look at me with more connection when they raise their heads from the trough' suggests that the [village people]'s gaze creates closer proximity between [Florens] and farm animals like swine than between [Florens] and the human beings around her. It also hints at the notion that keeping and feeding animals such as pigs is part of the labor that [Florens] is forced to do on the [Vaark] farm, thus adding to the variety of tasks that [Florens] needs to do. In other words, taking care of farm animals is an errand she has to run over the course of the day, just like being ordered to fetch [the blacksmith] to help cure her mistress is an errand. At this point, it is important to remind ourselves of Orlando Patterson's definition of slavery *not* as coerced labor but as social death (i.e., natal alienation, general dishonor, and openness to gratuitous violence), which post-slavery theoretical trajectories have taken up and expanded on by introducing such terms as *accumulation* and *fungibility* to describe the slave's violent positioning in the world (see Chapter 3). That is to say, forced labor was essential to the slave's experience on the plantation, in the field, and in the household, but it did not have positioning power over them, both structurally and ontologically speaking. With this in mind, I want to reiterate my claim that [Florens] needs to be understood to be positioned in her textual fragments *not* as someone who goes on a journey or quest for finding their agency. Her "being sent" through colonial Virginia's "wilderness" does *not* result from some sort of intra-human, interpersonal act between [Rebecca Vaark] and [Florence] (which would allow for notions of agency to surface). Instead, the text decidedly positions [Florens] as *fungible human property* for which the trek to [the blacksmith]'s dwelling is part of an errand on which they have been sent by their

---

[136] The scene of the auction block will continue to resonate in *A Mercy*. It will be echoed in the novel's final fragment, which is that of [Florens'] mother, the [*minha mãe*] (see Chapter 4.6). In it, she describes how, after her arrival in Barbados in the wake of her being shipped to the New World, she and the other enslaved and imprisoned African persons were made "to jump high, to bend over, to open our mouths. The children were best at this. Like grass trampled by elephants, they sprang up to life again. They had stopped weeping long ago. Now, eyes wide, they tried to please, to show their ability and therefore their living worth. How unlikely their survival" (*AM* 163).

mistress/owner. Such a positioning crucially allows me to read her [character] in the context of the formation of the New World slavery's calculus of property or "the master['s dream] of future increase" (Hartman, "Belly" 166). This enslaved girl's domestic labors on the [Vaark] farm thus need to be understood as constitutive of the master's/mistress's economic success. They also are indicative of the master's Human capacity and the powers they "have or lack [as subjects], the constituent elements of [their] structural position with which they are imbued or lack prior to [their] performance" (Wilderson, *Red* 8). This brings me back to the letter: The letter's close positioning to the scene of the auction block as well as to scenes or instances of forced labor in [Florens'] telling once again raises questions as to who will ultimately benefit from her successfully completing her errand. In some ways, it appears to grant [Florens] safe passage on her way to [the blacksmith]. For as [Florens] has it at the beginning of the novel, "[Lina and Mistress] tell me to hide the letter inside my stocking—no matter the itch of the sealing wax. I am lettered but I do not read what Mistress writes and Lina and Sorrow cannot. But I know what it means to say to any who stop me" (*AM* 2). Later at [Widow Ealing]'s house and after having examined [Florens], the [village people] take [Rebekka]'s letter with them. As a result, [Florens] is "hung with fear" as she waits for them to return. However, the reader never learns what the [village congregation] has decided because [Florens], aided by [Daughter Jane], escapes before they return and continues on her errand (112). Ultimately, and as previously suggested, her mistress will be the one who will profit from [Florens'] successful completion of this errand. Thinking about the letter, its positioning in the text, as well as about to whom it holds out what kind of a promise in this way returns me to post-slavery theoretical trajectories' questioning of narrative as a structure. That is, [Florens'] telling raises important questions as to whether narrative form can contain or account for someone like [Florens]. What does [Florens'] text in *A Mercy* tell us about the relationship between liberty and literacy that was so fundamental to slave narratives like Frederick Douglass's (cf. Monaghan)? How do [Florens'] textual fragments deconstruct, allegorize, and criticize this relationship? Again, what does it mean that Morrison allows a "thing" to tell its story and not a Human subject in a liberal individualist sense?

"Belonging": Part Two

The second scene under scrutiny here pushes the reader to navigate two different versions of belonging in the context of a massively violent fight that ensues between [Florens] and [the blacksmith]. This fight is the fifth fragment of [Florens'] telling. As suggested earlier, [Florens] desires [the blacksmith] from the moment that he sets foot on the [Vaark] farm: "You probably don't know anything at all about what your back looks like whatever the sky holds: sunlight, moonlight. I rest there. My hand, my eyes, my mouth. The first time I see it you are shaping fire with bellows. The shine of water runs down your spine and I have shock at myself for wanting to lick there" (*AM* 33).[137] For [Florens], in other words, her errand to [the blacksmith] encapsulates in her visceral desire to belong to and to be owned by him. We read:

> No one steals my warmth and shoes because I am small. Not one handles by backside. No one whinies like sheep or goat because I drop in fear and weakness. No one screams at the sight of me. No one watches my body for how it is unseemly. *With you my body is pleasure is safe is belonging. I can never not have you have me.* (*AM* 135; emphasis mine)

In the fight between these two [characters], [the blacksmith] embodies yet another version of belonging. As "a free black man," the text suggests, [the blacksmith] has the ability to possess his own self. "He had rights [...] and privileges, like Sir. He could marry, own things, travel, sell his own labor" (43). For some of the [Vaark] women, like [Lina and [Sorrow], [the blacksmith]'s ability to possess his own self signifies danger and uncertainty. As [Sorrow] asks, for instance, "Was he the danger Lina saw in him or was her fear mere jealousy? Was he Sir's perfect building partner or a curse on Florens, altering her behavior from open to furtive?" (123). Even though [the blacksmith], from an Afropessimist perspective, would need to be conceptualized as an abject sentient being, the novel's neat narrative ploy thus in fact aligns him with possessive individualist notions of liberal subjectivity, as embodied by [Jacob Vaark]. Indeed, the text suggests an alliance between these two [characters] multiple times and thus opens up the narrative possibility of Black (individual) liberty. [Lina] observes accordingly that

> Sir behaved as though the blacksmith was his brother. Lina had seen them bending their heads over lines drawn in the dirt. Another time she saw Sir slice a green apple, his left boot raised on a rock, his mouth working along with his hands; the smithy nodding, looking intently at his employer. Then Sir, as nonchalantly as you please, tipped a slice of apple on his knife and offered it to the blacksmith, who, just as nonchalantly, took it and put it in his mouth. (*AM* 58–59)

---

[137] As [Lina] observes, moreover: "Since his coming, there was an appetite in the girl that Lina recognized as once her own" (*AM* 58) and in [Sorrow]'s fragment, we learn that "[b]y the time Sorrow recovered [from the smallpox], Florens was struck down with another sickness much longer lasting and far more lethal" (125).

4.5 "I Am a Thing Apart": [Florens] and the Ruse of Belonging

In the scene, [Florens] finally has arrived at "the yard, the forge, the little cabin where you are" and she "lose[s] the fear that I may never again in this world know the sight of your welcoming smile or taste the sugar of your shoulder as you take me in your arms" (133). When she tells [the blacksmith] why she has come to his dwelling and that he is being summoned to the [Vaark] farm, he tells her to wait for him to return: "I am to wait here you say. I cannot join you because it is faster without me. And there is another reason, you say. You turn your head. My eyes follow where you look" (133). The reason is a little boy, a "foundling," called [Malaik], whom [the blacksmith] has taken in his care "until a future when a townsman or magistrate places him, which may be never because although the dead man's [the boy's presumed father] skin is rosy [but] the boy's is not. So maybe he is not a son at all" (134). For [Florens], the boy signifies the danger of her being "expelled" from [the blacksmith]'s dwelling. We continue reading: "I worry as the boy steps closer to you. *How you offer and he owns your forefinger.* As if he is your future. Not me" (134; emphasis mine). The text here establishes complex connections between the little boy and [Florens'] previous experiences of having been taken away from her mother and of the little girl at [Widow Ealing]'s house screaming at her. We continue reading: "The first time it is me peering around my mother's dress hoping for her hand that is only for her little boy. The second time it is a pointing screaming little girl hiding behind her mother and clinging to her skirts. Both times are full of danger and I am expel" (133–134). This time around, [Florens], through her desire, is determined not to go through such an experience again. As she tells us, "We talk of many things and I don't say what I am thinking. That I will stay. That when you return from healing Mistress whether she is live or no I am here with you always. Never without you. Here I am not the one to throw out. [...] I take of Sir's boots and lie on your cot trying to catch the fire smell of you" (134–135). The above phrase 'how you offer and he owns your forefinger' also exposes the reader to the notion that the only vocabulary [Florens] has got to navigate the world is one wedded to possession.

At the cabin, [Florens] and the little boy urgently await [the blacksmith]'s return. The text here, in building up to the fight between [Florens] and [the blacksmith], suggests that the encounter between [Malaik] and [Florens] and the aggression that he expresses towards her unleashes the violence that will erupt toward the end of the fragment. As [Florens] states, "He is silent but the hate in his eyes is loud. He wants my leaving. This cannot happen. I feel the clutch inside. This expel can never happen again" (*AM* 135). Indeed, the text here also suggests that in spite of the fact that [Florens] feels unsafe in this situation because she does not know whether [the blacksmith] will devote his attention entirely to her once he returns, she is willing and ready to take her place in [the blacksmith]'s household regardless of whether the little boy, too, is a member of this household.[138] This is best illustrated when she prepares breakfast for herself and [Malaik], even though she senses the boy's hostile feelings towards her. We read: "In the morning the boy is not here but I prepare porridge for us two. Again he is standing in the lane holding tight the corn-husk doll and looking toward where you ride away. [...] First I notice Sir's boots are gone. [...] I know he steals Sir's boots that belong to me" (136–137). It is when [Florens], not knowing how to navigate this situation and in trying to protect herself, accidentally injures the boy in her attempt to stop him from screaming, that [the blacksmith] returns and strikes her: "I am first to get the knocking away. The back of your hand strikes my face. I fall and curl up on the floor. Tight. No question. You choose the boy. You call his name first" (138). What follows is the fight between [Florens] and [the blacksmith], which I quote at some length here:

> Why are you killing me I ask you.
> I want you to go.
> Let me explain.
> No. Now.
> Why? Why?
> Because you are a slave.
> What?
> You heard me.
> Sir makes me that.
> I don't mean him.
> Then who?
> You.
> What is your meaning? I am a slave because Sir trades for me.
> No. You have become one.
> How?
> Your head is empty and your body is wild.
> I am adoring you.
> And a slave to that too.
> You alone own me.
> Own yourself, woman, and leave us be. You could have killed this child.
> No. Wait. You put me in misery.

---

138 Strehle discusses this as a "reconstitution of the Black family" (117) but I think that such a conceptualization fails to account for the ways in which the New World's grammar of property structures the slave's existence and the (im)possibility of vertical kinship formations. How can "the Black family" be reconstituted if it is perpetually open to gratuitous violence at any given moment, as illustrated by the novel's core scene in which [Florens] becomes the currency of a debt settlement between two slave traders?

4.5 "I Am a Thing Apart": [Florens] and the Ruse of Belonging — 207

> You are nothing but wilderness. No constraint. No mind.
> You shout the word – mind, mind, mind – over and over and then you laugh, saying as I live and breathe, a slave by choice.
> On my knees I reach for you. Crawl to you. You step back saying get away from me.
> I have shock. Are you meaning I am nothing to you? [...] Now I am living the dying inside. No. Not again. Not ever. Feathers lifting, I unfold. The claws scratch and scratch until the hammer is in my hand. (*AM* 139–140)

In this rapid exchange between [Florens] and [the blacksmith], the text situates [the blacksmith] in line with the grammar of the liberal Human. An outraged [blacksmith] tells [Florens] to leave him and the little boy alone and when she asks him 'why, why,' he responds, 'Because you are a slave' and, a few lines further, 'You have become one.' That is, he considers [Florens] to be a 'slave' to her emotions – her fear, her anger, her jealousy – and to be incapable of keeping these feelings under control. As he sees it, [Florens'] aggressive behavior and her violent attack on [Malaik] make her everything but a civilized and reasonable, thinking human being with good judgment. In his assessment of the situation, [Florens'] 'head is empty and [her] body is wild.' Indeed, he believes [Florens] to be a slave not only to her emotions in general but also to her sexual desires. When she tells [the blacksmith] that she is 'adoring' him, he sharply responds by saying, 'And a slave to that too,' referring to what he considers her being dependent on her sexual feelings. What the text asserts by the time we reach the end of the scene is that [the blacksmith] does not want to be for [Florens] what she desires him to be, namely the person who owns, and thus safeguards, her ('You alone own me'). Instead, he strongly insists on his independence, specifically his independence from the will and the desires of others and he thus echoes Locke's conceptions of the liberal subject. As he says, 'Own yourself, woman, and leave us be. [...] You are nothing but wilderness. No constraint. No mind.' To which [Florens] responds by 'reaching' for and by 'crawling' to him. That is to say, the text firmly positions [the blacksmith] within a microcosm of reason, possessive individualism, and freedom. It positions him over the desire that [Florens] represents in the scene. Again, standing shoulder-to-shoulder with [Jacob Vaark], the [character] of [the blacksmith] here mediates notions of Black male liberty and independence on the New World colonial scene, as it suggests the phantasma of Black freedom and individualism as one instance of "what could have been." *A Mercy* here aligns this [character] with liberal and possessive individualist discourses of freedom and thus tentatively opens up a narrative window of possibility for male Blackness to become a part of such discourses. However, *A Mercy* rejects and abandons this phantasma when [Florens] violently removes [the blacksmith] from the novel's narrative orbit shortly after he states with a bellow of rage: "You shout the word – mind, mind, mind – over and over and then you laugh, saying as I live an breathe, a slave by choice" (*AM* 139). Like [Vaark], that is, [the blacksmith]'s version of liberal possessive individualism is disnarrated from the novel when [Florens] violently attacks and possibility kills him (140, 155–156).

In the scene, the text juxtaposes this version of Black liberal, independent, and self-possessing manhood with [Florens'] wanting to be owned by [the blacksmith]. In many ways, this scene is an iteration of the novel's core scene of abandonment, in which [Florens] is taken away from her mother by [Jacob Vaark] and [Senhor D'Ortega] and which [Florens] perceives as her mother choosing her son over her daughter (see also Waegner 98).[139] In this second iteration, the scene is represented through her unfulfilled, greedy longing to be owned by someone else than her late master or, for that matter, her mistress [Rebekka]. It is in this context that I read [Florens'] attempt to explain to [the blacksmith] – in response to each 'cutting' word (*AM* 138) that he throws at her – that it is her master who has actually made her a slave. The phrases 'Sir makes me that' and 'I am a slave because Sir trades for me' illustrate this at the same time that they echo her previous assertion that she is "a thing apart." [Florens] knows that she is a slave, a "thing" for as well as in the eyes of others. With [Florens'] protestations and her trying to defend herself, the text offers a representation of belonging linked to emotion and it also links this representation to the body, to desire and to pleasure ('with you my body is pleasure is safe is belonging'). However, this is not an attempt at freeing herself from someone else's claim to ownership of her. [Florens] does not want to belong to her self ('own herself') in the same way that [the blacksmith] claims to be independent and to be possessing himself. Instead, she wants to be taken under [the blacksmith]'s care. [Florens] knows that he can protect her in the same way that her current master can because

> you say you are a free man from New Amsterdam and always are that. *Not like Will or Scully but like Sir. I don't know the feeling of or what it means, free and not free.* [...] Standing there between the beckoning wall of perfume and the stag I wonder what else the world may show me. It is as though I am loose to do what I choose, the stag, the wall of flowers. *I am a little scare of this looseness.* Is that how free feels? I don't like it. *I don't want to be free of you because I am live only with you.* [...] You are my shaper and my world as well. It is done. *No need to choose.* (*AM* 67–69; emphasis mine)

---

[139] In some ways, then, [the blacksmith] could perhaps be understood as figuring as a stand-in for her mother (see also Otten).

4.5 "I Am a Thing Apart": [Florens] and the Ruse of Belonging — 209

This is the moment when [Florens] explains that she does not know what it means to be 'free or not free' and that she does not want to be free in a liberal individualist sense. What she wants is to belong to [the blacksmith], as illustrated by the phrase 'I don't want to be free of you because I am live only with you.' [Florens] defies [the blacksmith]'s liberal discourse of "mind over desire" when she states, 'no need to choose.' What this phrase shows us is that [Florens] does not know what "choice" would be. As a sentient being and as human property, [Florens] cannot imagine desire as a kind of choosing agency of her own volition. The only desire she can know is to exist as a belonging thing as opposed to an abandoned thing. It is that kind of desire, which has made that choice for her, as the paragraph suggests. I read these lines as the text's meditation not only on proprietorial notions of belonging but also on the liberal conceptions of freedom that [the blacksmith] represents at this point in the novel. In other words, this is where *A Mercy* shows the reader that it cannot give Florence the vocabulary nor the psychic repertory of a Human and instead takes us very far into imagining a "thing apart" speaking for itself. For [Florens], a term like "dependence" does not have any distinctive signifying value because she does not exist in the liberal subject's orbit of a binary between self-determination and dependence. As soon as her body viscerally desires (for example when she wants 'to lick' [the blacksmith]'s skin), [Florens] can only represent that urge to be connected as a drive to be "owned" and to be owned in a way that leaves her "safe" (as opposed to abandoned). These words represent the only word choice her [character] can have for her interiority. [Florens'] own liberal vocabulary is very small at this point and she cannot counter the mass of "learned" words that [the blacksmith] throws at her. All the words and the juxtapositions he employs (slave–free, wild–civilized, mind–passion) mean nothing to her except they signify that she will be abandoned again. In other words, [Florens] cannot challenge [the blacksmith]'s words because his kind of Human(ist) liberal discourse is not at her disposal. What she can do, as we read, is to respond by physically attacking him.

At the end of her last fragment in *A Mercy* and after the fight with [the blacksmith], [Florens] returns to the [Vaark] farm. The two indentured servants who sometimes work on the farm, Willard and Scully, describe her upon her return as "the docile creature they knew had turned feral" (*AM* 144). Without her master's boots, which she had taken off at [the blacksmith]'s dwelling, [Florens] comes back to the farm barefoot, and without "anybody's shoes," in fact (2). In the next section of the chapter, I turn to her fragments' core metaphor of the shoes, which I reconstruct here in regard to this absence of shoes and I suggest that it signifies [Florens'] positioning outside of [the blacksmith]'s world of reason – her unbelonging, if you will.[140]

"I Have No Shoes": Unbelonging

> I ain't got no home, ain't got no shoes
> Ain't got no money, ain't got no class
> Ain't got no skirts, ain't got no sweater
> Ain't got no perfume, ain't got no bed
> Ain't got no man
>
> Ain't got no mother, ain't got no culture
> Ain't got no friends, ain't got no schoolin'
> Ain't got no love, ain't got no name
> Ain't got no ticket, ain't got no token
> Ain't got no god
>
> Hey, what have I got?
> Why am I alive, anyway?
> Yeah, what have I got
> Nobody can take away?
> — Nina Simone, "Ain't Got No / I Got Life."

> The history of blackness is testament to the fact that objects can and do resist. Blackness – the extended movement of a specific upheaval, an ongoing irruption that anarranges every line – is a strain that pressures the assumption of the equivalence of personhood and subjectivity. While subjectivity is defined by the subject's possession of itself and its objects, it is troubled by a dispossessive force objects exert such that the subject seems to be possessed – infused, deformed – by the object it possesses.
> — Fred Moten, *In the Break*

In the lyrics to the song "Ain't Got No / I Got Life" (1968), singer, pianist, and activist in the Civil Rights Movement Nina Simone (1933–2003) asks, "What have I got [that] nobody can take away?" Simone's words, which constitute the first epigraph to this section of the chapter, help me situate my reading of [Florens] after her struggle with [the blacksmith]. They speak to her devastation when she realizes that she has "no consequence in [the blacksmith]'s world"

---

140 For a reading of *A Mercy*'s "shoe and footstep imagery" as a way for Morrison to connote the historical event of Bacon's Rebellion and to "[stress] not only the possibility in Bacon's Rebellion for a cross-ethnic, cross-class coalition, however, but also the subsequent opportune 'divide and rule' strategy of the colonial governmental and economic leaders with its portentous judicial result: new laws were spawned which were directed against the Africans, serving to link slavery firmly to blackness," see Waegner (104, 103).

and that [the blacksmith] will not become for her what she wants him to be (*AM* 140). It remains unclear what happens to [the blacksmith] in the wake of the fight—at the end of which [Florens] takes a hammer into her hand (140). We continue reading: "Our clashing is long. I bare my teeth to bite you, to tear you open. Malaik is screaming. You pull my arms behind me. I twist away and escape you. The tongs are there, close by. Close by. I am swinging and swinging hard. Seeing you stagger and bleed I run" (155–156). Will [the blacksmith] survive and recover from his injuries so that Black male liberal freedom remains one possibility of proprietorial self-making within the novel's experimental representation of colonial Virginia? Or will he die, and with him, perhaps, the boy [Malaik]? And what will happen to [Florens] at the novel's New World colonial scene?

An outraged, grieving, and traumatized (though not by the fact that [the blacksmith] calls her a slave, but by the fact that he does not want to own her) [Florens] declares at the beginning of the sixth and last fragment of her telling, "What I read or cipher is useless now. Heads of dogs, garden snakes, all that is pointless. [...] *I have no shoes. I have no kicking heart no home no tomorrow*" (*AM* 155, 156; emphasis mine). [Florens'] words once again bring me back to the notion that [Florens'] trek to [the blacksmith] constitutes an errand, a task she is forced to complete. And this, precisely, is what she does when she returns to the [Vaark] patroonship and, therefore, to her "legal owner." With the phrase 'what I read or cipher is useless now,' the text suggests that it ultimately does not matter that she had wanted to stay with [the blacksmith]. The fact remains that she is and will be someone's property and that she will not be able to decide who this person is: 'all that is pointless.' Indeed, [Florens] knows that her mistress is "putting her up for sale. But not Lina. Sorrow she wants to give away but no one offers to take her." And while [Sorrow] plans to run away and wants [Florens] to escape with her, [Florens] decides to stay because she "has a thing to finish here" (157).

'I have no shoes. I have no kicking heart no home no tomorrow.' These words also return me to the beginning of the novel and to the first fragment of [Florens'] telling where she tells the reader that the "beginning begins with the shoes. When a child I am never able to abide being barefoot and always beg for shoes, anybody's shoes, even on the hottest days. [...] I am dangerous, [Florens' mother] says, and wild but she relents and lets me wear the throwaway shoes from Senhora's house, pointy-toe, one raised heel broke, the other worn and a buckle on top" (*AM* 2). [Florens'] words and the shoe imagery also return me to Hartman's observation in the first epigraph to the chapter as a whole that "trying to fit into the other's shoes becomes the very possibility of narration."[141] When [Florens] is sent to fetch [the blacksmith], she is wearing her master's boots, in which she hides her mistress's letter (2). As soon as she arrives at [the blacksmith]'s dwelling, she takes off her master's boots and the boy [Malaik] steals "Sir's boots" from her (137). Barefoot, [Florens] is "stepping through the cabin, the forge, in cinder and in pain of my tender feet. Bits of metal score and bite them" and she will "never find Sir's boots" again (137). If 'trying to fit in to the other's shoes becomes the very possibility of narration,' as Hartman has it, then what does it mean that [Florens] gets into the fight with [the blacksmith] "[o]n bleeding feet" (137)? What does it mean that she later makes her way back to the [Vaark] patroonship "[a]lone. It is hard without Sir's boots. Wearing them I could cross a stony riverbed" (155)? If [Florens] ultimately remains [Rebekka Vaark]'s (or someone else's) property, as the text suggests, then how does this relate to notions of narrative, self-making, and belonging and to what Hartman describes as "the limits of most available narratives to explain the position of the enslaved" (Hartman and Wilderson 184)? What does it mean for [Florens] to *have no shoes*?

Without her master's boots and after she has attacked [the blacksmith], [Florens] no longer wants to fit anybody's shoes, no longer imagines herself to be someone else anymore. We continue reading: "But *my way is clear after losing you* who I am thinking always as my life and my security from harm, from any who look closely at me only to throw me away. From all those who believe they have claim and rule over me" (*AM* 155; emphasis mine). That is, [Florens] does not seek to find the kinds of freedom that [Vaark] and [the blacksmith] represent and strive for within the novel's diegetic frame. With [Florens] no longer trying to use someone else's shoes, with her fighting [the blacksmith] and his liberal conceptions of freedom and of belonging as possessing one's own self, she positions her self outside of the nexus of mind over desire that he tries to impose on her. As [Florens] states, "See? You are correct. A minha mãe too. I am become wilderness but I am also Florens. In full. Unforgiven. Unforgiving. No ruth, my love. None. Hear me? Slave. Free. *I last*" (159; emphasis mine). If [Florens] defies notions of freedom as self-possession (as embodied by [the blacksmith] and [Jacob Vaark], respectively), then what does her [character] embody, what does she signify? What kind of a space (narrative, metaphorical, epistemic), or the possibility thereof, does [Florens] meditate on when she states that she "lasts"?

Even more so than deconstructing [Florens'] thwarted attempt to belong to someone other than her master/mistress, I suggest that the text here also uses

---

141 See Montgomery ("Traveling Shoes") and Waegner for general discussions of this trope.

4.5 "I Am a Thing Apart": [Florens] and the Ruse of Belonging

the trope of the shoes to comment on narrative and/as form itself. Three months after [Florens'] return to the [Vaark] patroonship, the farm is in complete disarray (*AM* 157), so much so that there "was so much to be done because, hardy as the women had always been, they seemed distracted, slower now" (143). [Sorrow], for example, attends to the needs of her newborn baby and would "interrupt any field chore if she heard a whimper from the infant always somewhere nearby" (143). And while [Florens] continues to "do chores. Chores that are making so sense" (156), she, too, begins to focus on something else, which is the scratching of her telling into the walls of her deceased master's unfinished mansion. [Florens] goes about her telling secretly because her "Mistress [...] forbade anyone to enter the new house" and she does so barefooted (131). As suggested earlier, reviewers and critics tend to read [Florens'] telling and her scratching her words into the wall and floor of a room in [Vaark]'s unfinished mansion as "a dramatic representation of her asserting her subjectivity through narrative, using the master's words to create a counter-narrative to the objectifying discourse of western mercantilism, and establishing her history and physical body as living testaments to her survival" (Bellamy 24; see also Müller, "Standing"). By contrast, I argue that what is being asserted in her text is not so much that her telling is a cathartic or recuperative act towards "giv[ing] voice to her own experiences" but instead a representation of the absence of transformative promises within the structure of the narrative of the liberal Human (Müller, "Standing" 82–83).[142] I quote her telling here at some length:

> If you are ever live or ever you heal you will have to bend down to read my telling, crawl perhaps in a few places. I apologize for the discomfort. Sometimes the tip of the nail skates away and the forming of words is disorderly. Reverend Father never likes that. He raps our fingers and makes us do it over. In the beginning when I come to this room I am certain the telling will give me the tears I never have. I am wrong. Eyes dry, I stop telling only when the lamp burns down. Then I sleep among my words. The telling goes on without dream and when I wake it takes time to pull away[.] (*AM* 156)

---

142 Müller writes, "By carving her tale into the walls and the floor of the upper storey room in the house her master has built – literally, her master's property – Florens has arrived on par with the blacksmith in a way the latter might not have anticipated. While he was generous with lessons about reading the world, his ability to shape the world [...] had not been within reach of Florens. Her act of carving, however, using a nail and thus by virtue of the material also a symbol of the blacksmith's skills, transforms the room and presents an act of agency comparable to the blacksmith's art. In this sense Florens moves from being shaped to being a shaper" ("Standing" 82).

The phrases 'sometimes the tip of the nail skates away' and 'the forming of the words is disorderly' here convey a sense of disarray and disruption, of the disorganizing of a particular structure or order. This is emphasized by the sentences that immediately follow these two phrases: 'Reverend Father [who taught [Florens] to read and write (*AM* 4)] never likes that. He raps our fingers and makes us do it over.' The text here signifies on the principles of the catechism as that which both creates and imposes a particular (social) order – a syntax one could say – in the context of which [Florens] learns not only to how write but also how to exist and to abide by on the New World colonial scene. As she states elsewhere, "Confession we tell not write as I am doing now. [...] I like talk. Lina talk, stone talk, even Sorrow talk. Best of all is your talk. At first I am brought here I don't talk any word. [...] Slowly a little talk is in my mouth and not on stone" (4). [Florens] here undercuts and erodes this order by drawing the reader's attention to the fact that her telling is 'in her mouth' and not written in or, rather, 'on stone.'

The first line of the paragraph – 'If you are ever live or ever you heal you will have to bend down to read my telling, crawl perhaps in a few places' – creates another intertextual connection with the literary tradition of the African American slave narrative. This time around, *A Mercy* references Harriet Jacobs's 1861 *Incidents in the Life of a Slave Girl. Written by Herself* with the phrase 'crawl perhaps in a few space.' In her narrative, Jacobs tells the reader how she attempts not only to escape enslavement but also her master's sexual use of her and that she escapes to the North after hiding in a crawl space above a storeroom in her grandmother's house. We read that Jacobs takes refuge in a small "garret [...] only nine feet long, and seven wide. The highest part was three feet high, and sloped down abruptly to the loose board floor. There was no admission for either light or air" for seven years before she is finally able to escape to freedom (95).[143] With this in mind, I connect [Florens'] "disorderly" words to Hartman's above observation on the impossibility of Black emplotment into liberal narratives of freedom as well as to Wilderson's argument that "social death interrogates narrative as a form" ("Aporia" 134–135) in order to think about the ways in which [Florens'] telling in *A Mercy* can be understood as a comment on narrative and on how, as a form, narrative perpetuates the New World grammar of Human freedom and (self-)possession. In the novel, that is, [Florens] continues to scratch her words into the wooden walls of [Jacob Vaark]'s abandoned mansion when she notes:

---

143 I will continue my discussion of the intertextuality between *A Mercy* and Jacobs's narrative in the next chapter of this study.

4.5 "I Am a Thing Apart": [Florens] and the Ruse of Belonging — 215

> There is no more room in this room. These words cover the floor. From now on you will stand to hear me. The walls make trouble because lamplight is too small to see by. I am holding light in one hand and carving letters with the other. My arms ache but I have need to tell you this. I cannot tell it to anyone but you. I am near the door and at the closing now. What will I do with my nights when the telling stops? [...] Sudden I am remembering. You won't read my telling. You read the world but not the letters of talk. (*AM* 158)

With the words filling up (the walls of) the room that she is in, [Florens] continues to talk to [the blacksmith] (as well as the reader?) even though her arms are hurting from the physical act of carving because she 'cannot tell it to anyone but you.' However, it also turns out that her words, spoken in the present tense, will not find the right addressee because the [the blacksmith]/the "you" 'read[s] the world but not the letters of talk.' This literary strategy of creating "failed scenes of address" throws into relief the different grammars at work in this scene (Best, "On Failing" 468–469; see also Wyatt). While the "you" reads/navigates the world of the Human, [Florens'] grammar, her 'letters of talk,' seems to speak in(to) a void—if by void we mean her being outside of narrative emplotment, outside of being recognized as human being by the Human.

That is also to say that unlike Harriet Jacobs's narrative, in which Jacobs escapes into freedom and thus finds some kind of (narrative) closure/freedom as she leaves her crawl space for a better future, [Florens'] telling does not offer resolution for this [character]. Instead, the novel lets [Florens'] words circulate around themselves, lets them 'talk to themselves' forever and in the now. The text here strategically emphasizes its circular form. The novel's internal formal structure with the six fragments of [Florens'] telling mirrors this when the reader links her last fragment back to the novel's beginning and [Florens'] opening sentence "Don't be afraid" (*AM* 1)—only for her telling to start all over again, infinitely. "If you never read this, no one will. These careful words, closed up and wide open, will talk to themselves. Round and round, side to side, bottom to top, top to bottom all across the room" (159). I suggest that it is this circularity of her telling – again, not writing in a Human sense but scratching in the wooden walls of her master's dilapidated, unfinished third house – which ultimately reflects her being positioned outside of the Human fold by the New World grammar property. On the level of form, it is the circular structure of her telling which represents this and makes visible how this enslaved girl is "a sentient being[, that needs to be read as an] *existence void of transformative promise*, which narrative holds out to human subjects[; this] is a painful lesson for the slave to inculcate, much less accept" (Wilderson, "Aporia" 139; emphasis mine). Morrison's novel directs its readers to this denial of transformative promise ("Aporia" 134) and thus of narration in a Human sense for the enslaved by way of letting [Florens] talk about and define her words as her "telling" from the very beginning of the text (*AM* 1). This reconfiguration of narrative as "telling" and of "the world" as "letters of talk" – which I call a strategy of anti-narration – thus tries to account for [Florens'] existence and to grasp her as that being who cannot be free and belong in the same way that [the blacksmith] demands her to be or, for that matter, as [Jacob Vaark] claims to be.

'I have no shoes. I have no kicking heart no home no tomorrow.' [Florens'] words, finally, also return me to post-slavery conceptualizations of social death as the absence of recognized kinship relations or natal alienation (see Chapter 3). With [Florens] not having shoes and being barefoot both at the beginning and the end of her fragmented text, *A Mercy* suggests that she does not have any relations that are recognized by the Human (Wilderson, "Aporia"). On a different, while related, note, this also means that such things as and conceptualizations of belonging are within the purview of the Human and thus necessarily part of the master's vocabulary. This vocabulary is structured by the Human's proprietorial grammar. Or, in the words of [Florens'] mistress [Rebekka Vaark], "She is owned by me and can be knowne by a burne mark in the palm of her left hand" (*AM* 110). [Florens] knows that she does not belong in the world of the liberal Human and she expresses her "unbelonging" at the very end of her telling when she states:

> Or. Or perhaps no. Perhaps these words need the air that is out in the world. Need to fly up then fall, fall like ash over acres of primrose and mallow. Over a turquoise lake, beyond the eternal hemlocks, through clouds but by rainbow and flavor the soil of the earth. Lina will help. She finds horror in this house and much as she needs to be Mistress's need I know she loves fire more. (*AM* 159)

[Florens'] telling here does not offer a compensatory strategy of dealing with, or perhaps undoing, her "unbelonging" at the New World colonial scene. Once again, *A Mercy* here wrestles with the impossibility of writing the enslaved into a narrative that does not at the same time obliterate them. In line with post-slavery theoretical trajectories, *A Mercy* does not offer a recuperative narrative for Blackness and it does not gesture towards "the germ of a new beginning if not a new world" (Wilderson, *Red* 337). Instead of offering "a roadmap to freedom so extensive it would free us from the epistemic air we breathe" (Wilderson, *Red* 338), [Florens'] telling here literally suggests to burn it (read: narrative) all down. "Perhaps these words need the air that is out in the world. Need to fly up then fall, fall like ash over acres of primrose and mallow" (*AM* 159). The fact that [Florens] is and remains the property of someone else as well as the fact that she will remain in her crawl space infinitely as represented by *A Mercy*'s circular narrative form, disrupt the assumptive logics of narrative writ large, disrupt an assumed line of flight from Human, to subject and [character]. Finally,

[Florens'] telling in this sense constitutes "an ongoing irruption that anarranges every line," as Fred Moten puts it in the second epigraph to this section of the chapter, of the liberal Human's narrative.

[Coda]: When 'Belonging' Becomes Unbelonging Becomes Lasting

Consider the very last lines of [Florens'] telling: "I will keep one sadness. That all this time I cannot know what my mother is telling me. Nor can she know what I am wanting to tell her. Mãe, you can have pleasure now because the soles of my feet are hard as cypress" (*AM* 159). The text here not only offers the third possibility of who might be the "you" that [Florens] addresses throughout, namely her mother, the [*minha mãe*]. It also suggests a different way of conceptualizing, a different vocabulary to talk about her existence in *A Mercy*. The soles of [Florens'] feet are 'hard as cypress,' she no longer wears her master's, or anybody's boots, for that matter. How does one address this, with which words that are not wedded to the grammar of the liberal property paradigm? To repeat one of my earlier questions, what does it mean for [Florens] to state that she "lasts," and which words can account for her?

Over the course of the preceding pages, I have tried to show that Morrison's novel meditates on different notions of belonging at the New World colonial scene with [Florens'] telling. The text's core metaphor of the shoes functions as a vehicle to negotiate such different versions of belonging and it ultimately exposes that belonging needs to be understood as being tied to claims to property. I suggested this in my reading of two scenes of [Florens'] texts in *A Mercy*: [Florens'] brief stay at [Widow Ealing]'s home, during which she is subjected to the dehumanizing gaze and practices of a [village congregation] hunting for witches, as well as the fight between her and [the blacksmith]. These scenes, in turn, evoke racial slavery's paradigmatic scenes of the auction block, the letter of sale, and the errand, and the ways they maintain and renew the property paradigm and the liberal subject's claim to freedom through (self-)ownership. I have also tried to show that a critique of the liberal subject and its assumption of freedom animates [Florens'] textual fragments in the novel. At issue in my reading of her texts are the ways in which [Florens] [character] defies this assumptive Human grammar. [Florens'] question is not a question about Human subjectivity or Human self-making but one about being and "lasting" in/as social death. Put another way, what her textual fragments elaborate on is a tension between notions of belonging bound by proprietorial notions. As such, they need to be understood as, on the one hand, being within the purview of the master/Human and, on the other, what appears to be [Florens'] desire to no longer belong to her mistress but to [the blacksmith]. For [Florens], belonging turns out to be a ruse because it only exists in the sense of being owned by someone and not in the sense of a relationality. For her, Human relationality does not exist. In [Florens'] fragments, belonging thus becomes unbelonging.

In Christina Sharpe's *In the Wake: On Blackness and Being*, Moten's "anarranging blackness" becomes "*anagrammatical* blackness that exists as an index of violability and also potentiality" (75). Sharpe writes:

> That is, we can see the moments when blackness opens up into the anagrammatical in the literal sense as when "a word, a phrase, or name is formed by rearranging the letters of another" [...]. We can also apprehend this in the metaphorical sense in how, regarding blackness, grammatical gender falls away and new meanings proliferate; how "the letters of a text are formed into a secret message by rearranging them" or a secret message is discovered through the rearranging of the letters of a text. [...] So, blackness anew, blackness as a/ temporal in and out of place and time putting pressure on meaning and that against which meaning is made. (*Wake* 76)

[Florens'] telling indexes what Sharpe calls the "violability of Blackness" for enslaved life in the New World and it does so by way of invoking core scenes and "dehumaning" mechanisms of slavery such as the errand, the letter/slave pass, the auction block (*Wake* 74). Her completing her errand and subsequent return to the [Vaark] patroonship as [Rebekka Vaark]'s property also function as an index of this violability.

Following Sharpe's logic, I want in closing also suggest that new meanings also proliferate in/with [Florens'] telling. Let me point you again to [Florens'] words, uttered as she is carving her telling into the wooden walls of the unfinished [Vaark] mansion: "I am become wilderness but I am also Florens. In full. Unforgiven. Unforgiving. No ruth, my love. None. Hear me? Slave. Free. I last" (*AM* 159). These words – both a pun of Martin Luther King's famous "I Have a Dream"- speech delivered at the 1963 March on Washington (M. King) and a reference to the African American spiritual tradition ("African American Spirituals") – once again directs the reader towards the literary tradition of the African American slave narrative. Unlike those narratives, however, her words push against meanings of resolution, emancipation, and liberty and instead resort to making the vague promise of [Florens'] enduring and survival: 'I last.' As I have argued, [Florens] will ultimately remain in the crawl space that is the telling of her words which, in turn, will not press for a resolution to this [character]'s situation. Like [Florens] herself, they will also just last. They will "talk to themselves" and, perhaps, they will even go up in the flames kindled by Lina (*AM* 159).

4.5 "I Am a Thing Apart": [Florens] and the Ruse of Belonging

[Florens] puts pressure on what it means to be a Human subject in a liberal, possessive individualist sense, challenging "the assumption of the equivalence of personhood and subjectivity" (Moten, *Break* 1) when she states that she has 'become wilderness but I am also Florens.' [Florens'] feathers have lifted, her claws have scratched [the blacksmith] and have torn him open (*AM* 140, 155). Unbelonging, for [Florens], becomes lasting. At the end of her fragment, she lasts as wilderness, with soles hard as cypress, refusing any version of being Human in a liberal, possessive individualist sense that the text confronts her with. Blackness, as embodied by [Florens'] [character], thus becomes a total refusal of the requirements set down by the property paradigm of being/becoming a Human being at the New World colonial scene. With [Florens'] texts, *A Mercy* fundamentally refuses the notion that a free(d) person would automatically belong to the Human fold (which is what the slave narratives by Mary Prince and Harriet Jacobs would strive for, see Chapter 4.6). In thus wrestling with the difficulty of emplotting a "thing apart" into the structure that is narrative, I suggest that *A Mercy*, with [Florens'] telling, pushes the reader to come up with new ways of thinking about the relationship between slavery, self-making, and narrative and perhaps even to find a new vocabulary that allows for accounting for the slave in narrative, although such hopeful thinking borders dangerously on the kind of critical engagements of the novel that I set out to confront. (Again, the vocabulary of the Human, just like its grammar and the form and structure of its narrative cannot account for the slave without obliterating them). We can recognize this when [Florens'] states, "You won't read my telling. You read the world but not the letters of talk" (*AM* 158). As suggested before, the "you" addressed in these sentences is most obviously [the blacksmith] but it may also refer to a readership and a critical environment that still ventures to offer reparative readings that fail to take the property paradigm into consideration, striving for resolution.

[Florens'] words are and remain a "story predicated on impossibility," a telling predicated on the impossibility of narrative to account for the gratuitous violence that makes her into someone else's property (Hartman, "Venus" 2). *A Mercy*'s circular telling of what happens to [Florens] on the North American colonial mainland, carved into the wooden walls of her deceased master's unfinished mansion by herself, offers one way of dealing with this impossibility on the literary level of representation. Her telling – written in the present tense – is forever in the now. With [Florens], grammatical narration falls away, if by the former ("grammar") we mean the structure and episteme of the liberal subject and by the latter ("narration") we mean the formation of this grammar. Without the assumed belonging to "anybody's shoes," finally, [Florens'] telling creates a Black anti-narrative in which a thing apart speaks, tells, lasts (*AM* 2).

4.6 "There is No Protection": The [*Minha Mãe*], Slave Narratives, and the Sexual Economies of Atlantic Slavery

[**Routing the Argument**] My overall argument in the chapter is that *A Mercy* brings Atlantic slavery and specifically the (im)possibility for vertical motherhood for enslaved women to its textual orbit with the fragment of the [*minha mãe*]. In other words, it brings chattel slavery's practices of making human property and the heritability of the "non/status" (Sharpe, "Black Studies" 62) of the enslaved to the fore. This, in turn, is reflected in both the novel's and the textual fragment's forms. My argument in the chapter needs to be understood as a continuation of my engagement with the slave narrative script in *A Mercy*, which I have previously touched upon in my reading of [Florens]. I argue that the function of the slave narrative script in the [*minha mãe*]'s fragment needs to be understood as the novel's insisting on the active afterlives of the slave past as its ethical frame of reference. (So, again, I argue contra Best's claim that *A Mercy* opens up a new paradigm within literary and cultural criticism with which to think about Blackness without taking the violence and loss generated by slavery as a point of departure). From this, I want to think about the (im)possibilities of motherhood under colonial Atlantic enslavement (following J. Morgan, Hartman, Sharpe, Spillers), away from romanticized narratives of individual development and (self-)emancipation prevalent in the critical discourse on the novel. This is also to argue that the novel needs to be taken seriously in its engagement with the slave past and that the [*minha mãe*] needs to be read in her own right, as it were.

> The inability to name these women is not just a problem of the colonial archive but is rather a problem embedded in our cultural grammar—it is an insurmountable reality that testifies to the ways this was not intended to be a story to pass on.
> — Jennifer L. Morgan, "*Partus*"

> Once in the water that thrown overboard person would have experienced the circular or bobbing motion of the wake and would have been carried by that wake at least for a short period of time. It is likely, though, that because many of those enslaved people were sick and were likely emaciated or close to it, they would have had very little body fat; their bodies would have been denser than seawater. It is likely, then, that those Africans, thrown overboard, would have floated just a short while, and only because of the shapes of their bodies. It is likely, too, that they would have sunk relatively quickly and drowned relatively quickly as well. And then there were the sharks that always traveled in the wake of slave ships.
> — Christina E. Sharpe, *In the Wake*

Introduction

The previous chapter dealt with the textual fragment of [Florens] and in it I argued that her [character] ultimately remains void of a transformative promise that narrative offers to the (liberal) Human. In this context, I examined how *belonging* as a calculus of ownership and property constitutes one of *A Mercy*'s critical themes. I argued that the novel navigates the liberal property paradigm's nexus of Human self-making and ownership in [Florens'] fragment through belonging in a proprietorial sense rather than through notions such as identity, female agency, or self-emancipation (from patriarchal formations of power). In this chapter, I turn to the [character] of the [*minha mãe*], whose text ends the novel and, as such, constitutes a kind of coda to *A Mercy*'s previous texts. As [Florens'] mother, the [*minha mãe*] is the only [character] in the novel who brings Atlantic chattel slavery to *Mercy*'s plotting in that she literally embodies the large-scale making of human property fueled by the Middle Passage. She represents what Spillers describes as

> [t]hose African persons in "Middle Passage" [who] were literally suspended in the "oceanic" if we think of the latter in its Freudian orientation as an analogy for undifferentiated identity: removed from the indigenous land and culture, and not-yet "American" either, these captive persons, without names that their captors would recognize, were in movement across the Atlantic, but they were also *nowhere* at all. ("Mama's Baby" 72)

In my reading of this [character] – and in contrast to the existing body of literature on *A Mercy*, which tends not to read this [character] in its own right – I argue that the novel takes up and allegorizes the script of the slave narrative (previously addressed in my discussion of [Florens'] telling) as well as some of its topoi in order to discuss Atlantic slavery's "sexual economies" (Davis, "Don't") and the (im)possibility of vertical kinship relations for the enslaved (Spillers, "Mama's Baby"). I also examine what this means for narrative, which is, after all, premised on a notion of generative power and emplot-ability of past, present, and future. If we follow Afropessimism's concerns about (the structure of) narrative, this notion becomes the slave narrative's inherent but always-already defeated promise. It is that which *A Mercy* interrogates with the [*minha mãe*].

The fragment of the [*minha mãe*] comes at the very end of the book and it is by far the shortest text in a novel that does not impress the reader by its length. Of a total of 165 pages in my edition of the book, the [*minha mãe*]'s textual fragment roughly spans four whole pages. With [Florens'] telling (the last segment of which immediately precedes that of the [*minha mãe*]), these four pages share the ambiguity of who speaks and who is being spoken to. The text here works with readers' assumptions of reading certain signals in the text almost by default as first-person slave narrative. In what follows, however, I argue that the [*minha mãe*]'s text cannot be read as such because it lacks narrative resolution, and that it therefore needs to be read as challenge to the slave narrative script and its liberatory narrative gestures. I also suggest that the fragment's positioning as a kind of coda in the novel serves a crucial function in the text's examination of the connections between private property and (the enforced absence of) subjectivity and personhood. That is, it critically speaks to the importance of chattel slavery and specifically of enslaved women's experiences during Atlantic slavery as part and parcel of this socio-cultural, political, and epistemic formation.

Black feminists and (cultural) historians of Atlantic and U.S. slavery like Saidiya V. Hartman, Jennifer L. Morgan, and Marisa J. Fuentes remind us that any attempt at reconstructing the perspectives of enslaved people from enslavement's archive is usually doomed to fail, since the existing documents generally only reflect the dominant narratives of the colonizers, enslavers, and slave owners (Fuentes; Hartman, "Venus," *Lose*; Morgan, "Partus," "Archives," *Laboring*). In *Dispossessed Lives: Enslaved Women, Violence, and the Archive*, for example, Fuentes writes that the

> same objectification [that made human beings into property] led to the violence in and of the archive. Enslaved women appear as historical subjects through the form and content of archival documents in the manner in which they lived: spectacularly violated, objectified, disposable, hypersexualized, and silenced. The violence is transferred from the enslaved bodies to the documents that count, condemn, assess, and evoke them, and we receive them in this condition. *Epistemic violence originates from the knowledge produced about enslaved women by white men and women in this society, and that knowledge is what survives in archival form.* (5; emphasis mine)

In a similar vein, Hartman in "Venus in Two Acts" addresses the epistemic violence of the archive of Atlantic enslavement as she elaborates on both the longing and the impossibility of recuperating the lives lost by this violence, specifically those of enslaved women and girls on the slave ships. Hartman explains:

> There are hundreds and thousands of other girls who share her circumstances and these circumstances have generated few stories. And the stories that exist are not about them, but rather about the violence, excess, mendacity, and reason that seized hold of their lives, transformed them into commodities and corpses, and identified them with names tossed-off as insults and crass jokes. The archive is, in this case, a death sentence, a tomb, a display of the violated body, and inventory of property, a medical treatise on gonorrhea, a few lines about a whore's life, an asterisk in the grand narrative of history. ("Venus" 2)

And Morgan, furthermore, addresses the "scandal and excess" (Hartman, "Venus" 5) of Atlantic slavery's archive in relation to questions on "the challenge of recovery" (J. Morgan, "Archives" 153) when she writes that the archive "might remain nothing more than the repository of testimony on the part of predominantly white witnesses to one-dimensional truth claims. And yet, to depend upon archival corroboration to rewrite the history of black life can route you back to the very negations at which you started" ("Archives" 156). In elaborating how she revisited the archive of colonial slave law in order to find "a new way of thinking about slavery, gender, and reproduction," Morgan reminds her readers of

> the promise and betrayals of the colonial record books. The archival echoes that are left of the material conditions of the seventeenth-century black Atlantic are meager indeed, and they whisper of unspeakable horror, of Atlantic crossings, rape, disorientation, and backbreaking forced labor. ("Archives" 158, 159)

Against this backdrop, what does it mean that *A Mercy* centers the historical fact and the legacies of chattel slavery as well as the difficulty of engaging with it from the perspective of the enslaved in the novel's plotting with the text of the [*minha mãe*] and thus from the perspective of someone who would not be represented in the archive of slavery? How does this relate to Afropessimism's questioning of the structure of narrative as that which cannot contain social death?

In addition to these questions concerning the archive of Atlantic slavery, notions of recuperation and questions about the representation of lives lost during Middle Passage, there is a different, while related, set of questions, which also animates my thinking in the chapter. To my knowledge at this point, critics, readers, and reviewers generally tend to read and discuss the [character] of the [*minha mãe*] only briefly, and in the context of their respective readings of [Florens], and in relation to themes such as mother-daughter relationships, motherloss, abandonment, and trauma. Put another way, while the archive of academic texts on *A Mercy* grows continuously and thickens accordingly, what one is able to read about the [*minha mãe*] in this archive are a few words about a [character] "most palpably present in a dispatch addressed to her estranged daughter that closes the novel" (Best, "On Failing" 468) as well as a few words about this [character]'s "maternal sacrifice" (Jennings 646; see also Morgenstern). This critical tendency comes with another problem: while critical discourse on the novel, on the one hand, often reads *A Mercy*, like its predecessor *Beloved*, into the genre of the neo-slave narrative (e. g., Mueller, "Standing"; Nehl), critics and reviewers, on the other hand, have neglected to comment on how the textual fragment of the [minha mãe] both invokes and rebukes central tropes of the literary tradition of the African American slave narrative. Put differently, the [minha mãe] is hardly ever dealt with in scholarship, which results in the subsequent erasure of Atlantic chattel slavery and its sexual economies as the novel's rather explicit frame of reference and signification.

Accordingly, the aim of the chapter is twofold: First, it is geared toward reading the [minha mãe]'s textual fragment in its own right, thereby pushing back against readings that tend to overlook or ignore this [character]. Second, by way my analysis of how this fragment continues to take up, allegorize, and challenge the redemptive logic of liberation inherent to the slave narrative script, I aim to show that *A Mercy* does not abandon its readers to a "more baffled, cut-off, foreclosed position with regard to the slave past" (Best, "Failing" 472) but that it puts that past at its narrative center with [characters] like the [minha mãe]. To that end, I will first juxtapose this narrative fragment with Olaudah Equiano's slave narrative (1789) and unpack the intertextual connections between *A Mercy* and Equiano's narrative.[144] I also extract the intertextuality between Equiano's text and Morrison's Pulitzer Prize-winning novel and neo-slave narrative *Beloved* (1987) to then demonstrate how *A Mercy* revisits those connections in its own narrative orbit. In a second step, I enter the [minha mãe]'s text into conversation with both Mary Prince's (1831) and Harriet Jacobs' (1861) respective woman-authored slave narratives in order to show how this [character] engages with the intricate connections between sexuality, race, and the marketplace during slavery and makes this a fundamental concern within *A Mercy*'s plotting. Lastly, I argue that the segment of the [minha mãe] ultimately needs to be read as Black anti-narration, that is, as an ongoing epistemic critique of modernity's calculus of property by way of its critical referral to slave narrative. As for my writing and analyses in this chapter, I also need to point out that in following *A Mercy*'s last fragment's form I will necessarily reproduce some of the text's very own repetitions.

Situating the [*Minha Mãe*]: Re-Visiting Slave Narratives

I open this study's final close reading of *A Mercy*'s [characters] with some additional observations about the textual fragment of the [minha mãe] in order to begin to articulate what its function is in *A Mercy*. As mentioned already, the fragment adds up to four pages in total and in these pages, the [minha mãe] ad-

---

[144] For a prominent conceptualization of intertextuality see Genette.

dresses a "you," which appears to be her daughter, [Florens]: "Neither one will want your brother. [...] But you wanted the shoes of a loose woman, and a cloth around your chest did no good" (*AM* 160, 164). These words come from the beginning and the near end of the fragment, respectively, and with them the [*minha mãe*] tries to explain to her daughter why she offers [Florens] to [Jacob Vaark] as the currency in his debt settlement with her master in her place. We continue reading: "I knew Senhor would not allow it. I said you. Take you, my daughter. Because I saw the tall man see you as a human child, not pieces of eight" (*AM* 164).[145] The beginning and the end of her text frame the words with which she tells [Florens] and, by extension, the reader (the "you") how she became a slave on Jublio, her master [Senhor D'Ortega]'s tobacco plantation in colonial Maryland. Here is what one learns about her: as a survivor of a long-standing and violent feud between "the king of we families and the king of others" (161), the [*minha mãe*] is taken captive and, together with many others, brought to the coast: "We increase in number or we decrease in number until maybe seven times ten or ten times ten of we are driven into a holding pen" (161–162). Shipped to Barbados by "whitened men" (162), she is sold to [Senhor D'Ortega] at a Barbadian slave market and then brought to his plantation in the Chesapeake. On the [D'Ortega] plantation, the [*minha mãe*] is immediately used by her master to reproduce new slave property (163–164). It is important to note in this context that the [*minha mãe*] is the only major [character] who does not have a proper name in *A Mercy*.[146] "Minha mãe" comes from the Portuguese and can be roughly translated into "my mother." Explicitly, this needs to be understood as a reference to the global dimensions of the transatlantic slave trade and specifically to the role of the Portuguese triangular trade between the African West coasts and South America and Brazil, which between 1560 and 1850

---

145 "Pieces of Eight" refers to silver coins fabricated during the Spanish empire. These silver coins "were the world's first global currency. As the coins of Spain they were used across the vast Spanish Empire, stretching from South America to the Philippines, but were also used outside the empire as well. In 1600 one coin would have been worth the equivalent of a modern £50 note. The front of the coin is decorated with the coat of arms of the Habsburgs, the rulers of Spain and the most powerful family in Europe" ("Pieces of Eight"). As I was working on this chapter, the phrase 'not pieces of eight' would for me also always evoke associations of the infamous "three-fifths clause" in the original draft of the American Constitution and, by extension, of the *Dred Scott v. Sanford* verdict (1857), in which the notorious Supreme Court of the United States Judge Roger B. Taney explained that, before the law, slaves were not considered part of the political community of personas (see e.g,. Weier, *Cyborg* 23; "The Thirteenth Amendment")
146 There are a number of minor [characters] in the novel such as the [village people] that [Florens] encounters during her errand to fetch to the [blacksmith]; her brother; and [Sorrow]'s child who do not have a name.

unswervingly was "the largest destination for slaves in the Americas" (*Transatlantic Slave Trade Database*). Apart from what we might want to call these "hard facts," the attentive reader also learns about the sexual violence that the [*minha mãe*] is subjected to not only by her master but also by her mistress. Indeed, the first two sentences of her text already point to this. We read: "Neither one will want you brother. I know their tastes. [...] It was as though you were hurrying up your breasts and hurrying also the lips of an old married couple" and a few lines further on, "I saw things in his eyes [Jacob Vaark] that said he did not trust Senhor, Senhora or their sons" (*AM* 160, 162). It is here – specifically in the close semantic proximity between 'Senhor, Senhora' as well as 'or their sons,' which suggests that the sons, too, will become masters who will use their enslaved female property for their profit and their pleasures – that the text addresses the "sexual violence and reproduction characteristic of enslaved women's experience." It is here that the text speaks to the ways the "reproduction of human property and the social relations of racial slavery were predicated upon the belly" (Hartman, "Belly" 167, 168). It is this understanding of the creation of the New World that the [*minha mãe*] seeks to pass on to her daughter. Her very last words attest to this: "Oh Florens. My love. Hear a tua mãe" (*AM* 165).[147] It is with these words, which will never reach her daughter, that both the [*minha mãe*]'s fragment and the novel end.

The following, interconnected questions arise in this context: What happens to this message after one finishes reading the book? What happens to the [*minha mãe*] after [Jacob Vaark] accepts her daughter [Florens] from [Senhor D'Ortega] as a partial debt settlement? Does she remain the property of the [D'Ortegas]? How long will she live or survive on the [D'Ortega] plantation? Does she find a way to resist "Senhor, Senhora or their sons" (*AM* 161)? Will she run? Where to? Will she be able to protect her son? What will happen to her son? Why does the reader never hear about him again? Will she bear more children, who will then become her master's property? Who is their father? How will she mourn the loss of her daughter? Will she ever get to name herself, just like [Sorrow] re-names herself "Complete" (132)?

* * *

In her widely circulated and cited 1987 essay "The Site of Memory," which was first published in the same year as her Pulitzer Prize-winning novel *Beloved*, Toni Morrison comments on the literary tradition and genre of the African American

---

[147] Here, the previously used construction of "the mina mãe" is inflected to signify "your mother"/"tua mãe."

slave narrative and on what she considers her role as a novelist to be when engaging with such narratives—narratives, which hardly mention the author's "interior life" (Morrison and Denard 70). To recall, slave narratives were autobiographical narratives about the emancipation of the formerly enslaved and, as such, they generally were "stories of spiritual as bodily captivity and liberation" (Gould 14). As a "generic field," slave narratives first emerged "during the 1770s and the 1780s" (Gould 11). They were written in the first person and would bear a preface attesting to the authenticity of the subject matter as narrated, which usually would include a foreword by a white amanuensis who testified to the author's credibility and "assure[d] the reader how little editing was needed" (Morrison and Denard 68–69). In general, slave narratives simultaneously argued for the humanity of the enslaved as well as for the abolition of slavery and, in doing so, they often "embrac[ed] distinctly American ideals and values – of Christian faith, of the centrality of the family, and of a notion of freedom that encompasses individualism and independence – that were rooted and central to the newly emerging Republic" (Fisch 2).[148] Moreover, antebellum slave narratives would come to include genres such as the "spiritual autobiography, the conversion narrative, the providential tale, criminal confession, Indian captivity narrative, sea adventure story, and the picaresque novel" (Gould 13). From the twentieth century, writers across the African diaspora have begun to deal with the histories and the afterlives of New World chattel slavery in what conventional wisdom would call the literary genre of the neo-slave narrative.[149] In her essay and thoughts on the early and antebellum slave narratives, Morrison reminds her readers that "whatever level of eloquence or the form, popular taste discouraged

---

[148] On the genre and for collections of early/antebellum slave narratives see generally Bontemps; Drake; Fisch; Gould; Osofsky; Reid-Pharr.

[149] The term "neo-slave narrative" first appears in Bernard B. Bell's 1987 study *The Afro-American Novel and Its Tradition* and has since been taken up by scholars such as Ashraf H. A. Rushdy and Elizabeth A. Beaulieu. Valerie Smith reminds us in this context that twentieth and twenty-first century neo-slave narratives, like their early and antebellum predecessors, "approach the institution of slavery from a myriad perspectives and embrace a variety of styles of writing: from realist novels grounded in historical research to speculative fiction, postmodern experiments, satire, and works that combine these diverse modes" (168). Unlike their forerunners, however, these texts "possess a measure of creative and rhetorical freedom unavailable to the freed and fugitive slaves who wrote narratives during the antebellum period" and they write from a perspective "informed and enriched by the study of slave narratives, the changing historiography of slavery, the complicated history of race and power relations in America and throughout the world during the twentieth century, and the rise of psychoanalysis and other theoretical frameworks" (V. Smith 169). See also Fulton; James; Keizer; Nehl; V. Smith, "Neo-Slave Narratives."

the writer from dwelling too long or too carefully on the more sordid details of their experience," including the sexual and reproductive violence that many enslaved women experienced routinely, which were often encoded in or "veiled" by the specific literary conventions of the genre (Morrison and Denard 69–70). For Morrison's own literary project, which is interested in precisely the formerly enslaved's unwritten interior lives, this means that that she has to rely on her own imagination to represent enslaved women's experiences of sexual and other violence. As she goes on to tell us: "My job becomes to rip that veil drawn over 'proceedings too terrible to relate'" (Morrison and Denard 70). *Beloved* would become Morrison's masterful attempt at "[m]oving that veil aside"—an endeavor that indeed requires her to "trust my own recollections. [...] But memories and recollections won't give me total access to the unwritten interior life of these people. Only the act of the imagination can help me (Morrison and Denard 71). Following Morrison's comments, critics have often read *Beloved* as a literary text offering to (partially) fill this gap that early and antebellum slave narratives had opened up strategically, classifying the novel as a neo-slave narrative (see, e.g., Patton; V. Smith, "Neo-Slave Narratives"; Spaulding).

Morrison's comments on "the American slaves' autobiographical narratives" (Morrison and Denard 67) and the role that the contemporary novelist's imagination plays in the context of attempting to represent aesthetically what the script of the slave narrative does not make explicit – the slave's "interior life" – help me situate my reading of the [*minha mãe*]'s textual fragment as a continuation of *A Mercy*'s previous intertextual engagement of the slave pass, which becomes [Rebekka Vaark]'s letter, as well as of Harriet Jacobs's trope of the crawl space, which in [Florens'] fragment becomes the room in which she does her "telling." That is, the [*minha mãe*]'s text, over the course of almost an entire page, metatextually signifies on early slave narratives in that it relates how the [*minha mãe*] is captured on the African continent, and it dwells on her Middle Passage to the New World. We read:

> We increase in number or we decrease in number until maybe seven times ten or ten times ten of we are driven into a holding pen. There we see *men we believe are ill or dead*. We soon learn they are neither. *Their skin was confusing*. The men guarding we and selling we are black. Two have hats and strange pieces of cloth at their throats. They assure we that *the whitened men do not want to eat we*. Still it is the continue of all misery. Sometimes we sang. Some of we fought. Mostly we slept or wept. Then the whitened men divided we and placed we in canoes. We come to *a house made to float on the sea*. Each water, river or sea, has sharks under. *The whitened ones guarding we like that as much as the sharks are happy to have a plentiful feeding place*. I welcomed the circling sharks but they avoided me as if knowing I preferred their teeth to the chains around my neck my waist my ankles. (*AM* 161–162; emphasis mine)

This is most obviously an intertextual moment with Olaudah Equiano's encounter with European slave traders and subsequent Middle Passage, as narrated in his famous *Interesting Narrative of the Life of Olaudah Equiano, or Gustavus Vassa, the African, Written by Himself* (1789):

> The first object which saluted my eyes when I arrived on the coast was the sea, and *a slave ship*, which was then riding at anchor, and waiting for its cargo.[150] These filled me with astonishment, which was soon converted into terror when I was carried on board. I was immediately handled and tossed up to see if I were found by some of the crew; and I was now persuaded that *I had gotten into a world of bad spirits, and that they were going to kill me.* Their complexions too differing so much from ours, their long hair, another language they spoke (which was very different from any I had ever heard) united to confirm me in this belief. [...] *I asked them if we were not to be eaten by those white men with horrible looks, red faces, and loose hair.* (205–206; emphasis mine)

In *A Mercy*, as the first paragraph shows, Equiano's slave ship becomes 'a house made to float on the sea'; the 'white men' and 'bad spirits' 'with horrible looks' in Equiano's narrative become 'the whitened ones guarding we,' 'men we believe are ill or dead' with 'confusing skin'; and in both texts the reader is confronted with the captive's fear of 'being eaten' by their white captors. *A Mercy* here evokes the image of the sharks traveling in the wake of the slave ship crossing the Atlantic Ocean from the West African coast to the shores of the New World that also often figures in slave narratives, as, for instance, in *A Narrative of the Lord's Wonderful Dealings with J. Marrant, a Black, Taken Down from His Own Relation* (1785) by John Marrant (Bontemps xii–xiv). Christina Sharpe echoes this in the first epigraph to the chapter when she writes, "And then there were the sharks that always traveled in the wake of slave ships" (*Wake* 40). Like Equiano, the [*minha mãe*] in her fragment in the novel tells us about her capture and subsequent transport to the West African coast, which results from a feud between her people and a different people. We read: "Insults had been moving back and forth to and fro for many seasons between the king of we families and the king of others. [...] Everything heats up and finally the

---

150 Before arriving at the coast, Equiano also travels in a canoe as part of his forced passage to the shores: "At last I came to the banks of a large river, which was covered with canoes, in which the people appeared to live with their household utensils and provisions of all kinds. I was beyond measure astonished at this, as I had never before seen any water larger than a pond or a rivulet: and my surprise was mingled with no small fear when I was put into one of these canoes, and we began to paddle and move along the river. [...] Thus I continued to travel, sometimes by land, sometimes by water, through different countries and various nations, till, at the end of six or seven months after I had been kidnapped, I arrived at the sea coast" (Equiano 205).

men of their families burn we house and collect those they cannot kill or find for trade" (*AM* 161). Like Equiano (207–209), the [*minha mãe*] then tells the reader about her Middle Passage to the New World and about how she arrives on Barbados, only to be shipped to the nascent American colonies in the Chesapeake.[151] "Barbados, I heard them say. After times and times of puzzle about why I could not die as others did. After pretending to so in order to get thrown overboard. [...] So it was as a black that I was purchased by Senhor, taken out of the cane and shipped north to his tobacco plants" (163). It is here also, at a Barbadian slave market, where "[o]ne by one we were made to jump high, to bend over, to open our mouths" that the [*minha mãe*] "learn[s] how I was not a person from my country, nor from my families. I was negrita" (163).[152] (In Equiano's narrative, we read: "Many merchants and planters now came on board, though it was in the evening. They put us in separate parcels, and examined us attentively. They also made us jump, and pointed to the land, signifying we were to go there" (209)).

In *Beloved*, we find a similar intertextual moment with Equiano's slave narrative. At the beginning of its second part (Morrison, *Beloved* 236–256), *Beloved*'s (neo-slave) narrative splits up into a series of fragmented monologues that follow after the character Beloved, the child that Sethe had murdered in order to protect her from enslavement, has returned to 124 Bluestone Road to haunt its living inhabitants Sethe and her daughter Denver: "Mixed in with the voices surrounding the house, recognizable und decipherable to Stamp Paid, were the thoughts of the women of 124, unspeakable thoughts, unspoken" (*Beloved* 235). These monologic thoughts by the different narrative voices of Sethe, Denver, and Beloved completely disrupt the novel's plotting; in them, all grammar, syntax, coherence, and semantics fall away as the text attempts to represent the haunting memories of the horrors of the Middle Passage and chattel slavery (Broeck, "Trauma"). "[C]haracters (Sethe's grandmother, Sethe's mother, Sethe, Beloved) are blurred and merged, one voice is speaking in various registers of personal memory at once, any time frame is abandoned" (Broeck, "Trauma" 9). In Beloved's narrative we read, for example:

> some who eat nasty themselves I do not eat the men without skin bring us their morning water to drink we have none if we had more to drink we could make tears we cannot make sweat or morning water so the men without skin bring us theirs [...] We are not crouching

---

[151] On the North American mainland, Equiano will arrive in Virginia, the [*minha mãe*] in "Mary's Land" (*AM* 4).
[152] Of course, this echoes [Florens'] experience of being examined by the [village people] (*AM* 111).

now we are standing but my legs are like my dead man's eyes I cannot fall because there is no room to the men without skin are making loud noises[.] (Morrison, *Beloved* 248–249).

Equiano's 'world of bad spirits' with 'white men with horrible looks, red faces, and loose hair' here become *Beloved*'s 'men without skin,' the captors and enslavers which bring the narrating 'I' of this text passage to the New World. While this narrating 'I' represents Beloved's voice, it also signifies on and references those generations of enslaved females, her ancestors "in 'Middle Passage'" (Spillers, "Mama's Baby" 72), that came before her, amalgamating their voices with hers. The words "All of it is now it is always now there will never be a time when I am not crouching and watching others who are crouching too" attest to this and they bring their experiences to the present moment (*Beloved* 248).

It is the intertextuality with Equiano's *Narrative* which helps shed light on another intertextual connection, namely that between *A Mercy* and *Beloved*. Consider the latter's "Middle Passage fragments," of which I quote the following:

> All of it is now it is always now *there will never be a time when I am not crouching and watching others who are crouching too* I am always crouching the man on my face is dead his face is not mine his mouth smells sweet but his eyes are locked [...] *small rats do not wait for us to sleep* someone is thrashing but there is no room to do it in if we had more to drink we could make tears [...] *we are all trying to leave our bodies behind* [...] *it is hard to make yourself die forever you sleep short and then return* in the beginning we could vomit now we do not [...] *in the beginning the women are away from the men and the men are away from the women the storms rock us and mix the men into the women and the women into the men* [...] the men without skin are making loud noises they push my own man through . . . They are not crouching now we are they are floating on the water they break up the little hill and push it through [...] the iron circle is around our neck[.] (*Beloved* 248, 249, 250; emphasis mine)

In *A Mercy*, phrases such as 'we are all trying to leave our bodies behind'; 'it is hard to make yourself die forever'; 'you sleep short and then return'; and 'in the beginning the women are away from the men and the men are away from the women the storms rock us and mix the men into the women and the women into the men' become:

> We are put into the house that floats on the sea and *we saw for the first time rats and it was hard to figure out how to die. Some of we tried; some of we did.* Refusing to eat that oiled yam. Strangling we throat. Offering we bodies to the sharks that follow all the way night and day. I know it was their pleasure to freshen us with a lash but I also saw it was their pleasure to lash their own. Unreason rules here. *Who lives who dies? Who could tell in that moaning and bellowing in the dark, in the awfulness? It is one matter to live in your own waste; it is another live in another's.* Barbados, I heard them say. After times

and times of puzzle about why I could not die as others did. After pretending to be so in order to get thrown overboard. (*AM* 162–163; emphasis mine)

These lines echo the rats that the narrating 'I' in *Beloved* describes, echo the captives' attempts to die in order to escape from the slave ship, echo the impossibility of telling the stacked cargo of human bodies apart. 'Who lives who dies? Who could tell in that moaning and bellowing in the dark?' Like *Beloved*'s narrating 'I,' the [*minha mãe*] describes her Middle Passage in the present tense before continuing to tell the reader that she finally arrived at Barbados, thus making the experience and the memory of the Middle Passage a matter both of the narrative's and the reader's present. By way of these intertextual connections, it seems that the [*minha mãe*] could easily be one of the women in *Beloved*'s broken narrative passages of the Middle Passage, and vice versa. She could be Sethe's mother just like [Florens] could be the granddaughter of any of *Beloved*'s blurred female voices in Middle Passage.[153]

To draw attention to these intertextual moments is not to say that *A Mercy* revisits and takes up (neo-) slave narrative scripts in order to replicate those exactly, of course. Unlike most early and antebellum slave narratives, for example, the fragment of the [*minha mãe*] does not have an amanuensis who vouches for the authenticity of her telling. Similarly, it remains unclear *who* is addressed in her fragment, as opposed to the abolitionist audiences that slave narratives like

---

[153] We can also trace this intertextuality through the trope of the face. In *Beloved*, this trope is repeated and modified throughout in Beloved's fragment: "I am Beloved and she is mine [...] how can I say things that are pictures I am not separate from her there is no place where I stop her face is my own and I want to be there in the place where her face it and to be looking at it too a hot thing [...] the man on my face is dead his face is not mine [...] I cannot lose her again my dead man was in the way like the noisy clouds when he dies on my face I can see hers she is going to smile at me [...] I see the dark face that is going to smile at me it is my dark face that is going to smile at me the iron circle is around our neck she goes in the water with my face [...] You are my face; I am you" (*Beloved*, 248, 250, 251, 256). Explicitly, *A Mercy* takes up *Beloved*'s concern with this image in [Florens'] fragment, where we read about a dream [Florens] has when she is at [the blacksmith]'s house: "I dream a dream that dreams back at me. I am on my knees in soft grass with white clover breaking through. There is a sweet smell and I lean close to get it. But the perfume goes away. I notice I am at the edge of a lake. The blue of it is more than sky, more than any blue I know. More than Lina's beads of the heads of chicory. I am loving it so, I can't stop. *I want to put my face deep there*. I want to. What is making me hesitate, making me not get the beautiful blue of what I want? I make me go nearer, lean over, clutching the grass for balance. Grass that is glossy, long and wet. *Right away I take fright when I see my face is not there. Where my face should be is nothing.* I put a finger in and watch the water circle. I put my mouth close enough to drink or kiss *but I am not even a shadow there*. Where it is hiding? Why is it?" (*AM* 135–136; emphasis mine).

Equiano's would conventionally target; again, the "you" that the [*minha mãe*] addresses, just like in [Florens'] telling, could be either [Florens] or the reader, respectively. And again, *A Mercy* does not repeat the slave narrative's redemptive logic of liberation. Whereas slave narratives would strive to show their author's literacy and, therefore, their humanity through their use of written language (L. Scott),[154] *A Mercy*, like Morrison's previous novels, makes use of language arising from African American oral traditions of storytelling.[155] To give another example, in contrast to Harriet Jacobs or Mary Prince, who in their paradigmatic narratives choose not to explicitly address the sexual violence that they were subjected to as slaves (see, for example, Jacobs, esp. pp. 47–51), the [*minha mãe*] explicitly tells the "you" about the two times that she was "broken into" by enslaved men who were ordered to do so by their master (*AM* 163–164). (I will return to this later in the chapter.) Rather, I draw attention to these intertextual moments because they help me situate *A Mercy*, once again, as a literary text that precisely takes the slave past and its calculus of ownership and possession as critical paradigm and makes the abandonment of narrative coherence its primary concern. *A Mercy*'s and, more specifically, the [*minha mãe*]'s text's rather explicit intertextual connections with its predecessor *Beloved*'s "Middle Passage narratives" speak to this. With the intertextuality between *A Mercy* and early/antebellum slave narratives as well as between *A Mercy* and *Beloved* in mind, the following interrelated ensemble of questions arises: If the "slave narrative is a text [and] a key artifact in the global campaign to end first the slave trade [...], then colonial slavery [...], and finally US slavery" (Fisch 2), and if the neo-slave narrative is fundamentally concerned with e. g. "the challenges of representing trauma and traumatic memories; the legacy of slavery [...] for subsequent generations; the interconnectedness of race and gender; [...] the commodification of black bodies; and the elusive nature of freedom" (V. Smith 168–169), then what does it mean that *A Mercy* so obviously comments as well as signifies on both these textual genres? What does this mean with respect to a) the fact

---

154 Morrison explains, "In addition to using their own lives to expose the horrors of slavery, they had a companion motive for their efforts. The prohibition against teaching a slave to read and write (which many Southern states carried severe punishment) and against a slave's leaning to read and write had to be scuttled at all costs. These writers knew that literacy was power. Voting, after all, was inextricably connected to the ability to read; literacy was a way of assuming and proving the 'humanity' that the Constitution denied them. That is why the narratives carry the subtitle 'written by himself,' or 'herself,' and include introductions and prefaces by white sympathizers to authenticate them" (Morrison and Denard 68).
155 See generally in this context Morrison's "Rootedness: The Ancestor as Foundation" (1984, reprinted in Morrison and Denard).

that most readers and reviewers tend to read [Florens'] telling as a neo-slave narrative but do not think about the [minha mãe] along such lines? And b) what does this mean in relation to Best's argument to read *Mercy* as "an appreciation of the slave past as [...] that which falls away," as a kind of antidote to *Beloved* ("On Failing" 466)?

In some ways, my answer(s) to these questions here comes as a repetition: *A Mercy*, in contrast to what many critics and readers have argued, does not offer a post-racial fantasy of American colonial beginnings but very consciously and deliberately pushes its readers to New World practices and economies of enslavement as both its aesthetic as well as its ethical frame of reference.[156] Put somewhat differently, what has struck me most in my engagement of the still growing (yet relatively small in relation to the plethora of scholarly works on, for example, *Beloved* as well as Morrison's other novels) body of reviews, scholarly articles, and book chapters on *A Mercy*, is that the [minha mãe] hardly ever appears in these writings—apart from, perhaps, in discussions focusing on, for example, [Florens'] feelings of abandonment or motherloss and the trauma this induces.[157] In "Venus in Two Acts" Hartman writes:

> There is not one extant autobiographical narrative of a female captive who survived the Middle Passage. This silence in the archive in combination with the robustness of the fort or barracoon, not as a holding cell or space of confinement but as an episteme, has for the most part focused the historiography of the slave trade on quantitative matters and on issues of markets and trade relations. Loss gives rise to longing, and in these circumstances, it would not be far-fetched to consider stories as a form of compensation or even as reparations, perhaps the only kind we will ever receive. ("Venus" 3–4)

'There is not one extant autobiographical narrative of a female captive who survived the Middle Passage.' My engagement with the slave narrative script in my reading of the [minha mãe], while not an attempt at compensation in Hartman's sense, is geared towards bringing to the fore, one, precisely the non-reading of the [minha mãe] in the novel's critical discourse and, two, the subsequent erasure of chattel slavery and slavery's economies of reproduction as the novel's rather explicit frame of signification.[158] To be clear, I do not simply argue that

---

156 And doing so against the backdrop of the historic election of Barack Obama as 44th and first Black President of the United States and an ensuing discourse of post-racialism (cf. Cho).
157 The novel's term for this is "[m]other hunger—to be one or have one" (*AM* 61). I discuss this in my chapter on [Lina].
158 As someone who is structurally positioned as a member of the Human fold (Wilderson, *Red*), I cannot argue for "reparations" or "compensation" in the same way that Hartman does. At best, my project and this study can be understood as an attempt at the white project

the [*minha mãe*] and her fragment in *A Mercy* should be read as a neo-slave narrative but that the re-staging of certain topoi/tropes prominent in slave narratives in *A Mercy* as well as the intertextual references to a neo-slave narrative like *Beloved* function to facilitate a discussion of the (im)possibility of motherhood as well as of vertical kinship relations under slavery in a novel that has been discussed within a post-racial paradigm. That is also to say that the [*minha mãe*] in fact does not write/tell a slave narrative, at least not in the sense that the previously mentioned definitions of the genre would have it. The function of this script cannot be to show the emancipation of the [*minha mãe*] because the novel's plotting does not allow for the [*minha mãe*] to become free. To the contrary, the text quite explicitly suggests that she remains someone's, possibly [Senhor/Senhora]'s, property.

With this in mind, I will in the next section of the chapter continue to think about the ways in which the [*minha mãe*]'s textual fragment in *A Mercy* explicitly engages early slave narratives. This time, I will focus on women-scripted narratives in order to discuss how *A Mercy* utilizes those narratives to make the sexual economy of slavery its explicit narrative concern. Following the fragment's form, I suggest that unlike those narratives, *A Mercy* does not offer a redemptive narrative of liberation for the [*minha mãe*] but one in which purposefully lacks narrative resolution.

Summoning Mary Prince and Harriet Jacobs

> In North America, the future of slavery depended upon black women's reproductive capacity as it did on the slave market. The reproduction of human property and the social relations of racial slavery were predicated upon the belly. Plainly put, subjection was anchored in black women's reproductive capacities.
> — Saidiya V. Hartman, "The Belly of the World"

> At length the vendue master, who was to offer us for sale like sheep or cattle, arrived, and asked my mother which was the eldest. She said nothing, but pointed to me. He took me by the hand, and led me out into the middle of the street, and, turning me slowly round,

---

of counter-history in Greenblatt and Gallagher's sense, namely as (the writing of) a history that "opposes itself not only to dominant narratives, but also to prevailing modes of historical thought and methods of research" (qtd. in Hartman, "Venus" 12–13)—an attempt located at the intersections of the study of literature, its discursive and historical context, as well as theoretical critique. In this context, Hartman goes on to explain that "the *history* of black counter-historical projects is one of failure, precisely because these accounts have never been able to install themselves as history, but rather are insurgent, disruptive narratives that are marginalized and derailed before they ever gain a footing" ("Venus" 13).

exposed me to the view of those who attended the vendue. I was soon surrounded by strange men, who examined and handled me in the same manner that a butcher would a calf or a lamb he was about to purchase, and who talked about my shape and size in like words—as if I could no more understand their meaning than the dumb beasts. I was then put up to sale. The bidding commenced at a few pounds, and gradually rose to fifty-seven, when I was knocked down to the highest bidder; and the people who stood by said that I had fetched a great sum for so young a slave. I then saw my sister led forth, and sold to different owners: so that we had not the sad satisfaction of being partners in bondage. When the sale was over, my mother hugged and kissed us, and mourned over us, begging of us to keep up a good heart, and do our duty to our new masters. It was a sad parting; one went one way, one another, and our poor mammy went home with nothing.
— Mary Prince, "History of Mary Prince"

The first few lines of the [*minha mãe*]'s text begin with her powerful account of how she tries to protect her girl child [Florens] from the sexual subjection – her being consumed by – their master and mistress.[159] The first paragraph of the fragment reads:

Neither one will want your brother. I know their tastes. Breasts provide the pleasure more than simpler things. Yours are rising too soon and are becoming irritated by the cloth covering your little girl chest. *And they see and I see them see. No good follows even if I offered you to one of the boys in the quarter.* Figo. You remember him. He was the gentle one with the horses and he played with you in the yard. I saved the rinds for him and sweet bread to take to the others. Bess, his mother, knew my mind and did not disagree. *She watched over her son like a hawk as I did over you. But it never does any lasting good, my love. There was no protection. None.* Certainly not with your vice for shoes. *It was as though you were hurrying up your breasts and hurrying also the lips of an old married couple.* (AM 160; emphasis mine)

These words, addressed to [Florens] as the reader will later understand, distinctly establish as the fragment's historical as well as epistemic frame of reference how chattel slavery "systematically expropriated black women's sexuality and reproductive capacity for white pleasure and profit" (Adrienne Davis, "Don't" 104). The phrases 'Neither one will want your brother. I know their tastes'; 'they see and I see them see'; as well as 'it was as though you were hurrying

---

[159] Sexual subjection committed by slave mistresses was part of the plantation economy. Some literary texts also address this. An example is Harriet Wilson *Our Nig: Sketches from the Live of a Free Black*, which was first published in 1859. For a discussion of "Black Women's Speech Acts that Expose Torture and Abuse by Slave Mistresses," see Fulton (41–60). A more recent, white-authored example is Valerie Martin's 2003 novel *Property*. For two excellent readings of this novel and the critical discourse on it, see Broeck, "Property," and Sharpe, "The Lie," respectively.

up your breasts and hurrying also the lips of an old married couple' point the reader to the sexual subjugation done by both slave and plantation masters and mistresses alike. In this context, the paragraph also puts on the table what Hartman in "The Belly of the World: A Note on Black Women's Labors" (2016) describes as the master's "dreams of future increase" (166), which is to say that it brings to the for the complex connections between enslaved women's reproductive capacity and the market. As Hartman writes:

> The mother's only claim—to transfer her dispossession to the child. The material relations of sexuality and reproduction defined black women's historical experiences as laborers and shaped the character of their refusal and resistance to slavery. The theft, regulation and destruction of black women's sexual and reproductive capacities would also define the afterlife of slavery. ("Belly" 166)

In the above paragraph in *A Mercy*, the text represents this with the phrases 'Bess, his mother, knew my mind and did not disagree. She watched over her son like a hawk as I did over you' and 'There was no protection. None.' These proleptic phrases prepare the reader for what the text will describe as the "breaking in" of the [*minha mãe*], [Bess], and another – unnamed – slave woman on subsequent pages (*AM* 161, 163–164).

That the above first paragraph of the [*minha mãe*]'s text sets the stage for the pages to follow is best illustrated by the sentence "There was no protection. None," around which her text is structured. This sentence will appear three more times over the following pages, albeit in modified versions. In its subsequent variations – "There was no protection and nothing in the catechism to tell them no"; "There is no protection. To be female in this place is to be an open wound that cannot heal. Even if scars form, the festering is ever below"; and "There is no protection but there is difference" (*AM* 160, 161, 164) – the text switches not only from past tense to present tense, thus indexing the ongoing danger/subjugation/subjection that enslaved women were exposed to at all times, including in the reader's moment. With each repetition, it also continuously provides more and more information about what it is that there is not protection against. "There was no protection and nothing in the catechism to tell them no" (160–161). In between this second and the third reiteration/variation of the sentence, the [*minha mãe*] tries to explain that she "hoped if we could learn letters somehow someday you could make your way" and that she "tried to tell Reverend Father" (161), who secretly teaches her, [Florens], and [Florens'] brother to read and write (4). However, [Reverend Father] does not seem to understand what the [*minha mãe*] is trying to tell him: that for enslaved girls and women, there was no protection against sexual and other violence. And that when [Jacob Vaark] arrives at [Senhor D'Ortega]'s plantation, she sees "things in his eyes that said he did not trust Senhor, Senhora or their sons. [...] He never looked at me the way Senhor does. He did not want" (161). It is here also, in the context of the third reiteration of the sentence, that the [*minha mãe*] begins to tell the "you" ([Florens]/the reader) about her Middle Passage to the New World; her arrival on Barbados; and her being sold to [Senhor D'Ortega] and subsequently transported to his tobacco plantation on the southern North American mainland, where she is forced to work in [D'Ortega]'s household (161–164). With this reiteration – "There is no protection. To be female in this place is to be an open wound that cannot heal. Even if scars form, the festering is ever below" (161) – the text, now in the present tense, addresses the "*reproductive calculus* of the institution. [T]he work of sex and procreation was the chief motor for reproducing the material, social, and symbolic relations of slavery" (Hartman, "Belly" 169; emphasis mine). Switching from what I tentatively call the micro level of the sexual subjugation of enslaved women (in this case, the sexual violence committed by 'Senhor, Senhora,' 'their tastes' and their 'lips' within the plantation household[160]) to the macro level (Saidiya Hartman's "reproductive calculus," Adrienne Davis's "sexual economy of slavery", or Christina Sharpe's *Monstrous Intimacies*), the text thus addresses the propertization of the captured and the enslaved, the making of human beings into shippable cargo or, to use Spillers's term again, into "flesh" ("Mama's Baby" 67). This unmaking of human beings into flesh is also represented by the fact that the [*minha mãe*] does not have a name.

Lastly, the fourth reiteration of the sentence – "There is no protection but there is difference" (*AM* 164) – is situated in the second of the last three paragraphs of the novel. In these paragraphs the [*minha mãe*] describes and explains how when [Jacob Vaark] came to the [D'Ortega] plantation to settle a debt between himself and [Senhor], she knew that there was only one "chance [to save her girl child from the [D'Ortegas]'s making use of her for their sexual pleasures]. [...] Because I saw the tall man see you as a human child, not pieces of eight" (164). The [*minha mãe*] then goes on to tell us that she "knelt before him. Hoping for a miracle. He said yes" (165). With the words that follow,[161]

---

160 For a historical study on antebellum Black and white southern women, class, race, and gender, see, e.g., Elisabeth Fox-Genovese's *Within the Plantation Household: Black and White Women of the Old South* and the more recent and brilliant monograph *They Were Her Property: White Women as Slave Owners in the American South* by Stephanie Jones-Rogers.
161 We continue reading: "I stayed on my knees. In the dust where my heart will remain each night and every day until you understand what I know and long to tell you: to be given dominion over another is a hard thing; to wrest dominion over another is a wrong thing; to give dominion

words that will never reach their addressee [Florens], the text once again switches to the present tense. The text here suggests that the [*minha mãe*]'s words will continuously be spoken again and anew within the novel's diegesis, hovering, repeating the words of the enslaved mother while haunting (in Avery F. Gordon's sense) every reading of the book.[162] That is, in *A Mercy* the [*minha mãe*]'s kneeling in the dust becomes the present moment, so that her message to her daughter becomes *Beloved*'s "always now." [163]

In the above epigraph, which comes from Mary Prince's slave narrative entitled *The History of Mary Prince, a West Indian Slave. Related by Herself. With a Supplement by the Editor. To Which Is Added, the Narrative of Asa-Asa, a Captured African* (1831), Prince tells the reader how, as a little girl, she was sold at a slave auction when her master needed to raise money for his wedding. She describes how she is being examined by potential buyers attending the auction and how she 'fetched a great sum for so young a slave.' Right after the sale, Prince is separated from her siblings, who are sold to different owners, and from her mother. We read: "When the sale was over, my mother hugged and kissed us, and mourned over us, begging of us to keep up a good heart, and do our duty to our new masters. It was a sad parting; one went one way, one another, and our poor mammy went home with nothing" (M. Prince 3–4).

Thirty years later, in 1861, Harriet Jacobs would publish her by now famous slave narrative *Incidents in the Life of a Slave Girl. Written by Herself.* In it, Jacobs tells the reader how she attempts not only to escape slavery but also her master's

---

of yourself to another is a wicked thing. Oh Florens. My love. Hear a *tua mãe*" (*AM* 165). I will return to these words at the end of the chapter.

162 In *Ghostly Matters*, Gordon thinks about haunting as "one way in which abusive systems of power make themselves known and their impacts felt in everyday life" (xvi). For Gordon, haunting "is not the same as being exploited, traumatized, or oppressed, although it usually involves these experiences or is produced by them. What's distinctive about haunting is that it is an animated state in which a repressed or unresolved social violence is making itself known, sometimes very directly, sometimes more obliquely" (xvi). Most importantly for my reading of the [*minha mãe*], haunting according to Gordon "alters the experience of being in time, the way we separate the past, the present, and the future" (xvi). In the context of this study's core argument that *A Mercy* opens the door to the notion that the grammar of the liberal property paradigm continues to structure the present political, cultural, and aesthetic moment, Gordon's conceptualization of haunting as that which changes the ways in which we think about events from/ within different temporalities helps me make sense of the ways in which *A Mercy*'s form conflates past, present, and future realities/readings.

163 In terms of form, this fragment mirrors the very last two pages of Morrison's *Beloved*. Those pages are structured by the repetition of the sentence "This was not a story to pass on" (twice) as well as its variation, "This is not a story to pass on," thus creating another intertextual connection of sorts between *A Mercy* and *Beloved* (Morrison, *Beloved* 323–324).

sexual use of her. Jacobs tells the reader how she escapes from her master's hold on her by hiding in a crawl space above a storeroom in her grandmother's house for seven years before finally fleeing to the North (see also Andrews). In Jacobs' narrative we read:

> A small shed had been added to my grandmother's house years ago. Some boards were laid across the joists at the top, and between these boards and the roof was a very small garret, never occupied by any thing but rats and mice. It was a pent roof, covered with nothing but shingles, according to the southern custom for such buildings. The garret was only nine feet long, and seven wide. The highest part was three feet high, and sloped down abruptly to the loose board floor. There was no admission for either light or air. My uncle Philip, who was a carpenter, had very skillfully made a concealed trap door, which communicated with the storeroom. He had been doing this while I was waiting in the swamp. The storeroom opened upon a piazza. To this hole I was conveyed as soon as I entered the house. The air was stifling; the darkness total. A bed had been spread on the floor. I could sleep quite comfortably on one side; but the slope was so sudden that I could not turn on the other without hitting the roof. The rats and mice ran over my bed; but I was weary, and I slept such sleep as the wretched may, when a tempest has passed over them. Morning came. I knew it only by the noises I heard; for in my small den day and night were all the same. (Jacobs 95–96)

I suggest that the [*minha mãe*]'s fragment in *A Mercy* allegorizes both these women-authored slave narratives. Indeed, Jacobs' crawl space already appears in [Florens'] text; it appears in the last fragment of [Florens'] telling, when she carves her words into the wooden walls of [Vaark]'s abandoned and unfinished mansion: "If you are live or ever you heal you will have to bend down to read my telling, *crawl perhaps in a few places*. I apologize for the discomfort" (*AM* 156; emphasis mine; see Chapter 4.5). In *A Mercy*, then, Jacobs' "loophole of retreat" (Jacobs 95) first becomes a room in which [Florens'] "careful words, closed up and wide open, will talk to themselves, round and round, side to side, bottom to top, top to bottom all across the room," a room in which "the walls make trouble because the lamplight is too small to see by" and in which [Florens], at the end of her telling, is "near the door and at the closing" while the "you" she addresses "will stand to hear me" (159, 158). Jacobs' 'small den' above her grandmother's storeroom, a space where she crawls rather than stand, here becomes a space crawling with words, a space in which the [blacksmith] as well as her reader-listeners – mark her words, "you will have to bend down to *read my telling*" (156; emphasis mine) – will need to 'crawl perhaps in a few places.' In the [*minha mãe*]'s coda, the text precisely takes up the image of the crawl space and of 'crawling in a few places.' It does so at the very end of the [*mina mãe*]'s words where she explains how she hoped for 'the tall man' to accept her daughter in her stead in order to settle [D'Ortega]'s debt: "I knelt before him. [...] I stayed on my knees. In the dust where my heart will remain each night and every day until you understand what I know and long to tell you" (164, 165). Here, [Florens'] crawling becomes the [*minha mãe*]'s kneeling in the dust, which becomes *A Mercy*'s metaphorical crawl space, if you will.

The above scene from Mary Prince's narrative represents what Spillers describes as "another instance of vestibular cultural formation where 'kinship' loses meaning, since it can be invaded at any given and arbitrary moment by the property relations" ("Mama's Baby" 74). In *A Mercy*, we encounter this through the [*minha mãe*]'s account of when [Senhor D'Ortega] gives [Florens] to [Jacob Vaark] as debt settlement:

> After the tall man dined and joined Senhor on a walk through the quarters, I was singing at the pump. A song about the green bird fighting then dying when the monkey steals her eggs. I heard their voices and gathered you and your brother to stand in their eyes. One chance, I thought. There is no protection but there is difference. You stood there in those shoes and the tall man laughed and said he would take me to close the debt. I knew Senhor would not allow it. I said you. Take you, my daughter. (*AM* 164)[164]

These lines suggest that the [*minha mãe*] offers her daughter to [Jacob Vaark] in a desperate attempt to protect her from her master/mistress/their sons, hoping that her daughter will be better off at a different plantation/household. However, I submit that this desire on the part of the [*minha mãe*] is disrupted by the fact that the [*minha mãe*] will forever continue to kneel in the dust, as suggested by the novel's very last lines:

> I stayed on my knees. In the dust where *my heart will remain each night and every day until you understand what I know and long to tell you:* to be given dominion over another is a hard thing; to wrest dominion over another is a wrong thing; to give dominion of yourself to another is a wicked thing.
> Oh Florens. My love. Hear a tua mãe. (*AM* 165; emphasis mine)

After the [*minha mãe*]'s last words, the reader is referred back to the beginning of the novel and, thus, to [Florens'] telling by the novel's circular, loop-like form, so that the text will be told infinitely and [Florens] will continue to *not* hear what her mother wants to tell her. The [*mina mãe*]'s words will remain where they are and her message to her daughter/the "you" will continue to float, haunt, and hover within as well as beyond the textual orbit of *A Mercy*. In this context

---

[164] In [Florens'] telling this scene is condensed into the image of her mother holding on to her little brother but giving her away. On the very first page of the novel we read, for example: "If a pea hen refuses to brood I read it quickly and, sure enough, that night I see a minha mãe standing hand in hand with her little boy, my shoes jamming the pocket of her apron" (*AM* 1).

and against the background of *A Mercy*'s form, what does it mean for the [*minha mãe*] to stay kneeling in the dust and, therefore, in the novel's metaphorical crawl space? What does it mean that *A Mercy* poses this question at the very end of the novel?

[Coda]: Fashioning Anti-Narrative

> Everybody knew what she was called, but nobody anywhere knew her name. Disremembered and unaccounted for, she cannot be lost because no one is looking for her, and even if they were, how can they call her if they don't know her name? Although she has claim, she is not claimed.
> — *Beloved*

Reading the [*minha mãe*]'s fragment in conversation with Mary Prince's and Harriet Jacobs' respective narratives, we can see how in the intertextual moments created by the juxtaposition of these texts *A Mercy* both takes up and changes the slave narrative scripts. In this way, this textual fragment of and coda to *A Mercy* can perhaps best be understood as a companion piece to [Florens'] fragment in that it continues the former's allegorical summoning of the literary tradition of the African American slave narrative at the same time that it confronts and changes those texts' redemptive narrative arcs of liberation. Against this backdrop, the [*minha mãe*] can best be read in terms of what Jennifer Morgan has described as "early theorists of power." In her discussion of colonial Virginia's *Partus Sequitur Ventrem* law (1662), which would tether notions of reproduction to questions of race, status, property, heredity, and descent (see Chapter 3), Morgan writes that for enslaved African women and women of African descent in colonial Virginia,

> the *partus* act was hardly theoretical. Instead, *it made African women read the social landscape, becoming early theorists of power.* Enslaved women would be the first to grapple with the ways the alienation of their children placed them at the crux of unprecedented individual and systemic violence, in service of extracting labor through a newly emerging language of race and racial hierarchy. [...] Centering the crucible of race, sex, and reproduction illustrates that rather than spaces of opportunity, the shifting terrains of colonial racial ideology were experienced as spaces of dread. ("*Partus*" 16; emphasis mine)

Morgan's above conceptualization of enslaved women in colonial Virginia as "early theorists of power" is productive for my argument because it directs the reader's attention to those historical and epistemic contexts that *A Mercy* both addresses and represents with the [*minha mãe*] (if one cares to read it for them, as it were). As I have tried to show in this chapter, the [*minha mãe*] decidedly brings questions about the (im)possibility of motherhood under slavery to *A Mercy*'s New World colonial scene, showing how 'giving away'/'offering' her child (because "Senhor would not allow" for herself to be the currency in his debt settlement with [Vaark] (*AM* 164)) into the hands of a complete stranger/new master appears as the only way out of the sexual subjection of the master/mistress/their sons. Moreover, the [*minha mãe*]'s text also speaks to the hegemonic archive of Atlantic enslavement and addresses issues concerning those who are not represented in, those who are eliminated or erased from this archive. With the text of the [*minha mãe*], that is, *A Mercy* laments the nameless girls and women that Hartman speaks of, laments those who came to the New World on board the slave ships (if they made it so far). To echo Hartman once more, "How can narrative embody life in words and at the same time respect what we cannot know? [...] How does one tell impossible stories?" ("Venus" 3). *A Mercy* links these issues to "the taking of me and Bess and one other to the curing shed" (*AM* 163) so that complex connections between the violence of the archive and the violence of the sexual economies of Atlantic slavery become visible.

However, reading the [*minha mãe*]'s textual fragment together with Mary Prince's and Harriet Jacobs' respective narratives also once again alerts us to the fact that, structurally speaking, Blackness is elaborated by Middle Passage and social death and, as such, will never return to "a prior meta-moment of plenitude, never Equilibrium: never a moment of social life" (Wilderson, "Aporia" 139). Again, Afropessimist interrogations of narrative – as a structure that cannot to account for Blackness and the violence of slavery – have argued that Black social death's narrative arc needs to be understood as "a flat line [...] that moves from disequilibrium to a moment in the narrative of faux-equilibrium, to disequilibrium restored and/or rearticulated" ("Aporia" 139). Unlike the narrative of the Human subject, which bears the capacity for change and transformation, it is the absence of such transformative promises, which structures the narrative of social death. Social death in fact bars the slave's access to narrative ("Aporia" 136). As I have tried to show, *A Mercy* adopts these theoretical concerns with the fragment of the [*minha mãe*], throwing into crisis conventional elements of storytelling and rupturing conventional (slave) narrative's time/space matrix (cf. Hartman, "Venus"; Wilderson, "Aporia") by way of the textual fragment's open-ended form and positioning at the end of *A Mercy*, which connects back to the beginning to the novel and thus infinitely continues [Florens'] circular telling. That is to say, the [*minha mãe*] will stay in her metaphorical "crawl space," will continue to kneel in the dust at the end of the novel as well as beyond its immediate frame. Unlike Jacobs and Prince, the [*minha mãe*] will not escape into freedom. (Neither will [Florens], for that matter; see Chapter 4.5). Unlike Jacobs in her narrative, the [*minha mãe*] does *not* succeed in running away from her master and she will not save her children from slavery, either. In *A Mercy*, the reader does not learn what will happen to the [*minha mãe*]'s little boy (he simply disappears from the narrative) nor, for that matter, do we learn what will happen to [Florens] after her mistress sells her. That is, while Prince's and Jacobs' texts are constructed as narratives of the liberation and emancipation of the (formerly) enslaved, the [*minha mãe*]'s fragment decidedly does not venture to become such a narrative. Hers is not the narrative of the creation/making of a free(d) subject but *an anti-narrative that theorizes power*. Put another way, I have tried to show that *A Mercy* questions narrative form and its ability to emplot and account for the enslaved by way of its intertextual allegorizing of the slave narrative script. With the text of the [*minha mãe*], that is, *A Mercy* both unsettles and refuses a narrative arc that strives towards liberation and instead scripts social death.

Let me turn once more to the last iteration of the one sentence that structures the whole text of the [*minha mãe*]: "There is no protection but there is difference" (*AM* 164). Again, this reiteration is not only written in the present tense but it also inscribes itself into the reader's present. What does it mean that the [*minha mãe*] is given these words at the very end of the novel? How does this confront the reader, how does this show that the reader might have erred in their hopes for difference? Why is the [*minha mãe*] suddenly given a voice within the novel's diegesis? Given that [Florens] will never hear what her mother wants to tell her, why does *A Mercy* leave the reader with the words that "to be given dominion over another is a hard thing; to wrest dominion over another is a wrong thing; to give dominion of yourself to another is a wicked thing" (*AM* 165)? What is the meaning of the notion of "difference" in this context? Against the backdrop of my discussion of this [character], I think that these questions and the sentence speak not so much what this difference could have looked like at the New World colonial scene or how this difference might have shown itself (was it simply the fact the "the tall man [saw] you as a human child, not pieces of eight" (164)?). As the text suggests, [Florens'] life at the [Vaark] farm is one in which she is, at least temporarily, spared slavery's sexual subjection. It is in this sense that she experiences "difference." What *A Mercy* draws our attention to throughout, as I have tried to show in this study, is that the property paradigm structured the New World and human existence – as (self-possessed) owners and as those who were owned by others – from its very inception. In this context, there was in fact no difference between being someone's property and being the property of someone else who was perhaps kinder or less cruel than others. With the property paradigm structuring existence at the New World colonial scene, in other words, there was no freedom to be gained and no Human subjectivity to be claimed for the slave. And thus, while difference was important, it was not redemptive. "There is no protection but there is difference" thus once more points us to the notion that there is no redemptive narrative arc for the enslaved. There is no freedom (within narrative and beyond), there is just "difference" (164).

Finally, *A Mercy*'s equivocations purposefully leave the reader with the text's refusal to create a redemptive narrative arc for the [*minha mãe*]. Many of the novel's critics, whether enthusiastic or not, however, have chosen not to confront *A Mercy*'s ambiguities and thus do not engage with a [character], who, as a fictional representation of those whose narratives and stories do not exist within or have been obliterated from the archive of slavery, is in fact structured by such archival lacunae. *A Mercy* does not tell a story with the coda-fragment of the [*minha mãe*]. Instead, it both fashions and becomes Black anti-narrative—an epistemic critique of modernity's calculus of property that is constantly being revisited, revised, and recalibrated, a critique that is ongoing and "always now."

5 Coda

5.1 Refusing Private Property or, On Telling Impossible Stories

> Don't be afraid. My telling can't hurt you in spite of what I have done and I promise to lie quietly in the dark – weeping perhaps or occasionally seeing the blood once more – but I will never again unfold my limbs to rise up and bare teeth. I explain. [...] Stranger things happen all the time everywhere. You know. I know you know. One question is who is responsible? Another is can you read?
> — *A Mercy*

> Slavery broke the world in half, it broke it in every way. It broke Europe. It made them into something else, it made them slave masters, it made them crazy. You can't do that for hundreds of years and not take a toll. They had to dehumanize, not just the slaves but themselves.
> — Toni Morrison in Gilroy, *The Black Atlantic*

> And how does one tell impossible stories?
> — Saidiya V. Hartman, "Venus"

I began this study with the observation that Toni Morrison's novel *A Mercy* fundamentally confronts its readers with a specific kind of negation advanced through what one critic has somewhat disparagingly alluded to as "insubstantial characters and this wisp of a narrative" (Mantel). I asserted that this negation needs to be theorized as a refusal – expressed and negotiated through the text's strategies of allegorical characterization – to reproduce liberal ideas of what it means to be a Human subject within *A Mercy*'s seventeenth-century plotting. Those ideas are inextricably bound by notions of ownership, (self-)possession, and private property. Within *A Mercy*'s diegesis, refusal in the form of allegorical characterization emerges as critical practice; it becomes a strategy to take up, to allegorize, and to confront a grammar, which structures being and self-making through private property. The novel's form itself thus ultimately becomes a means to interrogate and to (re-)configure the relation between literary narrative and a theoretical critique of early modern liberal self-making.

Slavery and freedom have been complicit from the very beginning and it is their "vexed genealogy" that continues to structure both the liberal imagination of personhood and Atlantic slavery's "afterlife" in the present moment (Hartman, *Scenes* 115; *Lose* 6). Throughout this study, I have argued that *A Mercy* urges us as readers to critically revisit our liberal Western legacies of colonialism and slavery – social, political, cultural, epistemic – by way of returning us to key historical moments of its emergence on the aesthetic level of literary representation. That is to say, the novel takes its readers back to the late seventeenth century when liberal ideas of individual rights, representational government, and Human emancipation from feudal rule, as well as claims to individual liberty and property, among others, were first articulated by early Enlightenment thinkers over and against the systems and practices of New World chattel slavery. In this context, I drew attention to an event that would come to define North American, and later U.S. beginnings and futures like no other: the arrival of the first enslaved Africans in colonial Virginia in 1619 (see Hannah-Jones). This event would come to mark one crucial site on which the interconnected concepts of freedom and individuality emerged and constituted themselves in parasitic relation to Atlantic slavery. I also suggested that in returning us to American beginnings in the colonial Chesapeake, Morrison's novel transports us back to a moment in time that historians have often read as a moment of pure possibility, a moment in which racial lines appear to have not been as rigid and rife as they would come to be in the eighteenth and nineteenth century, respectively. From the perspective of enslaved Black women, however, those lines were far from fluid, as Black feminist historians have taught us in their examination of the intricate connections between slavery, property, reproduction, and heritability in seventeenth-century Virginia. Entering into conversation with such Black feminist counter-historical projects, I have endeavored to illustrate how *A Mercy* pushes an interrogation of the ways in which a grammar of private property is already at work at a time commonly understood to be shaped by the (near) absence of inflexible racial lines. I have insisted throughout that *A Mercy*'s representation of late-seventeenth-century colonial Virginia returns us to a past in which chattel slavery and its regimes of violence and grammar of private property produce a subject, whose claims to freedom are both fueled by and grounded in the abjection of the enslaved, and I proposed to conceptualize this as the liberal *property paradigm*. The complex entanglements between individual liberty, slavery, and private property converge in a subject, which can be recognized in paradigmatic textual articulations and flourishing epistemic configurations from across the seventeenth-century English Atlantic, where it materializes in the early Enlightenment political philosophy of John Locke and in the Northern Quakers' concerns for their reputation, respectively. In different, while fundamentally related, ways, slavery and possession here emerge as a sounding board for the development and as a marker of white capacity and identity deliberations. The absence of captivity and bondage, freedom from social death and "violence's gratuitousness," as well as ownership of enslaved human beings become a metaphor for white liberty and being (Wilderson, *Red* 31).

A Mercy's experimental setup returns to the seventeenth-century slave past not to lay this past to rest, as some critics have suggested prominently (e. g., Best,

None, "On Failing"; Christiansë), but to draw attention to the ways the grammar of property planted on the North American mainland with the founding of the Virginia colony would structure social formations for centuries and its afterlives would continue to shape the present moment. *A Mercy*'s very first lines – "Don't be afraid. My telling can't hurt you in spite of what I have done" – are spoken by the slave girl [Florens], who introduces us in an anti-introductory fashion to *A Mercy*'s thought experiment that exposes us to slavery's multiple pasts at the New World colonial scene (*AM* 1; see also first epigraph). Through the text's creating of the different [characters] and the historical and mythical narratives that they evoke, we are by way of allegory confronted with the desire for freedom from feudal rule and religious doctrine that drove European settlers to embark on one of the countless oceangoing vessels. We are, in other words, exposed to those settlers' transatlantic passage to an environment, in which the fashioning of this new, liberal, and self-made self would come to be intricately connected to practices of enslavement. That white English women, too, would both struggle for and claim access to this new status of liberal subjectivity comes to the fore through the novel's gesturing towards the legal system of coverture as one site at which white married (or divorced) women in the seventeenth through the nineteenth century were legally dispossessed of their property. With the [character] of [Rebekka Vaark], the novel points the reader to the notion that white women would in fact "find ways to circumvent the constraints that coverture imposed" on them and that they would ultimately be able to claim mastery precisely of their (human) property (Jones-Rogers 27). Such claims to mastery, as we have seen with [Rebekka] as the novel's representation of late seventeenth-century formations of white womanhood, translate into claiming mastery of themselves as well as their slaves. *A Mercy* also exposes us to the histories of genocide and fundamental eradication of Indigenous life by European settlers and it does so by gesturing towards, for example, mythically rendered narratives of the distribution of smallpox infected blankets to Indigenous tribes or towards nineteenth-century futures of forced assimilation. I have suggested throughout that *A Mercy* constantly plays with the possibility of creating a version of the past that is not structured by the property paradigm at the same time that it confronts the reader with the histories and legacies of the Middle Passage and the archives of Atlantic slavery. In so doing, *A Mercy* draws attention not only to the conceptual conflation of enslavement, property, and reproductive capacity within the realm of colonial law but also to the continued disruption and denial of vertical and recognized kinship formations for the enslaved on the New World plantation. In this context, the examples of the [characters] of [Florens] and the [*minha mãe*] also expose the reader to a different realm of the past, namely that of literary history and ancestry, as it were. In its vision of a historical future yet to come, the text crucially enters into dialogue with its literary predecessors from the eighteenth, nineteenth, and twentieth centuries such as the African American slave narrative tradition and Morrison's prize-winning neo-slave narrative *Beloved*.

In the preceding chapters, *A Mercy*'s experimental setup and plotting animate an analytic aesthetic invested in "unimagining" the full scale "irreparable violence of the Atlantic slave trade" (Hartman, "Venus" 12), while facing the impasse that it is precisely this violence that is deeply woven into the social, political, and epistemic fabric of the present. From chattel slavery, Reconstruction, and Jim Crow over "police brutality, mass incarceration, segregated and substandard schools and housing, astronomical rates of HIV infection, and the threat of being turned away en masse at the polls," this violence "still constitute[s] the lived experience of Black life" today (Wilderson, *Red* 10). Thus revisiting the aesthetic territory of the New World colonial scene and straddling the line between unimagined pasts and devastating present-day futures, *A Mercy* shows us what could have been but what did not happen and what it is that we as readers should be grieving for. *A Mercy* imagines a history that could have taken a fundamentally different turn, creating a lost future in which conceptions of the (universal) self would not be intricately connected to proprietorial notions of being. By anticipating a historical future yet to come, that is, this experiment on the literary level of representation becomes a return to seventeenth-century American beginnings fueled by the dreadful knowledge that the workings of the liberal property paradigm are ubiquitous, past and present. This literary endeavor, I suggested, makes legible the positioning power and the violence of property in close dialogue with post-slavery Black Studies theoretical trajectories, which provide us with a set of analytical terms – *violence, dispossession, fungibility, abjection, reproduction, kinship, anticipatory wake* – as they ask us to scrutinize, intervene into, and deconstruct formations of private property. Put another way, the analyses of *A Mercy*'s [characters] have illustrated the manifold ways in which the property paradigm configures existence at the New World colonial scene in fundamentally different and often antagonistic ways. Throughout the study, I asserted that the novel attempts to account for the structural as well as structuring violence of slavery as that which places *A Mercy*'s [characters] in relation to one another. What this analytical focus on violence specifically brings to the fore is how the absence of relationality, that is social death, situates the slave woman [characters] outside of the realm of liberal self-making that the other [characters] are able to claim for themselves. This is also to say that I have rejected the critical discourse on *A Mercy* and its assumptions of "innocent" transformations fueled by the desire for property of quintessential Lockean subjectivity into morally corrupted and greedy versions of the liberal self, as is the case with

[Jacob Vaark]. I also have rejected this discourse's conceptualizations of cross-cultural, cross-racial, and cross-religious solidarity amongst *A Mercy*'s women [characters]; its embrace of concepts like dispossession as that which positions those [characters] in similar ways; as well as critics' politically salient, inclusionary gestures of universal female agency, identity deliberation, or self-emancipation from patriarchal formations of power.

All analyses of the novel's [characters] demonstrate, albeit to different degrees, that *A Mercy* establishes and reconfigures the relation between narrative form and theoretical critique. This relationship is often configured as distinct from the slave past in Black Studies' works that, like Stephen Best's, are committed to unearthing "a new set of relations between contemporary criticism and the black past on the basis of aesthetic values and sensibilities" (*None* 22). Here, rather than to theorize from the positioning power of gratuitous violence and the attendant structural ramifications for civil society as well as narrative, scholars like Best push towards disintegrating precisely this past and towards adopting a practice that will account for Black aesthetic and political articulation in abandonment of the slave past (Best, *None* 23). Rather than reproduce the "abandonment aesthetic" promoted by such disintegrative critical approaches to engaging with narrative and other works of art, I have argued that *A Mercy* both heightens and insistently questions narrative's embeddedness within the fold of the Human and points us to literary narrative's inability to account for Black social death. *A Mercy* thus opens up a path for us to think about the ways literary narrative and narrative form may allow for a new kind of "potentiality" (Sharpe, *Wake*) that devises new means for interrogating Atlantic chattel slavery's epistemic formations on the literary level of representation. This kind of "formal agency" (Best, *Fugitive*) is fundamentally different from widespread theorizations of agency, which frequently come into view as enunciations of capacity—to be gained or to be held by the liberal subject or by literary character as this subject's representation within narrative, respectively. Instead, it needs to be conceptualized as the insistent questioning of the relationship between the violence of social death, literary narrative, and form. This is what I have conceptualized throughout as allegorical anti-narration.

As strategy, allegorical anti-narration is not only wedded to Black Studies' critical inquiry of "speaking from the standpoint of the slave in a slave society" (to echo Jared Sexton). It also is fundamentally committed to notions of ambiguity, non-resolution, and contradiction. As an ongoing epistemic critique of modernity's calculus of property, allegorical anti-narration challenges (the structure of) narrative, which is to say that it challenges narrative's striving for resolution and its restorative desires (Wilderson, "Aporia"; Spillers, "Mama's Baby"). In *A Mercy*, we can recognize this confrontation in, for example, [Florens'] reconfiguration of narrative as "telling" or in [Sorrow]'s fugitively generative (racial) ambiguity, which opens up an utopian moment for the possibility of making Black generations beyond the property paradigm. *A Mercy*'s lingering with ambiguity, then, becomes a way to address the absence of narrative, which is to say social death, within a narrative text.

In the face of the ongoing abjection of Black life, antiblackness, and the longue durée of Atlantic slavery, Christina Sharpe – citing Hartman and Wilderson – conceptualizes the project of Black Studies as the "continued imagining of the unimaginable: [the] continued theorizing from the 'position of the unthought'" ("Black Studies" 59). As an antidote to literary narratives (and, by extension, the critical discourses accompanying those narratives) that neglect to account for and to reckon with Atlantic slavery's structuring violence, *A Mercy* heeds Sharpe's conceptualization. That is, the novel both portrays and theorizes the unimaginable at the same time that it attends to "the ongoing state of emergency in which black life remains in peril" (Hartman, "Venus" 13). In this way, *A Mercy* crucially becomes a sanctuary for Black critical thinking and fabulation.[165] *A Mercy* is history: Its representation of late seventeenth-century Virginia's colonial historical landscapes makes visible the fundamental differences between what it means to be dispossessed and what it means to be fungible property. *A Mercy* is theory: In and through its insistent questioning, we can recognize that the gratuitous violence of social death is "prelogical to narrative construction" (Wilderson, private conversation) and that the absence of narrative is social death (Wilderson, "Aporia"). *A Mercy* is anti-narrative: Not only does its representation of North American beginnings in colonial Virginia expose us to a version of the past that confronts us with and makes us work through a set of questions such as, *What else could have happened? What could have been had history not taken us down the devastating path of the epistemic, philosophical, social, political, and cultural entanglements of slavery and freedom fueling the liberal imagination of self?* This representation also fundamentally exposes us as readers to the im-

---

[165] I use the term "fabulation" here with reference to Tavia Nyong'o's theory of *Afro-Fabulations: The Queer Drama of Black Life*. In his eponymous study, Nyong'o conceptualizes afro-fabulation as "black feminist and posthumanist acts of speculation [that] are never simply a matter of inventing all tales from whole cloth. More nearly, they are the tactical fictionalizing of a world that is, from the point of view of black social life, already false. It is an insurgent movement – in the face of an intransigent and ever-mutating anti-blackness – toward something else, something other, something more. While moments of afro-fabulation are indeed often ephemeral and fleeting [...] they may also be monumental and enduring" (6). The connections and incongruences between Hartman's and Wilderson's concerns about the emplot-ability of social death and the practice and "insurgent movement" of afro-fabulation need to be examined more closely in a different project.

portance of language, of narrative construction and fabulation, as well as the necessity to recount the impossible to paraphrase Hartman from the third epigraph of this section ("Venus" 10).

A Mercy's groundbreaking refusal, which lingers and sprouts in the combination of theoretical intervention with literary/narrative form, raises important questions that take us beyond the novel's immediate diegetic frame. In other words, the larger implications of this project as the first book-length in-depth study of Morrison's novel beyond the realm of close analysis and literary criticism can be delineated along the following lines: First, novels like *A Mercy* alert us to the notion that language and words matter and are crucial to any engagement of Western modernity's genealogies grounded in Atlantic chattel slavery. And while language and words matter and often are the only vehicles available for any such engagement, *A Mercy* also confronts us with the limits as to what we can know, understand, and address through them. It points us to the limitations of what words can actually tell. That is to say, we (as literary scholars, critics, and readers of literary narratives) need a new vocabulary, new words, new terms that bring us closer to being able to describe and to account for the structures, patterns, mechanisms, and systems of white liberal self-making over and against private property, fungibility, and the abjection of Blackness. Black feminist thinkers like Christina Sharpe have recently introduced us to notions such as the "Black anagrammatical" or "Black annotation and redaction," which delineate Blackness and its repertoires of aesthetic, epistemic, and representational articulation as "exist[ing] as an index of violability and also potentiality" (*Wake* 75; see also Moten, *Break*).[166] Like [Florens] in *A Mercy*, who exposes us to her "telling," we as readers and critics need words and concepts that are

---

[166] The struggle to find the "right" and adequate vocabulary has always been an essential part of the groundbreaking work of Black feminist thinkers. As Hortense Spillers reminds us, her pioneering work endeavored to "find a vocabulary that would make it possible, and not all by myself, to make a contribution to a larger project. I was looking for my generation of black women who were so active in other ways, to open a conversation with feminists. Because my idea about where we found ourselves in the late 1970s and the mid-1980s, was that we were really out of the conversation that we had, in some ways, historically initiated. In other words, the women's movement and the black movement have always been in tandem, but what I saw happening was black people being treated as a kind of raw material. That the history of black people was something you could use as a note of inspiration but it was never anything that had anything to do with you—you could never use it to explain something in theoretical terms. There was no discourse that it generated, in terms of the mainstream academy that gave it a kind of recognition. And so my idea was to try to generate a discourse, or a vocabulary that would not just make it desirable, but would necessitate that black women be in the conversation" (Spillers et al. 300).

able to account for the ambiguity, the contradictions, and the absence of resolution that literary texts like *A Mercy* confront us with—texts that unsettle and decenter white knowledge formations in fundamental ways. Put somewhat differently, literary critics need a new language in narrative theory to speak about the limits of narrative with respect to social death.

Within the realm of the (narratological) study of literary character, and second, there also is a need for a re-examination of the concepts that we use to address fictional entities within literary narrative. This re-examination needs to follow in the steps of post-slavery interventions, like Afropessimism, which profoundly question the assumptive logics of critical theory's "ensembles of questions dedicated to the status of the subject as a relational being" (Douglass and Wilderson 117). Throughout the study, I have opted to use square brackets as a way of connoting the various demands or claims to New World subject-making and property, or the structurally induced absence thereof, that *A Mercy*'s fictional entities make in their respective textual fragments. While bracketing subject-making and the concept of fictional character in this way has helped me grapple with the aporia of not having adequate terms to account for the notion that not all "beings are on the same side of social life," both in narrative and beyond (Wilderson, "Aporia" 141), this orthographical shortcut demands further scrutiny and elaboration, as do other narratological terms, concepts, and methods, such as "close reading." Most often and to recall Bal's suggestions for cultural analysis and use of "travelling" concepts (see Chapters 1 and 4), existing narratological methodologies are bound by a premise of "intersubjective" exchange or conversation (e.g., Bal, *Traveling* 11). At the heart of this premise seems to be an assumption of a universal subject and relational being. If we follow post-slavery theoretical thinker's pressing interventions, as this study has done throughout, such an assumption needs to be thoroughly scrutinized, challenged, and dismantled.

Third, novels like *A Mercy* force us not to impose, consciously or unconsciously, a white hermeneutic reading and inflection on such Black-authored texts, the "ventriloquism or unbidden translation" of which will only reproduce and maintain the violence of white knowledge formations and assumptions of universalism (Broeck, *Gender* 11; see also Mills, "White Ignorance"). The push here must be to destabilize such white reading practices and to distort white coherence. Recall how [Jacob Vaark] as a paradigmatic example of the New World liberal subject is so forcefully disnarrated from the novel and how *A Mercy* refuses to reinscribe white subjective coherence within its diegesis in this way. By referring the reader, scholar, student, and critic back to questions concerning ethics and methodology, in other words, *A Mercy* opens up avenues towards a radical pedagogy committed to lingering with ambiguity and non-resolution.

This, and fourth, becomes highly relevant with respect to reading and studying *A Mercy* in the context of American Studies in Germany. As a discipline that continues to wrestle not only with what I have called the "challenge of remove" in the introduction to this study but also with the notion that it is a predominantly white(-authored) field, following a pedagogy and a practice of lingering with ambiguity or with what Frank Wilderson calls "pyrotechnics" (*Red* 337) seems more than urgent. Put somewhat differently, *A Mercy*'s call for critical vigilance holds true for those of us who, like myself, belong to the "social formation of contemporaries who do not magnetize bullets" (Wilderson, "Prison Slave" 20).

Finally, there is a fundamental difference between a grammar and its critique, on the one hand, and literary representation *as the critique of* this grammar, on the other. *A Mercy* embodies the task and challenge of becoming the means by which to provide the critique of the New World grammar of private property and Atlantic slavery. As such, it offers a response to Hartman's question and imperative not only to tell that which is impossible to tell but to also remain alert to "the incommensurability between the experience of the enslaved and the fictions of history, by which I mean the requirements of narrative, the stuff of subjects and plots and ends" ("Venus" 10). Within the realm of literary representation, *A Mercy* insists on and allows for these contradictions to coexist and to disseminate. It is in the imperious space of Toni Morrison's writing (to paraphrase Namwali Serpell), in other words, that theoretical critique, historical writing, and narrative form converge and thrive. It is here that the reader is cautioned/reassured "not to be afraid" of the enormous challenge they will meet on the pages that follow. It is here that *A Mercy*'s form becomes both the refusal and the argument.

Works Cited

Adams, Tim. "Return of the Visionary." Review of *A Mercy*, by Toni Morrison. *The Guardian*, 26 Oct. 2008, www.guardian.co.uk/books/2008/oct/26/mercy-toni-morrison. Accessed 27 Aug. 2011.

Adusei-Poku, Nana. "Everyone Has to Learn Everything or Emotional Labor Rewind." *Allianzen/Alliances: Kritische Praxis an Weißen Institutionen*, edited by Julian Warner, Elisa Liepsch, and Matthias Pees, Bielefeld, transcript, 2018, pp. 34–49.

"African American Spirituals." *Library of Congress*, www.loc.gov/item/ihas.200197495/. Accessed 21 Dec. 2019.

"Afro-Pessimism: An Introduction." *Racked & Dispatched*, February 2017, rackedanddispatched.noblogs.org/files/2017/01/Afro-Pessimism2.pdf.

Ahmed, Sara. "Melancholic Universalism." 15 Dec. 2015. *Feministkilljoys Blog*, www.feministkilljoys.com/2015/12/15/melancholic-universalism/. Accessed 22 Nov. 2019.

Ahmed, Sara. *The Cultural Politics of Emotion*. Edinburgh UP, 2014.

Ahmed, Sara. "A Phenomenology of Whiteness." *Feminist Studies*, vol. 8, no. 2, 2007, pp. 149–168.

Alber, Jan, and Monika Fludernik. *Postclassical Narratology: Approaches and Analyses*. The Ohio State UP, Columbus, 2010.

Allen, Paula Gunn. "Pocahontas to Her English Husband, John Rolfe." *Life Is a Fatal Disease: Collected Poems 1962–1995*. West End Press, 1997.

Anderson, Benedict R. *Imagined Communities*. Verso, 1990.

Andres, Emmanuelle. "Reading/Writing "The Most Wretched Business": Toni Morrison's *A Mercy*." *Black Studies Papers*, vol. 1, no. 1, 2014, pp. 91–104.

Andrews, William L. "Harriet A. Jacobs (Harriet Ann), 1813–1897." University of Chapel Hill, 2000. *Documenting the American South*, www.docsouth.unc.edu/fpn/jacobs/bio.html. Accessed 5 Jan. 2020.

"anticipate." *New Oxford American Dictionary*, edited by Angus Stevenson and Christine A. Lindberg. 3rd ed., Oxford UP, 2010, p. 68.

"anticipate, v." *Oxford English Dictionary*, Oxford UP, 2019, www-1oed-1com-10066f0dh00e2.erf.sbb.spk-berlin.de/view/Entry/8552?rskey=FxzIQW&result=3&isAdvanced=false#eid. Accessed 10 Jan. 2019.

Anolik, Ruth B. "Haunting Voices, Haunted Text: Toni Morrison's *A Mercy*." *21st-Century Gothic: Great Gothic Novels since 2000*, edited by Daniel Olson, Scarecrow, 2011, pp. 418–433.

Applebaum, Barbara. "Critical Whiteness Studies." *Oxford Research Encyclopedia of Education*, doi:10.1093/acrefore/9780190264093.013.5. Accessed 06 Aug. 2018.

Aristotle. "Poetics." *The Norton Anthology of Theory and Criticism*, edited by Vincent B. Leitch. Norton, 2001, pp. 88–118.

Armitage, David. "John Locke, Carolina, and the 'Two Treatises of Government.'" *Political Theory*, vol. 32, no. 5, 2004, pp. 602–627. *JSTOR*, www.jstor.org/stable/4148117. 18 Jan. 2019.

Arneil, Barbara. *John Locke and America: The Defence of English Colonialism*. Claredon Press, 1998.

Arneil, Barbara. "Trade, Plantations, and Property: John Locke and the Economic Defense of Colonialism." *Journal of the History of Ideas*, vol. 55, no. 4, 1994, pp. 591–609.

Ashcraft, Richard. *Revolutionary Politics & Locke's "Two Treatises of Government"*. Princeton UP, 1986.

Babb, Valerie M. "*E Pluribus Unum?:* The American Origins Narrative in Toni Morrison's *A Mercy*." *MELUS: Multi-Ethnic Literature of the U.S.*, vol. 36, no. 2, 2011, pp. 147–164.

Bachmann-Medick, Doris. "From Hybridity to Translation: Reflections on Travelling Concepts." *The Trans/National Study of Culture: A Translational Perspective*, edited by Doris Bachmann-Medick, De Gruyter, 2014, pp. 119–136.

Baillie, Justine J. *Toni Morrison and Literary Tradition: The Invention of an Aesthetic*. Bloomsbury Academic, 2013.

Baker, Houston A., Jr. and K. Merinda Simmons, editors. *The Trouble with Post-Blackness*. Columbia UP, 2015.

Bal, Mieke. *Travelling Concepts in the Humanities: A Rough Guide*. U of Toronto P, 2012.

Bal, Mieke. "From Cultural Studies to Cultural Analysis." *Kritische Berichte: Zeitschrift Für Kunst- Und Kulturwissenschaften*, vol. 35, no. 2, 2007, pp. 33–44.

Banner, Stuart. *American Property: A History of How, Why, and What We Own*. Harvard UP, 2011.

Bartley, Aryn. "'My Telling Can't Hurt You'": Teaching Toni Morrison's *A Mercy* in a Survey of American Literature." *Teaching American Literature: A Journal of Theory and Practice*, vol. 5, no. 3/4, 2012, pp. 16–28.

Bassard, Katherine C. "'And the Greatest of These': Toni Morrison, the Bible, Love." *Toni Morrison: Memory and Meaning*, edited by Adrienne L. Seward, and Justine Tally, UP of Mississippi, 2014, pp. 119–131.

Bear, Charla. "American Indian Boarding Schools Haunt Many." NPR Morning Edition, 12 May 2008. *NPR*, www.npr.org/templates/story/story.php?storyId=16516865&t=1576773536099. Accessed 19 Dec. 2019.

Beaulieu, Elizabeth A. *Black Women Writers and the American Neo-Slave Narrative: Femininity Unfettered*. Vol. 192, Greenwood Press, 1999.

Beckles, Hilary M. *A History of Barbados: From Amerindian Settlement to Caribbean Single Market*. Cambridge UP, 2006.

Bell, Bernard W. *The Afro-American Novel and its Tradition*. U of Massachusetts P, 1987.

Bell, Derrick A. J. "Property Rights in Whiteness: Their Legal Legacy, their Economic Costs." *Critical Race Theory: The Cutting Edge*, edited by Richard Delgado and Jean Stefancic, Temple UP, 2013, pp. 63–70.

Bellamy, Maria R. "'These Careful Words ... Will Talk to Themselves': Textual Remains and Reader Responsibility in Toni Morrison's *A Mercy*." *Contested Boundaries: New Critical Essays on the Fiction of Toni Morrison*, edited by Maxine L. Montgomery, Cambridge Scholars, 2013, pp. 14–32.

"belonging, n." *Oxford English Dictionary*, Oxford UP, 2018, www.oed.com.331745941.erf.sbb.spk-berlin.de/view/Entry/17508?rskey=VDVjWb&result=2&isAdvanced=false#eid. Accessed 24 Jan. 2018.

Benjamin, Walter. *The Origin of German Tragic Drama*. 1928. Verso, 2009.

Bennett, Joshua. *Being Property Once Myself: Blackness and the End of Man*. The Belknap P of Harverd UP, 2020.

Berg, Manfred. "Disenfranchisement: The Political System of White Supremacy." *American Studies Journal*, vol. 45, 2000, pp. 10–17.

Bernasconi, Robert, and Sybol Cook. *Race and Racism in Continental Philosophy*. Indiana UP, 2003.
Bernasconi, Robert, and Anika M. Mann. "The Contradictions of Racism: Locke, Slavery, and the *Two Treatises*." *Race and Racism in Modern Philosophy*, edited by Andrew Valls, Cornell UP, 2005, pp. 89–107.
Best, Stephen M. *None Like Us*. Duke UP, 2018.
Best, Stephen M. "On Failing to Make the Past Present." *Modern Language Quarterly*, vol. 73, no. 3, 2012, pp. 453–474.
Best, Stephen M. *The Fugitive's Properties: Law and the Poetics of Possession*. U of Chicago P, 2004.
Bever, Lindsey. "I'm just a Sociopath," Dylann Roof Declared After Deadly Church Shooting Rampage, Court Records Say." The Washington Post, 17 May 2017. *Washington Post*, www.washingtonpost.com/news/post-nation/wp/2017/05/17/im-just-a-sociopath-dylann-roof-declared-after-deadly-church-shooting-spree-court-records-say/. Accessed 9 Dec. 2019.
Bhandar, Brenna. *Colonial Lives of Property*. Duke UP, 2018.
Bhandar, Brenna. "Critical Legal Studies and the Politics of Property." *Property Law Review*, vol. 3, 2014, pp. 186–194.
Bhandar, Brenna. "Disassembling Legal Form: Ownership and the Racial Body." *New Critical Legal Thinking: Law and the Political*, edited by Matthew Stone, Illan R. Wall, and Costas Douzinas, Routledge, 2012, pp. 112–127.
Bhandar, Brenna, and Davina Bhandar. "Cultures of Dispossession: Critical Reflections on Rights, Status and Identities." *Darkmatter: In the Ruins of Imperial Culture*, vol. 14, 2016, www.darkmatter101.org/site/2016/05/16/cultures-of-dispossession/. Accessed 5 Jan. 2020.
Birk, Hanne, and Birgit Neumann. "Go-between: Postkoloniale Erzähltheorie." *Neue Ansätze in der Erzähltheorie*, edited by Ansgar Nünning, and Vera Nünning, WVT, Trier, 2002, pp. 115–152.
Blackburn, Robin. *The Making of New World Slavery: From the Baroque to the Modern 1492–1800*. Verso, 1997.
Blumberg, Jess. "A Brief History of the Salem Witch Trials." 23 Oct. 2007. *Smithsonianmag*, www.smithsonianmag.com/history/a-brief-history-of-the-salem-witch-trials-175162489/. Accessed 5 Jan. 2020.
Bobo, Jacqueline, Cynthia Hudley, and Claudine Michel, editors. *The Black Studies Reader*. Routledge, 2004.
Boesenberg, Eva. "Reconstructing "America": The Development of African American Studies in the Federal Republic of Germany." *Germans and African Americans*, edited by Larry A. Greene, and Anke Ortlepp, UP of Mississippi, 2011, pp. 218–230.
Bok, Sissela. *Secrets: On the Ethics of Concealment and Revelation*. Pantheon Books, 1982.
Bontemps, Arna W. *Great Slave Narratives*. Beacon, 1972.
Bradford, William. *Of Plymouth Plantation (1620–1647)*. Mcgraw Hill Book Co, 1981.
Brand, Dionne. *Love Enough*. Vintage Canada, 2014.
Broeck, Sabine. *Gender and the Abjection of Blackness*. SUNY P, 2018.
Broeck, Sabine. "Legacies of Enslavism and White Abjectorship." *Postcoloniality—Decoloniality—Black Critique: Joints and Fissures*, edited by Sabine Broeck, and Carsten Junker, Campus Verlag, 2014, pp. 109–128.

Broeck, Sabine. "The Challenge of Black Feminist Desire: Abolish Property." *Black Intersectionalities: A Critique for the 21st Century*, edited by Monica Michlin, and Jean-Paul Rocchi, Liverpool UP, 2013, pp. 211–224.

Broeck, Sabine. "Lessons for A-Disciplinarity—Some Notes on What Happens to an Americanist When She Takes Slavery Seriously." *Postcolonial Studies Across the Disciplines*, edited by Jana Gohrisch, and Ellen Grünkemeier, Amsterdam, Rodopi, 2013, pp. 349–357.

Broeck, Sabine. "Property: White Gender and Slavery." *Gender Forum: An Internet Journal for Gender Studies*, vol. 14, 2006, genderforum.org. Accessed 15 Dec. 2019.

Broeck, Sabine. "Trauma, Agency, Kitsch and the Excesses of the Real: Beloved within the Field of Critical Response." *America in the Course of Human Events*, edited by Joself Jarab, Marcel Arbeit, and Jenel Virden, VU UP, Amsterdam, 2006, pp. 201–215.

Broeck, Sabine. "Never Shall We *Be* Slaves: Locke's Treatises, Slavery, and Early European Modernity." *Blackening Europe: The African American Presence*, edited by Heike Raphael-Hernández, Routledge, 2004, pp. 235–248.

Brooks, Peter. *Troubling Confessions: Speaking Guilt in Law and Literature*. U of Chicago P, 2000.

Brooks, Peter, and Paul Gerwitz. *Law's Stories: Narrative and Rhetoric in the Law*, Yale UP, 1998.

Bross, Kristina. "Florens in Salem." *Early American Literature*, vol. 48, no. 1, 2013, pp. 183–188. *Project MUSE*, doi:10.1353/eal.2013.0014. Accessed 15 Jan. 2019.

Buchanan, Ian. *A Dictionary of Critical Theory*. Oxford UP 2010.

Buck-Morss, Susan. "Hegel and Haiti." *Critical Inquiry*, vol. 26, no. 4, 2000, pp. 821–865.

Buckle, Stephen. "Tully, Locke and America." *British Journal for the History of Philosophy*, vol. 9, no. 2, 2001, pp. 245–281.

Buinicki, Martin T. *Negotiating Copyright: Authorship and the Discourse of Literary Property Rights in Nineteenth-Century America*. Routledge, 2006.

Butler, Judith, and Athēna Athanasiou. *Dispossession: The Performative in the Political*. Polity Press, 2013.

Campt, Tina M. *Listening to Images*. Duke UP, 2017.

Campt, Tina M. "Introduction to In the Wake: A Salon in Honor of Christina Sharpe." *YouTube*, uploaded by Barnard Center for Research on Women, 7 Feb 2017, www.youtube.com/watch?v=DGE9oiZr3VM. Accessed 17 Mar. 2018.

Campt, Tina M. *Image Matters*. Duke UP, 2012.

Cantiello, Jessica W. "From Pre-Racial to Post-Racial?: Reading and Reviewing *A Mercy* in the Age of Obama." *MELUS: Multi-Ethnic Literature of the U.S.*, vol. 36, no. 2, 2011, pp. 165–183.

Carby, Hazel V. "White Woman Listen! Black Feminism and the Boundaries of Sisterhood." *Black British Cultural Studies: A Reader*, edited by Houston A. Baker Jr., Manthia Diawara, and Ruth H. Lindeborg, The U of Chicago P, 1996, pp. 110–128.

Carey, Barbara. "Morrison's Call to Conscience." Review of *A Mercy*, by Toni Morrison. 16 Nov. 2008. *The Star*, www.thestar.com/article/537678. Accessed 27 Aug. 2011.

Carey, Brycchan. *From Peace to Freedom: Quaker Rhetoric and the Birth of American Antislavery, 1657–1761*. Yale UP, 2012.

Carey, Brycchan. "Inventing a Culture of Anti-Slavery: Pennsylvanian Quakers and the Germantown Protest of 1688." *Imagining Transatlantic Slavery*, edited by Cora Kaplan, and John R. Oldfield, Palgrave Macmillan, 2010, pp. 17–32.

Carlacio, Jami. "Narrative Epistemology: Storytelling as Agency in *A Mercy*." *Toni Morrison: Paradise, Love, A Mercy*, edited by Lucille P. Fultz, Bloomsbury, 2013, pp. 129–146.

Carretta, Vincent. "Olaudah Equiano: African British Abolitionist and Founder of the African American Slave Narrative." *The Cambridge Companion to the African American Slave Narrative*, edited by Audrey A. Fisch, Cambridge UP, 2007, pp. 44–60.

Chappell, Vere C. *The Cambridge Companion to Locke*. Cambridge UP, 1994.

Charles, Ron. "Souls in Chains." Review of *A Mercy*, by Toni Morrison. 9 Nov. 2008. *The Washington Post*, www.washingtonpost.com/wp-dyn/content/article/2008/11/06/AR2008110602817_pf.html. Accessed 27 Aug. 2011.

Chloé Taylor. *The Culture of Confession from Augustine to Foucault*. Routledge, 2009.

Cho, Sumi. "Post-Racialism." *Iowa Law Review*, vol. 94, no. 5, 2009, pp. 1589–1649.

Cholant, Gonçalo. "Americanity and Resistance in *A Mercy*, by Toni Morrison." *Op. Cit.: A Journal of Anglo-American Studies*, vol. 2, no. 2, 2016, pp. 1–13.

Christian, Barbara. "The Race for Theory." *Cultural Critique*, vol. 6, 1987, pp. 51–63.

Christiansë, Yvette. *Toni Morrison: An Ethical Poetics*. Fordham UP, 2013.

Cillerai, Chiara. "One Question is Who is Responsible? Another is can You Read?" Reading and Responding to Seventeenth-Century Texts using Toni Morrison's Historical Reconstructions in *A Mercy*." *Early American Literature*, vol. 48, no. 1, 2013, pp. 178–183. *Project MUSE*, doi:10.1353/eal.2013.0010. Accessed 4 Apr. 2017.

Cillerai, Chiara. "Introduction." *Early American Literature*, vol. 48, no. 1, 2013, pp. 177–178. *Project MUSE*, doi:10.1353/eal.2013.0006. Accessed 4 Apr. 2017.

Clifford, James. "Traveling Cultures." *Routes: Travel and Translation in the Late Twentieth Century*. Harvard UP, 1997, pp. 17–46.

Clymer, Jeffory. *Family Money: Property, Race, and Literature in the Nineteenth Century*. Oxford UP, 2012.

Conner, Marc C. ""What Lay Beneath the Names": The Language and Landscapes of *A Mercy*." *Toni Morrison: Paradise, Love, A Mercy*, edited by Lucille P. Fultz, Bloomsbury, London, 2013, pp. 147–165.

Conner, Marc C. "Modernity and the Homeless: Toni Morrison and the Fictions of Modernism." *Toni Morrison: Memory and Meaning*, edited by Adrienne L. Seward, and Justine Tally, UP of Mississippi, 2014, pp. 19–32.

Coombe, Rosemary J. *Cultural Life of Intellectual Properties: Authorship, Appropriation and the Law*. Duke UP, 1998.

Cox, Sandra. ""Mother Hunger": Trauma, Intra-Feminine Identification, and Women's Communities in Toni Morrison's *Beloved*, *Paradise*, and *A Mercy*." *Contested Boundaries: New Critical Essays on the Fiction of Toni Morrison*, edited by Maxine L. Montgomery, Cambridge Scholars, 2013, pp. 96–125.

Curtis, Susan. "History, Fiction, Imagination, and *A Mercy*." *Early American Literature*, vol. 48, no. 1, 2013, pp. 188–193. *JSTOR*, www.jstor.org/stable/24476313. Accessed 4 Apr. 2017.

Davies, Margaret M. *Property: Meanings, Histories, Theories*. Routledge-Cavendish, 2007.

Davis, Adrienne D. "Slavery and the Roots of Sexual Harassment." *Directions in Sexual Harassment Law*, edited by Catharine A. MacKinnon, and Reva B. Siegel, Yale UP, 2004, pp. 457–477.

Davis, Adrienne D. "Don't Let Nobody Bother Yo' Principle": The Sexual Economy of American Slavery." *Sister Circle: Black Women and Work*, edited by Sharon Harley, New Rutgers UP, 2002, pp. 103–127.

Davis, Adrienne D. "The Private Law of Race and Sex: An Antebellum Perspective." *Stanford Law Review*, vol. 51, no. 2, Jan. 1999, pp. 221–288.

Davis, Angela. "Reflections on the Black Woman's Role in the Community of Slaves." *The Black Scholar*, vol. 3, no. 4, 1971, pp. 2–15.

Davis, David Brion. *The Problem of Slavery in Western Culture*. Cornell UP, 1970.

Dayan, Colin. "Legal Terrors." *Representations*, vol. 92, no. 1, 2005, pp. 42–80.

Deane, Charles. Preface. *A True Relation of Virginia,* by Smith. Boston, Wiggin and Lunt, 1866, pp. ix–xlvii. *Hathitrust*, babel.hathitrust.org/cgi/pt?id=loc.ark:/13960/t6n01rk1x&view=1up&seq=13. Accessed 4 Apr. 2017.

de Groot, Jerome. *The Historical Novel*. Routledge, 2010.

Dimock, Wai C. *Residues of Justice: Literature, Law, Philosophy*. U of California P, 1996.

"dispossession, n." *Oxford English Dictionary*, Oxford UP, 2019, www-1oed-1com-10066f0dh00f3.erf.sbb.spk-berlin.de/view/Entry/55133?redirectedFrom=dispossession#eid. Accessed 9 Dec. 2019.

"Doc Netzwerke." *Universität Bremen*, 2019, www.uni-bremen.de/byrd/promovierende/doc-netzwerke/. Accessed 15 Dec. 2019.

Dolin, Kieran, editor. *Law and Literature*. Cambridge UP, 2018.

Donahue, Deirdre. "Slavery of a Different Sort Toils in Toni Morrison's 'A Mercy'." Review of *A Mercy*, by Toni Morrison. 14 Nov. 2008. *USA Today*, www.usatoday.com/life/books/reviews/2008–11–12-morrison-mercy_N.htm#. Accessed 27 Aug. 2011.

Douglass, Frederick. "Narrative of the Life of Frederick Douglass, an American Slave, Written by Himself." *The Norton Anthology of African American* Literature, edited by Henry Louis Gates, Jr. and Nellie Y. McKay, Norton, 2004, pp. 385–482.

Douglass, Patrice D. "Black Feminist Theory for the Dead and Dying." *Theory and Event*, vol. 21, no. 1, 2018, pp. 106–123.

Douglass, Patrice D. "The Claim of Right to Property: Social Violence and Political Right." *Zeitschrift für Anglistik und Amerikanistik*, vol. 65, no. 2, 2017, pp. 145–159.

Douglass, Patrice D., and Frank B. Wilderson III. "The Violence of Presence." *The Black Scholar*, vol. 43, no. 4, 2013, pp. 117–123.

Drake, Kimberly. "Rewriting the American Self: Race, Gender, and Identity in the Autobiographies of Frederick Douglass and Harriet Jacobs." *Critical Insights: The Slave Narrative*, edited by Kimberly Drake, Grey House, 2014, pp. 43–64.

DuCille, Ann. "Of Race, Gender, and the Novel; or, Where in the World Is Toni Morrison?" *Novel: A Forum on Fiction*, vol. 50, no. 3, 2017, pp. 357–378.

Dunn, John. *The Political Thought of John Locke: An Historical Account of the Argument of the 'Two Treatises of Government.'* Cambridge UP, 1969.

Dussel, Enrique, and Eduardo Mendieta. *The Underside of Modernity: Apel, Ricoeur, Rorty, Taylor, and the Philosophy of Liberation*. Humanity Books, 1999.

Eaton, Alice. "Becoming a She Lion: Sexual Agency in Toni Morrison's Beloved and *A Mercy*." *Contested Boundaries: New Critical Essays on the Fiction of Toni Morrison*, edited by Maxine L. Montgomery, Cambridge Scholars, 2013, pp. 53–66.

Eder, Jens, Fotis Jannidis, and Ralf Schneider. "Characters in Fictional Worlds: An Introduction." *Revisionen: Characters in Fictional Worlds: Understanding Imaginary*

Beings in Literature, Film, and Other Media, edited by Jens Eder, Fotis Jannidis, and Ralf Schneider, De Gruyter, 2011, pp. 3–66.
Ely, James W. *The Guardian of Every Other Right*. 3rd ed. Oxford UP, 2008.
Emerson, Cheryl A. "My Skin is Black Upon Me": Toni Morrison's *A Mercy* and the Question of a Female Job." *South Atlantic Review*, vol. 82, no. 2, Summer, 2017, pp. 12–23.
Equiano, Olaudah. "The Interesting Narrative of the Life of Olaudah Equiano, Or Gustavus Vassa, the African, Written by Himself." *The Norton Anthology of African American Literature*, edited by Henry Louis Gates, Jr. and Nellie Y. McKay, Norton, 2004, pp. 189–212.
Essi, Cedric, Stephen Koetzing, Paula von Gleich, Samira Spatzek, and Gesine Wegner. "*COPAS* at Twenty: Interrogating White Supremacy in the United States and Beyond." *COPAS*, vol. 20, no. 2, 2019: pp. 1–17.
Euchner, Walter. *John Locke Zur Einführung*. Junius, Hamburg, 2011.
Eze, Emmanuel C. *Race and the Enlightenment: A Reader*. Blackwell, 1997.
Fanon, Frantz. *The Wretched of the Earth*. Translated by Richard Philcox, Grove Press, 2008.
Fanon, Frantz. *Black Skin, White Masks*. Translated by Richard Philcox, Grove Press, 2008.
Farr, James. "So Vile and Miserable an Estate": The Problem of Slavery in Locke's Political Thought." *Political Theory*, vol. 14, no. 2, 1986, pp. 263–290.
Federici, Silvia. *Caliban and the Witch*. Autonomedia, New York, 2009.
Fields, Barbara J. "Slavery, Race, and Ideology in the United States of America." *Racecraft: The Soul of Inequality in American Life*, edited by Karen E. Fields, and Barbara J. Fields, Verso, 2012, pp. 111–148.
Fisch, Audrey A., editor. *The Cambridge Companion to the African American Slave Narrative*. Cambridge UP, 2007.
Fischer, Sibylle. "Atlantic Ontologies: On Violence and being Human." *E-Misférica: Carribean Rasanblaj*, vol. 12, no. 1., 2015. hemisphericinstitute.org/hemi/es/emisferica-121-caribbean-rasanblaj/fischer. Accessed 16 Nov. 2016.
Fludernik, Monika. "The Genderization of Narrative." *GRAAT 21: Recent Trends in Narratological Research: Papers from the Narratology Round Table*, edited by John Pier, Publications des Groupes de Recherches Anglo-Américaines de l'Université François Rabelais de Tours, 1999, pp. 153–175.
Fludernik, Monika. "Histories of Narrative Theory (II): From Structuralism to the Present." *A Companion to Narrative Theory*, edited by James Phelan, and Peter J. Rabinowitz, Wiley-Blackwell, 2008, pp. 36–59.
Forster, E. M. *Aspects of the Novel and Related Writings*. Arnold, London, 1974.
Foster, Dennis A. *Confession and Complicity in Narrative*. Cambridge UP, 1987.
Fox-Genovese, Elizabeth. *Within the Plantation Household: Black and White Women of the Old South*. U of North Carolina P, 2006.
Franklin, Ruth. "Enslavements." Review of *A Mercy*, by Toni Morrison. 24 Dec. 2008. *New Republic*, newrepublic.com/article/62123/enslavements. Accessed 14 Nov. 2017.
Freeman, Judith. "Souls in Search of Freedom." Review of *A Mercy*, by Toni Morrison. 16 Nov. 2008, *Los Angeles Times*, articles.latimes.com/print/2008/nov/16/entertainment/ca-toni-morrison16. Accessed 27 Aug. 2011.
Fuentes, Marisa J. *Dispossessed Lives: Enslaved Women, Violence, and the Archive*. U of Pennsylvania P, 2016.

Fulton, DoVeanna S. *Speaking Power: Black Feminist Orality in Women's Narratives of Slavery*. SUNY P, 2006.

Fultz, Lucille P. *Toni Morrison: Paradise, Love, A Mercy*. Bloomsbury, 2013.

Gallagher, Catherine, and Stephen Greenblatt. "Counterhistory and the Anecdote." *Practicing New Historicism*, U Chicago P, 2001.

Gallego-Durán, Mar. "Newness Trembles Me"? Representations of White Masculinity in Toni Morrison's *A Mercy*." *Toni Morrison: Memory and Meaning*, edited by Adrienne L. Seward, and Justine Tally, UP of Mississippi, 2014, pp. 243–254.

Gallego-Durán, Mar. "Nobody Teaches You to Be a Woman": Female Identity, Community and Motherhood in Toni Morrison's *A Mercy*." *Toni Morrison's A Mercy: Critical Approaches*, edited by Shirley A. Stave and Justine Tally. Cambridge Scholars, 2011, pp. 103–118.

Gates, David. "Original Sins." Review of *A Mercy*, by Toni Morrison. 28 Nov. 2008. *The New York Times*, www.nytimes.com/2008/11/30/books/review/Gates-t.html?pagewanted=all&_r=0. Accessed 8 Jan. 2014.

Gates, Henry L., and Jennifer Burton, editors. *Call and Response: Key Debates in African American Studies*. Norton, 2010.

Genette, Gérard. *Palimpseste: Die Literatur auf Zweiter Stufe*. Translated by Wolfram Bayer, and Dieter Hornig, Frankfurt am Main, Suhrkamp, 2004.

"Germantown Friends' Protest Against Slavery, 1688." *The Quaker Origins of Antislavery*, edited by J. W. Frost. Norwood Editions, 1980.

Gillespie, Carmen. *Toni Morrison: Forty Years in the Clearing*. Bucknell UP, 2015.

Gilroy, Paul. *The Black Atlantic: Modernity and Double Consciousness*. Verso, 1999.

Glausser, Wayne. "Three Approaches to Locke and the Slave Trade." *Journal of the History of Ideas*, vol. 51, no. 2, 1990, pp. 199–216.

Goad, Jill. "Enslaved by Mother and Lover: Florens' Impossible Search for Self-Love in Toni Morrison's *A Mercy*." *New Academia: An International Journal of English Language Literature and Literary Theory*, vol. 3, no. 2, 2014, pp. 1–6.

Gordon, Avery F. *Ghostly Matters: Haunting and the Sociological Imagination*. U of Minnesota P, 2008.

Gordon, Jane A., and Lewis Gordon, editors. *A Companion to African-American Studies*. Vol. 11, Wiley-Blackwell, 2008.

Gordon, Jane A., and Lewis Gordon, editors. "Introduction: On Working Through a Most Difficult Terrain." *A Companion to African-American Studies*, edited by Lewis R. Gordon, and Anna J. Gordon, Vol. 11, Wiley-Blackwell, 2008, xx–xxxiv.

Gould, Philip. "The Rise, Development, and Circulation of the Slave Narrative." *The Cambridge Companion to the African American Slave Narrative*, edited by Audrey A. Fisch, Cambridge UP, 2007, pp. 11–27.

Graeber, David. "Manners, Deference, and Private Property: Or, Elements for a General Theory of Hierarchy." *Possibilities: Essays on Hierarchy, Rebellion, and Desire*, AK Press, 2007, 13–55.

Greene, Jack P. *Pursuits of Happiness: The Social Development of Early Modern British Colonies and the Formation of American Culture*. U of North Carolina P, 1988.

Greeson, Jennifer R. "American Enlightenment: The New World and Modern Western Thought." *American Literary History*, vol. 25, no. 1, 2013, pp. 6–17. *Project MUSE*, muse.jhu.edu/article/496738. Accessed 2 Jan. 2014.

Greeson, Jennifer R. "The Prehistory of Possessive Individualism." *PMLA*, vol. 127, no. 4, 2012, pp. 918–924.

Grewal, Gurleen. "*A Mercy* (Review)." *MELUS: Multi-Ethnic Literature of the U.S.*, vol. 36, no. 2, 2011, pp. 191–193. *JSTOR*, http://www.jstor.org/stable/23035289. Accessed 2 Jan. 2014.

Grosfoguel, Ramón. "The Structure of Knowledge in Westernized Universities, epistemic Racism/Sexism and the Four Genocides/Epistemicides of the Long 16th Century." *Human Architecture: Journal of the Sociology of Self-Knowledge*, vol. XI, no. 1, 2013, pp. 73–90.

Guasco, Michael. "The Fallacy of 1619: Rethinking the History of Africans in Early America." *AAIHS*, 4 Sep. 2017, www.aaihs.org/the-fallacy-of-1619-rethinking-the-history-of-africans-in-early-america/. Accessed 2 Feb. 2019.

Gura, Philip F. "Puritan Origins." *A Concise Companion to American Studies*, edited by John C. Rowe, Wiley-Blackwell, 2010, pp. 19–35.

Gustafson, Sandra M., and Gordon Hutner. "Projecting Early American Literary Studies: Introduction." *American Literary History*, vol. 22, no. 2, 2010, pp. 211–216. *JSTOR*, www.jstor.org/stable/27856614. Accessed 2 Jan. 2014.

Hahn, Johannes. *Der Begriff des "Property" bei John Locke: Zu den Grundlagen Seiner Politischen Philosophie*. Peter Lang, 1984.

Hall, Catherine. "Gendering Property, Racing Capital." *History Workshop Journal*, vol. 78, 2014, pp. 22–38. *Project MUSE*, doi:10.1093/hwj/dbu024. Accessed 30 Oct. 2016.

Hallet, Wolfgang. "Methoden Kulturwissenschaftlicher Ansätze: *Close Reading* und *Wide Reading*." *Methoden Der Literatur- Und Kulturwissenschaftlichen Textanalyse: Ansätze—Grundlagen—Modellanalysen*, edited by Bauder-Begerow, Irina, Vera Nünning, and Ansgar Nünning, Metzler, Stuttgart, 2010, pp. 293–315.

Hannah-Jones, Nikole. "Our Founding Ideals of Liberty and Equality Were False When They Were Written." *The 1619 Project*, edited by Nikole Hannah-Jones. Special issue of *The New York Times Magazine*, 18 Aug. 2019, pp. 14–26.

Hannah-Jones, Nikole, editor. *The 1619 Project*, special issue of *The New York Times Magazine*, 18 Aug. 2019.

Harley, Sharon, and Rosalyn Terborg-Penn. *The Afro-American Woman: Struggles and Images*. Black Classic Press, Baltimore, 1978.

Harpham, Edward J. "Locke's Two Treatises in Perspective." *John Locke's Two Treatises of Government: New Interpretations*, edited by Edward J. Harpham, UP of Kansas, 1992, pp. 1–13.

Harris, Cheryl I. "The Afterlife of Slavery: Markets, Property and Race." *YouTube*, uploaded by Artists Space, 2 Mar. 2016, www.youtube.com/watch?v=dQQGndN3BvY.

Harris, Cheryl I. "Finding Sojourner's Truth: Race, Gender, and the Institution of Property." *Cardozo Law Review*, vol. 18, no. 309, 1997, pp. 309–409.

Harris, Cheryl I. "Whiteness as Property." *Harvard Law Review*, vol. 106, no. 8, 1993, pp. 1707–1791.

Hartman, Saidiya V. "The Belly of the World: A Note on Black Women's Labors." *Souls: A Critical Journal of Black Politics, Culture, and Society*, vol. 18, no. 1, 2016, pp. 166–173.

Hartman, Saidiya V. "Venus in Two Acts." *Small Axe*, vol. 12, no. 2, 2008, pp. 1–14.

Hartman, Saidiya V. *Lose Your Mother: A Journey Along the Atlantic Slave Route*. Farrar, Straus and Giroux, 2007.

Hartman, Saidiya V. *Scenes of Subjection: Terror, Slavery, and Self-Making in Nineteenth-Century America*. Oxford UP, 1997.
Hartman, Saidiya V., and Frank B. Wilderson, III. "The Position of the Unthought." *Qui Parle*, vol. 13, no. 2, 2003, pp. 183–201.
Haselstein, Ulla. "Die Gegen-Öffentlichkeit der Allegorie. Einleitung." *Allegorie. DFG-Symposium 2014*, edited by Ulla Haselstein, De Gruyter, 2016, pp. 335–353.
Haselstein, Ulla. "Vorbemerkungen der Herausgeberin." *Allegorie. DFG-Symposium 2014*, edited by Ulla Haselstein, De Gruyter, 2016, pp. ix–xvi.
Hejinian, Lyn. "Wild Captioning." *Qui Parle: Critical Humanities and Social Sciences*, vol. 20, no. 1, 2011, pp. 279–299.
Held, Susanne. *Eigentum und Herrschaft bei John Locke Und Immanuel Kant: Ein Ideengeschichtlicher Vergleich*, Lit, Berlin, 2006.
Heltzel, Ellen Emry. "Toni Morrison's Powerful New Novel 'A Mercy' Tracks, Examines Forces of Slavery." Review of *A Mercy*, by Toni Morrison. 06 Nov 2008. *The Seattle Times*, seattletimes.nwsource.com/html/books/2008355833_br09morrison.html. Accessed 27 Aug. 2011.
Herman, David. "Histories of Narrative Theory (I): A Genealogy of Early Developments." *A Companion to Narrative Theory*, edited by James Phelan, and Peter J. Rabinowitz, Wiley-Blackwell, 2008, pp. 19–35.
Herman, David. *Narratologies: New Perspectives on Narrative Analysis*. The Ohio State UP, 1999.
Hesse, Clara. "The Rise of Intellectual Property, 700 B.C.-A.D. 2002: An Idea in the Balance." *Daedalus—Journal of the American Academy of Arts and Sciences*, vol. 131, no. 2, 2002, pp. 26–45.
Hine, Darlene C. "Female Slave Resistance: The Economics of Sex." *Hine Sight: Black Women and the Re-Construction of American History*. Carlson, 1994, pp. 27–36.
Hine, Darlene C. "Rape and the Inner Lives of Black Women in the Middle West: Preliminary Thoughts on the Culture of Dissemblance." *Signs: Journal of Women in Culture and Society*, vol. 14, no. 4, 1989, pp. 912–920.
Hinshelwood, Brad. "The Carolinian Context of John Locke's Theory of Slavery." *Political Theory*, vol. 41, no. 3, 2013, pp. 562–590.
Hirschmann, Nancy J., and Kirstie M. MacClure. *Feminist Interpretations of John Locke*. Pennsylvania State UP, 2007.
Hohfeld, Wesley N. "Some Fundamental Legal Conceptions as Applied in Judicial Reasoning." *The Yale Law Journal*, vol. 23, no. 1, 1913, pp. 16–59.
Homestead, Melissa J. *American Woman Authors and Literary Property, 1822–1869*. Cambridge UP, 2005.
Hong, Grace K. "Property." *Keywords for American Cultural Studies*, edited by Bruce Burgett, and Glenn Hendle, New York UP, 2007, pp. 180–182.
Hull, Gloria T., Patricia B. Scott, and Barbara Smith. *All the Women are White, all the Blacks are Men, But some of Us are Brave: Black Women's Studies*. Feminist Press, Old Westbury, NY, 1982.
Hulme, Peter. "The Spontaneous Hand of Nature: Savagery, Colonialism, and the Enlightenment." *The Enlightenment and its Shadows*, edited by Peter Hulme, and Ludmilla J. Jordanova, Routledge, 1990, pp. 16–34.

Hulme, Peter. *Colonial Encounters: Europe and the Native Caribbean, 1492–1797*. Methuen, 1986.
"indenture, n." *Oxford English Dictionary*, Oxford UP, 2019, www-1oed-1com-10066f0nr0117.erf.sbb.spk-berlin.de/view/Entry/94314?rskey=vnNztW&result=1&isAdvanced=false#eid. Accessed 5 Jan. 2020.
Irr, Caren. "Literature as Proleptic Globalization, or a Prehistory of the New Intellectual Property." *The South Atlantic Quarterly*, vol. 100, no.3, 2001, pp. 773–802.
Jackson, Zakiyyah I. "Waking Nightmares: Zakiyyah Iman Jackson on David Marriott." *GLQ: A Journal of Lesbian and Gay Studies*, vol. 17, no. 2–3, 2011, pp. 357–363. *Project MUSE*, muse.jhu.edu/article/437423. Accessed 24 June 2019.
Jacobs, Harriet. *Incidents in the Life of a Slave Girl*. Dover Publications, 2001.
Jahn, Manfred. *Narratology: A Guide to the Theory of Narrative*. English Department, University of Cologne, 2017, www.uni-koeln.de/~ame02/pppn.htm#N7. Accessed 5 Jan. 2020.
James, Joy, editor. *The New Abolitionists: (Neo) Slave Narratives and Contemporary Prison Writings*. SUNY P, 2005.
Jannidis, Fotis. *Figur und Person: Beitrag zu Einer Historischen Narratologie*. De Gruyter, 2004.
Jehlen, Myra, and Michael Warner. *The English Literatures of America: 1500–1800*. Routledge, 1997.
Jennings, La Vinia D. "*A Mercy:* Toni Morrison Plots the Formation of Racial Slavery in Seventeenth-Century America." *Callaloo*, vol. 32, no. 2, 2009, pp. 645–649.
Jimenez, Teresa G. "They Hatch Alone: The Alienation of the Colonial Subject in Toni Morrison's *A Mercy.*" *Berkeley Undergraduate Journal*, vol. 22, no. 2, 2010, pp. 1–11.
Jones, Dalton A., et al. editors. *Black Holes: Afro-Pessimism, Blackness and the Discourses of Modernity*, special issue of *Rhizomes: Cultural Studies in Emerging Knowledge*, vol. 29, 2016. doi:10.20415/rhiz/029/. Accessed 4 Sep. 2017.
Jones-Rogers, Stephanie E. *They Were Her Property: White Women as Slave Owners in the American South*. Yale UP, 2019.
Kakutani, Michiko. "Bonds That Seem Cruel Can Be Kind." Review of *A Mercy*, by Toni Morrison. 4 Nov. 2008. *The New York Times*, www.nytimes.com/2008/11/04/books/04kaku.html?pagewanted=pr int. Accessed 27 Aug. 2011.
Karavanta, Mina. "Toni Morrison's *A Mercy* and the Counterwriting of Negative Communities: A Postnational Novel." *MFS Modern Fiction Studies*, vol. 58, no. 4, 2012, pp. 723–746. *Project MUSE*, doi:10.1353/mfs.2012.0068. Accessed 27 June 2014.
Kawash, Samira. "Freedom and Fugitivity: The Subject of Slave Narrative." *Dislocating the Color Line: Identity, Hybridity, and Singularity in African American Narrative*. Stanford UP, 1997, pp. 23–84.
Keizer, Lerne R. *Black Subjects: Identity Formation in the Contemporary Narrative of Slavery*. Cornell UP, 2004.
Kelleter, Frank. "American Literary History—Early American Literature". *English and American Studies: Theory and Practice*, edited by Martin Middeke et al., Verlag J.B. Metzler, 2012, pp. 101–110.
Kelly, Paul J. *Locke's Second Treatise of Government: A Reader's Guide*. Continuum, 2007.
Kelly, Paul J. "Reception and Influence." *Locke's Second Treatise of Government: A Reader's Guide*. Continuum, 2007, pp. 138–152.

King, Lovalerie. "Property and American Identity in Toni Morrison's *Beloved*." *Toni Morrison: Memory and Meaning*, edited by Adrienne L. Seward, and Justine Tally, UP of Mississippi, 2014, pp. 159–171.

King, Lovalerie. *Race, Theft, and Ethics: Property Matters in African American Literature*. Louisiana State UP, 2007.

King, Martin Luther, Jr. "I Have a Dream." *National Archives*, www.archives.gov/files/press/exhibits/dream-speech.pdf. Accessed 21 Dec. 2019.

Kirby, Sean M. "Naming and Identity in Toni Morrison's *Beloved* and *Song of Solomon*." *Inquiries Journal/Student Pulse*, vol. 6, no. 6, 2014, www.inquiriesjournal.com/a?id=904. Accessed 4 Sep. 2017.

Knopf, Kerstin. "A Benighted America?: Slavery, Convict Labor, and Chain Gangs in American Prison Literature." *Rural America*, edited by Antje Kley, and Heike Paul, Heidelberg, Winter, 2015, pp. 95–110.

Lanser, Susan S. *Fictions of Authority: Women Writers and Narrative Voice*. Cornell UP, 1992.

Lanser, Susan S. "Sexing Narratology: Toward a Gendered Poetics of Narrative Voice." *Grenzüberschreitungen: Narratologie im Kontext*, edited by Andreas Solbach, and Walter Grünzweig, Tübingen, Narr, 1999, pp. 167–184.

Lanser, Susan S. "Sexing the Narrative: Propriety, Desire, and the Engendering of Narratology." *Narrative*, vol. 3, no. 1, 1995, pp. 85–94.

Larkins, Sharon. "Using Trade Books to Teach About Pocahontas." *Georgia Social Sciences Journal*, vol. 19, no.1, 1988, pp. 21–25.

Laslett, Peter. "Introduction." *Two Treatises of Government*, edited by Peter Laslett, Cambridge UP, 1988, pp. 3–136.

Leisy, Ernest E. *The American Historical Novel*. U of Oklahoma P, 1970.

Levitas, Ruth. *The Concept of Utopia*. Philip Alan, 1990.

Lipsitz, George. *The Possessive Investment in Whiteness: How White People Profit from Identity Politics*. Temple UP, 1998.

Locke, John. *Two Treatises of Government*, edited by Peter Laslett, Cambridge UP, 1988.

Logan, Lisa M. "Thinking with Toni Morrison's *A Mercy* (A Response to "Remembering the Past: Toni Morrison's Seventeenth Century in Today's Classroom")." *Early American Literature*, vol. 48, no. 1, 2013, pp. 193–199. *Project MUSE*, doi:10.1353/eal.2013.0020. Accessed 4 Apr. 2017.

Lowe, Lisa. *The Intimacies of Four Continents*. Duke UP, 2015.

Luck, Chad. *The Body of Property: Antebellum American Fiction and the Phenomenology of Possession*. Fordham UP, 2014.

Lyles-Scott, Cynthia. "A Slave by Any Other Name: Names and Identity in Toni Morrison's *Beloved*." *Names: A Journal of Onomastics*, vol. 56, no. 1, March 2008, pp. 23–28.

Mackenthun, Gesa. *Metaphors of Dispossession: American Beginnings and the Translation of Empire, 1492–1637*. U of Oklahoma P, 1997.

Macpherson, Crawford B. *Property: Mainstream and Critical Positions*. Blackwell, 1978.

Macpherson, Crawford B. *The Political Theory of Possessive Individualism: Hobbes to Locke*. Clarendon Press, 1962.

Maier, Pauline. *Inventing America: A History of the United States*. Norton, 2006.

Malaklou, M. Shadee, and Tiffany Willoughby-Herard, editors. *Afro-Pessimism and Black Feminism*, special issue of *Theory & Event*, vol. 21, no. 1, 2018.

Malaklou, M. Shadee, and Tiffany Willoughby-Herard, editors. "Notes from the Kitchen, the Crossroads, and Everywhere Else, too: Ruptures of Thought, Word, and Deed from the "Arbiters of Blackness Itself." *Afro-Pessimism and Black Feminism*, special issue of *Theory & Event*, vol. 21, no. 1, 2018, pp. 2–67.

Mantel, Hilary. "How Sorrow Became Complete." Review of *A Mercy*, by Toni Morrison. 08 Nov. 2008. *The Guardian*, www.guardian.co.uk/books/2008/nov/08/a-mercy-toni-morrison. Accessed 21 Feb. 2012.

Marable, Manning, editor. *The New Black Renaissance: The Souls Anthology of Critical African American Studies*. Paradigm, 2005.

Margolin, Uri. "Character." *The Cambridge Companion to Narrative*, edited by David Herman, Cambridge UP, 2007, pp. 66–79.

Marriott, David. *On Black Men*, Edinburgh UP, 2000.

Martin, Valerie. *Property*. Vintage Contemporaries, 2003.

Mayberry, Susan N. "Visions and Revisions of American Masculinity in *A Mercy*." *Toni Morrison: Paradise, Love, A Mercy*, edited by Lucille P. Fultz, Bloomsbury, 2013, pp. 166–184.

McCartney, Martha. "John Smith (bap. 1580–1631)." *Encyclopediavirginia*, www.encyclopediavirginia.org/Smith_John_bap_1580–1631. Accessed 19 Dec. 2019.

McClintock, Anne. *Imperial Leather: Race, Gender and Sexuality in the Colonial Contest*. Routledge, 1995.

McDougall, T. Mars. "The Water is Waiting": Water, Tidalectics, and Materiality." *Liquid Blackness Journal*, vol. 6, no. 3, 2016, pp. 50–63.

McHale, Brian. "Ghosts and Monsters: On the (Im)Possibility of Narrating the History of Narrative Theory." *A Companion to Narrative Theory*, edited by Peter J. Rabinowitz, and James Phelan, Wiley-Blackwell, 2008, pp. 60–72.

Melton, Gene, II. "The (Neo)Slave Narrative in Black and White: Toni Morrison's Re-Envisioning of Masculinity in *A Mercy*." *Contested Boundaries: New Critical Essays on the Fiction of Toni Morrison*, edited by Maxine L. Montgomery, Cambridge Scholars, 2013, pp. 34–52.

Michlin, Monica. "Writing/Reading Slavery as Trauma: Othering, Resistance, and the Haunting Use of Voice in Toni Morrison's *A Mercy*." *Black Studies Papers*, vol. 1, no. 1, 2014, pp. 105–123.

Miller, Cheryl. "Mine, Mine, Mine." Review of *A Mercy*, by Toni Morrison. 3 Jan. 2009. *Commentary Magazine*, www.theatlantic.com/magazine/print/2009/01/mercy/7223/. Accessed 8 Nov. 2011.

Miller, Lee. *Roanoke: Solving the Mystery of the Lost Colony*. Penguin, 2000.

Miller, Perry. *Errand into the Wilderness*. Harvard UP, 1956.

Mills, Charles W. *Black Rights / White Wrongs: The Critique of Racial Liberalism*. Oxford UP, 2017.

Mills, Charles W. "White Ignorance." *Race and Epistemologies of Ignorance*, edited by Shannon Sullivan, and Nancy Tuana, State U of New York P, 2007, pp. 11–38.

Mills, Charles W. *The Racial Contract*. Cornell UP, 1997.

"minion, n." *Oxford English Dictionary*, Oxford UP, 2019, www-1oed-1com-10066f09l00a7.erf.sbb.spk-berlin.de/view/Entry/118859?rskey=auKkhf&result=1&isAdvanced=false#eid. Accessed 21 Dec. 2019.

Mishra, Pramid K. "[A]ll the World was America": The Transatlantic (Post)Coloniality of John Locke, William Bartram, and the Declaration of Independence." *The New Centennial Review*, vol. 2, no. 1, 2002, pp. 213–258.

Mitchell, Angelyn. *The Freedom to Remember: Narrative, Slavery, and Gender in Contemporary Black Women's Fiction*. Rutgers UP, 2002.

Monaghan, E. Jennifer. "Reading for the Enslaved, Writing for the Free: Reflections on Liberty and Literacy." American Antiquarian Society. *Proceedings of the American Antiquarian Society Volumes 108–111*. Vol. 108, Part 2, October 1998, www.americanantiquarian.org/proceedings/44525153.pdf.

Montgomery, Maxine L. *Contested Boundaries: New Critical Essays on the Fiction of Toni Morrison*. Cambridge Scholars, 2013.

Montgomery, Maxine L. "Got on My Traveling Shoes: Migration, Exile, and Home in Toni Morrison's *A Mercy*." *Journal of Black Studies*, vol. 42, no. 4, 2011, pp. 627–637.

Moore, Caroline. Review of *A Mercy*, by Toni Morrison. 14 Nov. 2008. *The Telegraph*, www.telegraph.co.uk/culture/books/fictionreviews/3563259/A-Mercy-by-Toni-Morrison-review.html. Accessed 8 Jan. 2014.

Moore, Geneva C. "A Demonic Parody: Toni Morrison's *A Mercy*." *Southern Literary Journal*, vol. 44, no. 1, 2011, pp. 1–18. Project MUSE, doi:10.1353/slj.2011.0014. Accessed 18 Apr. 2017.

Moore, Lindsay. "Women and Property Litigation in Seventeenth-Century England and North America." *Married Women and the Law: Coverture in England and the Common Law World*, edited by Tim Stretton, and Krista J. Kesselring, Queen's UP, 2013, pp. 113–138.

Morgan, Edmund S. *American Slavery, American Freedom: The Ordeal of Colonial Virginia*. Norton, 2003.

Morgan, Jennifer L. "Partus Sequitur Ventrem: Law, Race, and Reproduction in Colonial Slavery." *Small Axe*, vol. 55, 2018, pp. 1–17.

Morgan, Jennifer L. "Partus Sequitur Ventrem": Considering Slave Law and Re/Production for Enslaved Women." *YouTube*, uploaded by Mothering Slaves, 21 Apr. 2015, www.youtube.com/watch?v=YGEARpgIzEY.

Morgan, Jennifer L. "Archives and Histories of Racial Capitalism." *Social Text* 125, vol. 33, no. 4, 2015, pp. 153–161.

Morgan, Jennifer L. *Laboring Women: Reproduction and Gender in New World Slavery*. U of Pennsylvania P, 2004.

Morgan, Philip D. *Slave Counterpoint: Black Culture in the Eighteenth-Century Chesapeake and Lowcountry*, 1st ed, U of North Carolina P, 1998.

Morgenstern, Naomi. "Maternal Love/Maternal Violence: Inventing Ethics in Toni Morrison's *A Mercy*." *MELUS*, vol. 39, no. 1, Spring 2014, pp. 7–29. JSTOR, www.jstor.org/stable/24569889. Accessed 4 Apr. 2017.

Morrison, Toni. *A Mercy*. Chatto & Windus, 2008.

Morrison, Toni. *Playing in the Dark: Whiteness and the Literary Imagination*. Harvard UP, 1992.

Morrison, Toni. *Beloved*. 1987. Vintage, 2007.

Morrison, Toni. *Sula*. 1973. Vintage, 2005.

Morrison, Toni, and Carolyn C. Denard. *What Moves at the Margin: Selected Nonfiction*. UP of Mississippi, 2008.

Morrison, Toni. Interview by Michelle Norris and Robert Siegel. *All Things Considered*, 27 Oct. 2008. *NPR*, www.npr.org/templates/transcript/transcript.php?storyId=96118766. Accessed 5 Dec. 2017.

Morrison, Toni. "Toni Morrison On Bondage and A Post-Racial Age." Interview by Michel Martin. *Tell Me More*, 10 Dec. 2008. *NPR*, www.npr.org/templates/story/story.php?storyId=98072491&t=1577729926602. Accessed 30 Dec. 2019.

Moten, Fred. "Blackness and Nothingness (Mysticism in the Flesh)." *The South Atlantic Quarterly*, vol. 112, no. 4, 2013, pp. 737–780.

Moten, Fred. "The Case of Blackness." *Criticism*, vol. 50, no. 2, 2008, pp. 177–218.

Moten, Fred. "Black Op." *PMLA*, vol. 123, no. 5, 2008, pp. 1743–1747.

Moten, Fred. *In the Break: The Aesthetics of the Black Radical Tradition*. U of Minnesota P, 2003.

Moten, Fred, and Stefano Harney. *A Poetics of the Undercommons*. Sputnik & Fizzle, 2016.

Moya, Paula L. *The Social Imperative: Race, Close Reading, and Contemporary Literary Criticism*. Stanford UP, 2015.

Mudgett, Kathryn. "The Natural and the Legal Geographies of the Body: Law's Corpus Written on the Lives of Florens and Sethe." *Contested Boundaries: New Critical Essays on the Fiction of Toni Morrison*, edited by Maxine L. Montgomery, Cambridge Scholars, 2013, pp. 67–81.

Müller, Stefanie. *The Presence of the Past in the Novels of Toni Morrison*. Heidelberg, Winter, 2013.

Müller, Stefanie. "Standing Up to Words: Writing and Resistance in Toni Morrison's *A Mercy*." *Black Studies Papers*, vol. 1, no. 1, 2014, pp. 73–89.

Mustakeem, Sowande' M. *Slavery at Sea: Terror, Sex, and Sickness in the Middle Passage*. U of Illinois P, 2016.

Myers, B. R. "Mercy!" Review of *A Mercy*, by Toni Morrison. January/February 2009. *The Atlantic*, www.theatlantic.com/magazine/print/2009/01/mercy/7223/. Accessed 8 Nov. 2011.

"Naming Practices." *Whitneyplantation*, www.whitneyplantation.com/education/louisiana-history/slavery-in-louisiana/slave-trade-in-louisiana/naming-practices/. Accessed 12 Dec. 2019.

Neal, Marc A. "#BlackTwitter, #Hashtag Politics, and the New Paradigm of Black Protest." 1 Sep. 2014. *NewBlackManInExile*, www.newblackmaninexile.net. Accessed 25 Jan. 2016.

Neary, Lynn, and Toni Morrison. "Toni Morrison Discusses *A Mercy*." *Youtube*, uploaded by NPR, 20 Oct 2008, www.youtube.com/watch?v=7IZvMhQ2LIU.

Nehl, Markus. *Transnational Black Dialogues: Re-Imagining Slavery in the Twenty-First Century*. Bielefeld, transcript, 2016.

Neumann, Birgit, and Ansgar Nünning. "Travelling Concepts as a Model for the Study of Culture." *Travelling Concepts for the Study of Culture*, edited by Birgit Neumann, Ansgar Nünning, and Mirjam Horn, De Gruyter, 2012, pp. 1–22.

New Netherland Institute: Exploring America's Dutch Heritage. www.newnetherlandinstitute.org. Accessed 15 Dec. 2019.

Newman, Judie, et al. "Round Table – Saidiya Hartman, Lose Your Mother: A Journey Along the Atlantic Slave Route." *Journal of American Studies*, vol. 44, no. 1, 2010. doi:10.1017/S0021875810000071.

Nicol, Kathryn, and Jennifer Terry. "Guest Editors' Introduction: Toni Morrison: New Directions." *MELUS: Multi-Ethnic Literature of the U.S.*, vol. 36, no. 2, 2011, pp. 7–12. Project MUSE, doi:10.1353/mel.2011.0022. Accessed 25 Mar. 2015.

"ninny, n." *Oxford English Dictionary*, Oxford UP, 2019, www-1oed-1com-10066f0690229.erf.sbb.spk-berlin.de/view/Entry/127209?rskey=rqymWL&result=1&isAdvanced=false#eid. Accessed 20 Dec. 2019.

Norment, Nathaniel, editor. *The African American Studies Reader*. Carolina Academic Press, 2001.

Nünning, Ansgar. "Kulturwissenschaft(en)—Cultural Studies—Travelling Concepts: Kritische Bestandsaufnahme und Entwicklungsperspektiven." *Theorie Ohne Praxis—Praxis Ohne Theorie?*, edited by Jürgen Joachimsthaler, and Eugen Kotte, Peter Lang, 2009, pp. 21–43.

Nünning, Ansgar. "Narratology Or Narratologies? Taking Stock of Recent Developments, Critique and Modest Proposals for Future Usages of the Term." *What is Narratology? Questions and Answers Regarding the Status of a Theory*, edited by Tom Kindt, and Hans-Harald Müller, De Gruyter, 2003, pp. 239–276.

Nyong'o, Tavia. *Afro-Fabulations: The Queer Drama of Black Life*. New York UP, 2019.

Nyong'o, Tavia. "Barack Hussein Obama, Or, the Name of the Father." *The Scholar and Feminism Online*, vol. 7, no. 2, 2009, pp. 116–132.

Omry, Keren. "Salt Roads to Mercy." *Toni Morrison's A Mercy: Critical Approaches*, edited by Shirley A. Stave, and Justine Tally, Cambridge Scholars, 2011, pp. 85–102.

Osofsky, Gilbert, editor. *Puttin' on Ole Massa: The Slave Narrative of Henry Bibb, William Wells Brown, and Solomon Northrup*. Harper & Row, 1969.

Otten, Terry. "To Be One or To Have One": "Motherlove" in the Fiction of Toni Morrison." *Contested Boundaries: New Critical Essays on the Fiction of Toni Morrison*, edited by Maxine L. Montgomery, Cambridge Scholars, 2013, pp. 82–95.

Owens, Craig. "The Allegorical Impulse: Toward a Theory of Postmodernism." *October*, vol. 12, 1980, pp. 67–86.

Owens, Craig. "The Allegorical Impulse: Toward a Theory of Postmodernism Part 2." *October*, vol. 13, 1980, pp. 58–80.

Painter, Nell Irvin. *Southern History Across the Color Line*. U of North Carolina P, 2002.

Palladino, Mariangela. *Ethics and Aesthetics in Toni Morrison's Fiction*. BRILL, 2018.

Patterson, Orlando. *Slavery and Social Death: A Comparative Study*. Harvard UP, 1982.

Patton, Venetria K. "Black Subjects Re-Forming the Past Through the Neo-Slave Narrative Tradition." *MFS Modern Fiction Studies*, vol. 54, no. 4, Winter 2008, pp. 877–883.

Paul, Heike. *The Myths that Made America: An Introduction to American Studies*. Bielefeld, transcript, 2014.

Pease, Donald E. *The New American Exceptionalism*. U of Minnesota P, 2009.

Peterson, James B. "Eco-Critical Focal Points: Narrative Structure and Environmentalist Perspectives in Morrison's *A Mercy*." *Toni Morrison's A Mercy: Critical Approaches*, edited by Shirley A. Stave, and Justine Tally, Cambridge Scholars, 2011, pp. 9–22.

"Pieces of Eight." BBC. *A History of the World*, www.bbc.co.uk/ahistoryoftheworld/objects/JO391t6cRtGxstjbE4EEmg. Accessed 23 Dec. 2019.

Piesche, Peggy. "Towards a Future African Diasporic Theory: Black Collective Narratives Changing the Epistemic Map." 2016, www.academia.edu/23234061/Towards_a_Future_Af

rican_Diasporic_Theory_Black_Collective_Narratives_Changing_the_Epistemic_Map. Accessed 19 Oct. 2019.

Pocock, J. G. A. "The Myth of John Locke and the Obsession with Liberalism." *John Locke. Papers Read at a Clark Library Seminar 10 December 1977*, edited by J. G. A. Pocock, and Richard Ashcraft, 1980.

Pratt, Mary L. "Arts of the Contact Zone." *Profession*, 1991, pp. 33–40.

Prince, Gerald. "On a Postcolonial Narratology." *A Companion to Narrative Theory*, edited by James Phelan, and Peter J. Rabinowitz, Wiley-Blackwell, 2008, pp. 372–381.

Prince, Gerald. "The Disnarrated." *Style*, vol. 22, no. 1, Spring 1988, pp. 1–8.

Prince, Mary. *The History of Mary Prince, a West Indian Slave. Related by Herself. with a Supplement by the Editor. to which is Added, the Narrative of Asa-Asa, a Captured African (1831)*. University of Chapel Hill, 2000. *Documenting the American South*, docsouth.unc.edu/neh/prince/prince.html. Accessed 5 Jan. 2019.

"Property Law." *Library of Congress Research Guides, American Women: Resources from the Law Library*, guides.loc.gov/american-women-law/state-laws. Accessed 5 Jan. 2020.

Putnam, Amanda. "Mothering Violence: Ferocious Female Resistance in Toni Morrison's *the Bluest Eye, Sula, Beloved*, and *A Mercy*." *Black Women, Gender & Families*, vol. 5, no. 2, 2011, pp. 25–43.

Quanquin, Hélène. "There are Two Great Oceans": The Slavery Metaphor in the Antebellum Women's Rights Discourse as Redescription of Race and Gender." *Interconnections: Gender and Race in American History*, edited by Carol Faulkner, and Alison M. Parker, U of Rochester P, 2012, pp. 75–104.

Quijano, Aníbal. "Coloniality and Modernity/Rationality." *Cultural Studies*, vol. 21, no. 2, 2007, pp. 168–178.

Radin, Margaret J. "Property and Personhood." *Stanford Law Review*, vol. 34, no. 5, 1982, pp. 957–1015.

Randall, Alice. *The Wind Done Gone*. Houghton Mifflin, 2001.

Raynaud, Claudine. "Memory Work." *Black Studies Papers*, vol. 1, no. 1, 2014, pp. 29–36.

Raynor, Deirdre. J., and Johnella E. Butler. "Morrison and The Critical Community." *The Cambridge Companion to Toni Morrison*, edited by Justine Tally, Cambridge UP, 2007, pp. 175–184.

"refusal, n." *Oxford English Dictionary*, Oxford UP, 2019, www.oed.com.33 1745941.erf.sbb.spk-berlin.de/viewdictionaryentry/Entry/161137?print. Accessed 9 Dec. 2019.

Reid-Pharr, Robert F. "The Slave Narrative and Early Black American Literature." *The Cambridge Companion to the African American Slave Narrative*, edited by Audrey A. Fisch, Cambridge UP, 2007, pp. 137–149.

Renov, Michael. *The Subject of Documentary*. U of Minnesota P, 2004.

Rice, James D. "Bacon's Rebellion (1676–1677)." 3 Oct. 2014. *Encyclopediavirginia*, www.encyclopediavirginia.org/bacon_s_rebellion_1676-1677#start_entry. Accessed 12 Dec. 2019.

Rimmon-Kenan, Shlomith. *Narrative Fiction: Contemporary Poetics*. Routledge, 2002.

Rogers, G. A. J. "The Influence of Locke's Philosophy in the Eighteenth Century: Epistemology and Politics." *The Continuum Companion to Locke*, edited S.-J. Savonius-Wroth, Paul Schuurman, and Jonathan Walmsley, Continuum, 2010, pp. 281–291.

Rojas, Fabio. *From Black Power to Black Studies: How a Radical Social Movement Became an Academic Discipline*. Johns Hopkins UP, 2007.

Rose, Carol M. *Property and Persuasion: Essays on the History, Theory, and Rhetoric of Ownership*. Westview Press, 1994.

Roye, Susmita. "Toni Morrison's Disrupted Girls and their Disturbed Girlhoods: *The Bluest Eye* and *A Mercy*." *Callaloo*, vol. 35, no. 1, 2012, pp. 212–227. Project MUSE, doi:10.1353/cal.2012.0013. Accessed 27 May 2015.

Roynon, Tessa. *The Cambridge Introduction to Toni Morrison*. Cambridge UP, 2013.

Roynon, Tessa. "Her Dark Materials: John Milton, Toni Morrison, and Concepts of "Dominion" in *A Mercy*." *African American Review*, vol. 44, no. 4, 2011, pp. 593–606.

Roynon, Tessa. "Miltonic Journeys in *A Mercy*." *Toni Morrison's A Mercy: Critical Approaches*, edited by Shirley A. Stave, and Justine Tally, Cambridge Scholars, 2011, pp. 45–62.

Rushdy, Ashraf H. A. *Neo-Slave Narratives: Studies in the Social Logic of a Literary Form*. Oxford UP, 1999.

Ryan, Judylyn S. "Language and Narrative Technique in Toni Morrison's Novels." *The Cambridge Companion to Toni Morrison*, edited by Justine Tally, Cambridge UP, 2007, pp. 151–161.

Said, Edward W. "Traveling Theory (1982)." *The World, the Text, and the Critic*. Harvard UP, 1983, pp. 226–247.

Salmon, Emily Jones. "John Rolfe (d. 1622)." *Encyclopediavirginia*, www.encyclopediavirginia.org/Rolfe_John_d_1622. Accessed 19 Dec. 2019.

Sandy, Mark. "Cut by Rainbow": Tales, Tellers, and Reimagining Wordsworth's Pastoral Poetics in Toni Morrison's *Beloved* and *A Mercy*." *MELUS: Multi-Ethnic Literature of the U.S.*, vol. 36, no. 2, 2011, pp. 35–51.

Schneck, Peter. "Who Owns *Uncle Tom's Cabin*? Literature as Cultural Property." *Copyrightig Creativity: Creative Values, Cultural Heritage Institutions and Systems of Intellectual Property*, edited by Helle Porsdam, Ashgate, 2015, pp. 129–150.

Schreiber, Evelyn J. "Echoes of 'The Foreigner's Home' in *A Mercy*." *Race, Trauma, and Home in the Novels of Toni Morrison*. Louisiana State UP, 2010, pp. 157–176.

Schreiber, Evelyn J. "Personal and Cultural Memory in *A Mercy*." *Toni Morrison: Memory and Meaning*, edited by Adrienne L. Seward, and Justine Tally, UP of Mississippi, 2014, pp. 80–92.

Scott, Anne F., and Susanne Lebsock. "Excerpts from *Virginia Women: The First Two Hundred Years*." 2017, www.history.org/history/teaching/enewsletter/volume4/february%2006/virginiawomen.cfm. Accessed 28 Jan. 2018.

Scott, Darieck. *Extravagant Abjection: Blackness, Power, and Sexuality in the African American Literary Imagination*. NYU P, 2010.

Scott, Lynn O. "Autobiography: Slave Narratives." *Oxford Research Encyclopedia of Literature*, doi:10.1093/acrefore/9780190201098.013.658. Accessed 24 July 2019.

Sedlmeier, Florian. "Die Allegorie in der Postkolonialen Literatur und Literaturtheorie." *Allegorie. DFG-Symposium 2014*, edited by Ulla Haselstein, De Gruyter, 2016, pp. 528–556.

Serpell, Namwali. "On Black Difficulty. Toni Morrison and the Thrill of Imperiousness." 26 Mar. 2019. *Slate*, slate.com/culture/2019/03/toni-morrison-difficulty-black-women.html. Accessed 9 Aug. 2019.

Seward, Adrienne L., and Justine Tally. *Toni Morrison: Memory and Meaning*. UP of Mississippi, 2014.

Seward, Adrienne L., and Justine Tally. "Introduction." *Toni Morrison: Memory and Meaning*, edited by Adrienne L. Seward, and Justine Tally, UP of Mississippi, 2014, pp. xv–xxv.

Sexton, Jared. "Afro-Pessimism: The Unclear Word." *Rhizomes: Cultural Studies in Emerging Knowledge*, vol. 29, 2016.

Sexton, Jared. "Unbearable Blackness." *Cultural Critique*, vol. 90, 2015, pp. 159–178.

Sexton, Jared. "The Social Life of Social Death: On Afro-Pessimism and Black Optimism." *Intensions Journal*, vol. 5, 2011, pp. 1–47.

Sexton, Jared. "African American Studies." *A Concise Companion to American Studies*, edited by John C. Rowe. Wiley-Blackwell, 2010. doi:10.1002/9781444319071.ch10. Accessed 29 Oct. 2018.

Sexton, Jared. "People-of-Color-Blindness: Notes on the Afterlife of Slavery." *Social Text* 103, vol. 28, no. 2, 2010, pp. 31–56.

Sexton, Jared. *Amalgamation Schemes*. U of Minnesota P, 2008.

Sharpe, Christina E. *In the Wake: On Blackness and Being*. Duke UP, 2016.

Sharpe, Christina E. "Lose Your Kin." *New Inquiry*, 16 Nov. 2016, thenewinquiry.com/essays/lose-your-kin. Accessed 5 Jan. 2020.

Sharpe, Christina E. "Black Studies: In the Wake." *The Black Scholar*, vol. 44, no. 2, 2014, pp. 59–69.

Sharpe, Christina E. "The Lie at the Center of Everything." *Black Studies Papers*, vol. 1, no. 1, 2014, pp. 189–214.

Sharpe, Christina E. "Response to 'Ante-Anti-Blackness.'" *Lateral*, vol. 1, 2012. doi:10.25158/L1.1.17. Accessed 10 Dec. 2019

Sharpe, Christina E. *Monstrous Intimacies: Making Post-Slavery Subjects*. Duke UP, 2010.

Shields, E. T., Jr. "The Genres of Exploration and Conquest Literatures." *A Companion to the Literatures of Colonial America*, edited by Susan Castillo, and Ivy Schweitzer, Blackwell, 2005, pp. 353–368.

Shilliam, Robbie. "Colonial Architecture Or Relatable Hinterlands? Locke, Nandy, Fanon, and the Bandung Spirit." robbieshilliam.wordpress.com/writings/. Accessed 20 Sep. 2017

Shook, Lauren. "[L]ooking at Me My Body Across Distances": Toni Morrison's *A Mercy* and Seventeenth-Century European Religious Concepts of Race." *Early Modern Black Diaspora Studies*, edited by C. L. Smith et al., Palgrave Macmillan, 2018, pp. 157–173.

Shorto, Russell. *The Island at the Center of the World: The Epic Story of Dutch Manhattan, the Forgotten Colony that Shaped America*. Doubleday, 2004.

Simone, Nina. "Ain't Got No / I Got Life." *Ain't Got No / I Got Life*, Digimode, 1996.

Sirvent, Roberto. "BAR Book Forum: Julia Jordan-Zachery's 'Shadow Bodies' and Sabine Broeck's 'Gender and the Abjection of Blackness.'" 19 Dec 2018. *Black Agenda Report*, www.blackagendareport.com/bar-book-forum-julia-jordan-zacherysshadow-bodiesand-sabine-broecks-gender-and-abjection-blackness. Accessed 21 July 2019.

"Slavery, Institutional Racism, and the Development of State Surveillance as a Response to Resistance." 29 July 2014. *PrivacySOS*, https://privacysos.org/blog/slavery-institutional-racism-and-the-development-of-state-surveillance-as-a-response-to-resistance/. Accessed 12 Dec. 2019.

"Slavery and Indentured Servants." *Library of Congress Research Guides, American Women: Resources from the Law Library*, guides.loc.gov/american-women-law/state-laws. Accessed 5 Jan. 2020.

Smallwood, Christine. "Back Talk: Toni Morrison." *The Nation*, vol. 287, no. 19, Dec 8, 2008, p. 37.

Smallwood, Stephanie E. *Saltwater Slavery: A Middle Passage from Africa to American Diaspora*. Harvard UP, 2007.

Smith, John. *A True Relation of Virginia*. Boston, Wiggin and Lunt, 1866. *Hathitrust*, https://babel.hathitrust.org/cgi/pt?id=loc.ark:/13960/t6n01rk1x&view=1up&seq=13. Accessed 5 Jan. 2019.

Smith, John. *The Generall History of Virginia*. London, printed by I.D. and I.H. for Michael Sparkes, 1624. University of Chapel Hill, 2000. *Documenting the American South*, docsouth.unc.edu/southlit/smith/smith.html. Accessed 5 Jan. 2019.

Smith, Valerie. *Toni Morrison: Writing the Moral Imagination*. Wiley-Blackwell, 2012.

Smith, Valerie. "Neo-Slave Narratives." *The Cambridge Companion to the African American Slave Narrative*, edited by Audrey A. Fisch, Cambridge UP, 2007, pp. 168–186.

"solidarity, n." *Oxford English Dictionary*, Oxford UP, 2019, www.oed.com/view/Entry/184237?redirectedFrom=solidarity#eid. Accessed 15 Mar. 2022.

Spanos, William V. *American Exceptionalism in the Age of Globalization: The Specter of Vietnam*. SUNY P, 2008.

Spatzek, Samira. "'Own Yourself, Woman': Toni Morrison's *A Mercy*, Early Modernity, and Property." *Black Studies Papers*, vol. 1, no. 1, 2014, pp. 57–71.

Spaulding, A. Timothy. *Re-Forming the Past: History, the Fantastic, and the Postmodern Slave Narrative*. Ohio State UP, 2005.

Spillers, Hortense J. "The Idea of Black Culture." *CR: The New Centennial Review*, vol. 6, no. 3, 2006, pp. 7–28.

Spillers, Hortense J. *Black, White, and in Color: Essays in American Literature and Culture*. The U of Chicago P, 2003.

Spillers, Hortense J. "Mama's Baby, Papa's Maybe: An American Grammar Book." *Diacritics*, vol. 17, no. 2, 1987, pp. 64–81.

Spillers, Hortense, et al. "'Whatcha Gonna do?': Revisiting 'Mama's Baby, Papa's Maybe: An American Grammar Book': A Conversation with Hortense Spillers, Saidiya Hartman, Farah Jasmine Griffin, Shelly Eversley, & Jennifer L. Morgan." *Women's Studies Quarterly*, vol. 35, no. 1/2, 2007, pp. 299–309.

Squadrito, Kathy. "Locke and the Dispossession of the American Indian." *Philosophers on Race: Critical Essays*, edited by Julie K. Ward, and Tommy L. Lott, Blackwell, 2002, pp. 101–125.

Staff and Agencies. "Dylann Roof Sentenced to Death for the Murders of Nine Black Church Members." 10 Jan. 2017. *The Guardian*, www.theguardian.com/us-news/2017/jan/10/dylann-roof-sentenced-to-death-charleston-church-shooting. Accessed 9 Dec. 2019.

Stanton, Timothy. "The Reception of Locke in England in the Early Eighteenth Century: Metaphysics, Religion and the State." *The Continuum Companion to Locke*, edited by S.-J. Savonius-Wroth, Paul Schuurman, and Jonathan Walmsley, Continuum, 2010, pp. 292–301.

Stapelbroek, Koen. "Property." *The Continuum Companion to Locke*, edited by S.-J. Savonius-Wroth, Paul Schuurman, and Jonathan Walmsley, Continuum, 2010, pp. 201–203.

Stave, Shirley A. "'More Sinned Against than Sinning': Redefining Sin and Redemption in *Beloved* and *A Mercy*." *Contested Boundaries: New Critical Essays on the Fiction of Toni Morrison*, edited by Maxine L. Montgomery, Cambridge Scholars, 2013, pp. 126–143.

Stave, Shirley A. "Across Distances without Recognition: Misrecognition in Toni Morrison's *A Mercy*." *Toni Morrison's A Mercy: Critical Approaches*, edited by Shirley A. Stave, and Tally Justine, Cambridge Scholars, 2011, pp. 137–150.
Stave, Shirley A., and Justine Tally. "Introduction." *Toni Morrison's A Mercy: Critical Approaches*, edited by Shirley A. Stave, and Justine Tally, Cambridge Scholars, 2011, pp. 1–8.
Stave, Shirley A., and Justine Tally. *Toni Morrison's A Mercy: Critical Approaches*. Cambridge Scholars, 2011.
Stevenson, Bryan. "Slavery Gave America a Fear of Black People and a Taste for Violent Punishment. Both Still Define our Criminal-Justice System." *The 1619 Project*, edited by Nikole Hannah-Jones. Special issue of *The New York Times Magazine*, 18 Aug. 2019, pp. 80–81.
Strehle, Susan. "I Am a Thing Apart": Toni Morrison, *A Mercy*, and American Exceptionalism." *Critique: Studies in Contemporary Fiction*, vol. 54, no. 2, 2013, pp. 109–123.
Stretton, Tim, and Krista J. Kesselring. "Introduction: Coverture and Continuity." *Married Women and the Law: Coverture in England and the Common Law World*, editors Tim Stretton, and Krista J. Kesselring, McGill-Queen's UP, 2013, pp. 3–23.
Tally, Justine. "Contextualizing Toni Morrison's Ninth Novel: What Mercy? Why Now?" *Toni Morrison's A Mercy: Critical Approaches*, edited by Shirley A. Stave, and Justine Tally, Cambridge Scholars, 2011, pp. 63–84.
Tally, Justine. "Palimpsest: Reading John Winthrop through the Morrison Trilogy." *Toni Morrison: Memory and Meaning*, edited by Adrienne L. Seward, and Justine Tally, UP of Mississippi, 2014, pp. 132–143.
Tambling, Jeremy. *Allegory*. Routledge, 2010.
Taylor, Alan. *American Colonies: The Settling of North America*. Penguin, 2002.
Taylor, Chloë. *The Culture of Confession from Augustine to Foucault: A Genealogy of the 'Confessing Animal.'* Routledge, 2009.
Tedder, Charles. "Post Racialism and its Discontents: The Pre-National Scene in Toni Morrison's *A Mercy*." *Contested Boundaries: New Critical Essays on the Fiction of Toni Morrison*, edited by Maxine L. Montgomery, Cambridge Scholars, 2013, pp. 144–159.
Teele, Elinor. Review of *A Mercy*, by Toni Morrison. 16 Dec. 2008. *California Literary Review*, calitreview.com/1895/a-mercy-by-toni-morrison/. Accessed 8 Jan. 2014.
Teller, Katalin. "Mieke Bals Ansatz zur Kulturellen Analyse und Ihre Theorie der Wandernden Begriffe im Lichte einer Möglichen Interkulturellen Narratologie." *Narratologie Interkulturell: Entwicklungen—Theorien*, edited by Magdolna Orosz, and Jörg Schönert, Peter Lang, 2004, pp. 167–178.
"The Thirteenth Amendment." *Exploring Constitutional Conflicts*, law2.umkc.edu/faculty/projects/ftrials/conlaw/thirteenthamendment.html. Accessed 23 Dec. 2019.
Thomas, Brook. *Cross-Examinations of Law and Literature*. Cambridge UP, 2009.
Thomas, Brook. *Law and Literature*. Tübingen, Narr, 2001.
Todaro, Lenora. "Toni Morrison's *A Mercy*: Racism Creation Myth." Review of *A Mercy*, by Toni Morrison. 19 Nov. 2008. *The Village Voice*, www.villagevoice.com/2008/11/19/toni-morrisons-a-mercy-racism-creation-myth/. Accessed 12 Dec. 2019.
Tompkins, Jane. *Sensational Designs: The Cultural Work of American Fiction 1790–1860*. Oxford UP, 1985.

Touré, and Michael E. Dyson. *Who's Afraid of Post-Blackness? A Look at What it Means to Be Black Now*. Free Press, 2011.

Transatlantic Slave Trade Database. Slave Voyages Database at Emory U, V. 2.2.7. www.slavevoyages.org. Accessed 15 Dec. 2019.

Tully, James. *An Approach to Political Philosophy: Locke in Contexts*. Cambridge UP, 1993.

Tully, James. *A Discourse on Property: John Locke and His Adversaries*. Cambridge UP, 1980.

Updike, John. "Dreamy Wilderness: Unmastered Women in Colonial Virginia." Review of *A Mercy*, by Toni Morrison. 3 Nov. 2008. *The New Yorker*, www.newyorker.com/arts/critics/books/2008/11/03/081103crbo_books_updike?printable=true. Accessed 8 Aug. 2011.

Uzgalis, William. "An Inconsistency Not to Be Excused": On Locke and Racism." *Philosophers on Race: Critical Essays*, edited by Julie K. Ward, and Tommy L. Lott, Blackwell, 2002, pp. 81–100.

Uzgalis, William. ". . . The Same Tyrannical Principle": Locke's Legacy on Slavery." *Subjugation and Bondage: Critical Essays on Slavery and Social Philosophy*, edited by Tommy L. Lott, Rowman & Littlefield, 1998, pp. 49–77.

Vega-González, Susana. "Orphanhood in Toni Morrison's *A Mercy*." *Toni Morrison's A Mercy: Critical Approaches*, edited by Shirley A. Stave, and Justine Tally, Cambridge Scholars, 2011, 119–136.

Verheul, Jaap. "Introduction: Utopia and Dystopia in American Culture." *Dreams of Paradise, Visions of Apocalypse: Utopia and Dystopia in American Culture*, edited by Jaap Verheul, VU UP, 2004, pp. 1–12.

Vieira, Fátima. "The Concept of Utopia." *The Cambridge Companion to Utopian Literature*, edited by Gregory Claeys, Cambridge UP, 2010, pp. 3–27.

"Virginia Company." *Historicjamestowne*, historicjamestowne.org/history/virginia-company/. Accessed 16 Dec. 2019.

Wadud, Imani. "Free but Not Equal": In the Wake of Trayvon Martin—American Anger and Visual Activism." *COPAS–Current Objectives of Postgraduate American Studies*, vol. 14, no. 2, 2013, pp. 1–17, copas.uni-regensburg.de/article/view/198/225. Accessed 25 Sep. 2018.

Waegner, Cathy C. "Ruthless Epic Footsteps: Shoes, Migrants, and the Settlement of the Americas in Toni Morrison's *A Mercy*." *Post-National Enquiries: Essays on Ethnic and Racial Border Crossings*, edited by Jopi Nyman, Cambridge Scholars, 2009, pp. 91–112.

Wagner-Martin, Linda. *Toni Morrison: A Literary Life*. Palgrave Macmillan, 2015.

Walcott, Rinaldo. *On Property: Policing, Prisons, and the Call for Abolition*. Biblioasis, 2021.

Walcott, Rinaldo. "The Problem of the Human: Black Ontologies and 'the Coloniality of our Being.'" *Postcoloniality—Decoloniality—Black Critique: Joints and Fissures*, edited by Sabine Broeck, and Carsten Junker, Campus, 2014, pp. 93–108.

Ward, Julie K., and Tommy L. Lott, editors. *Philosophers on Race: Critical Essays*. Blackwell, 2002.

Wardi, Anissa. "The Politics of "Home" in *A Mercy*." *Toni Morrison's A Mercy: Critical Approaches*, edited by Shirley A. Stave, and Justine Tally, Cambridge Scholars, 2011, pp. 23–44.

Warren, Calvin L. *Ontological Terror: Blackness, Nihilism, and Emancipation*. Duke UP, 2018.

Warren, Kenneth W. *What was African American Literature?* Harvard UP, 2012.

Weheliye, Alexander G. "After Man." *American Literary History*, 2008, pp. 321–36. *Project MUSE*, muse.jhu.edu/article/233035. Accessed 14 Feb. 2014.

Weheliye, Alexander G. *Habeas Viscus*. Duke UP, 2014.
Weier, Sebastian. *Cyborg Black Studies: Tracing the Impact of Technological Change on the Constitution of Blackness*. 2015. University of Bremen, PhD Dissertation, elib.suub.uni-bremen.de/edocs/00104897-1.pdf.
Weier, Sebastian. "Consider Afro-Pessimism." *Amerikastudien/American Studies*, vol. 59, no. 3, 2014, pp. 419-433.
Weinbaum, Alys E. *The Afterlife of Reproductive Slavery*. Duke UP, 2019.
Weinbaum, Alys E. *Wayward Reproductions: Genealogies of Race and Nation in Transatlantic Modern Thought*. Duke UP, 2004.
Weisberg, Richard H. *The Failure of the Word: The Protagonist as Lawyer in Modern Fiction*. Yale UP, 1984.
Weisberg, Richard H. *Poetics and Other Strategies of Law and Literature*. Columbia UP, 1992.
Welchman, Jennifer. "Locke on Slavery and Inalienable Rights." *Canadian Journal of Philosophy*, vol. 25, no. 1, 1995, pp. 67-81.
White, Deborah G. *Ar'n't I a Woman?: Female Slaves in the Plantation South*. Norton, 1985.
Wilderson, Frank B., III. *Afropessimism*. Liveright, 2020.
Wilderson, Frank B., III. "Social Death and Narrative Aporia in 12 Years a Slave." *Black Camera*, vol. 7, no. 1, 2015, pp. 134-149.
Wilderson, Frank B., III. "The Black Liberation Army and the Paradox of Political Engagement." *Postcoloniality— Decoloniality—Black Critique: Joints and Fissures*, edited by Sabine Broeck, and Carsten Junker, Campus, 2014, pp. 175-210.
Wilderson, Frank B., III. *Red, White & Black: Cinema and the Structure of U.S. Antagonisms*. Duke UP, 2010.
Wilderson, Frank B. III. "Grammar & Ghosts: The Performative Limits of African Freedom." *Theatre Survey*, vol. 50, no. 1, 2009, pp. 119-125.
Wilderson, Frank B., III. *Incognegro: A Memoir of Exile and Apartheid*. South End Press, 2008.
Wilderson, Frank B., III. "The Prison Slave as Hegemony's (Silent) Scandal." *Social Justice*, vol. 30, no. 2, 2003, pp. 18-27.
Wilderson, Frank B., III. "Gramsci's Black Marx: Whither the Slave in Civil Society?" *Social Identities*, vol. 9, no. 2, 2003, pp. 225-240.
Wilderson, Frank B., III., Samira Spatzek, and Paula von Gleich. "The Inside-Outside of Civil Society": An Interview with Frank B. Wilderson, III." *Black Studies Papers*, vol. 2, no. 1, 2016, pp. 4-22.
Williams, Eric E. *Capitalism and Slavery*. Williams Press, 2008.
Williams, Patricia J. *The Alchemy of Race and Rights: Diary of a Law Professor*. Harvard UP, 1991.
Wilson, Harriet E. *Our Nig: Sketches from the Live of a Free Black*, edited by Henry Louis Gates, Jr., and Richard J. Ellis, Vintage, 2011.
Winthrop, John. "A Modell of Christian Charity, 1630." *The English Literatures of America, 1500-1800*, edited by Myra Jehlen, and Michael Warner, Routledge, 1997, pp. 151-159.
Woolhouse, Roger, and Timothy Stanton. "Contemporary Locke Scholarship." *The Continuum Companion to Locke*, edited by S.-J. Savonius-Wroth, Paul Schuurman, and Jonathan Walmsley, Continuum, 2010, 314-320.
Woubshet, Dagmawi. "Introduction for Hortense Spillers." *Callaloo*, vol. 35, no. 4, 2012, pp. 925-928.

Wright, Kai, editor. *The African American Experience: Black History and Culture through Speeches, Letters, Editorials, Poems, Songs, and Stories*. Black Dog & Leventhal, 2011.

Wyatt, Jean. "Failed Messages, Maternal Loss, and Narrative Form in Toni Morrison's *A Mercy*." *MFS Modern Fiction Studies*, vol. 58, no. 1, 2012, pp. 128–151. *CrossRef*, doi:10.1353/mfs.2012.0006. Accessed 29 Nov. 2015.

Wynter, Sylvia. "Unsettling the Coloniality of Being/Truth/Power/Freedom: Towards the Human, After Man, Its Overrepresentation—An Argument." *CR: The New Centennial Review*, vol.3, no. 3, 2003, pp. 237–337.

Wynter, Sylvia. "1492: A New World View." *Race, Discourse, and the Origin of the Americas: A New World View*, edited by Vera Lawrence Hyatt, and Rex Nettleford, Smithsonian Institution Press, 1995, pp. 5–57.

Wynter, Sylvia. "No Humans Involved": An Open Letter to My Colleagues." *Forum N. H. I.: Knowledge for the 21st Century*, vol. 1, no. 1, 1994, pp. 42–73.

Yancy, George. "Whiteness as Ambush and the Transformative Power of Vigilance." *Black Bodies, White Gazes: The Continuing Significance of Race*, edited by George Yancy, Rowman & Littlefield, 2008, pp. 227–250.

Zauditu-Selassie, K. *African Spiritual Traditions in the Novels of Toni Morrison*. UP of Florida, 2009.

Zug, Marcia. "The Mail-Order Brides of Jamestown, Virginia." 31 Aug. 2016. *The Atlantic*, theatlantic.com/business/archive/2016/08/the-mail-order-brides-of-jamestown-virginia/4980 83/. Accessed 25 Aug. 2017.

Index

abjection 24, 54, 56, 63, 70, 75 f., 78, 81, 88, 94 f., 97 f., 247, 249, 251 f.
abjectorship 24, 54, 63, 75 f., 98
abolition 65, 89, 92, 227
accumulation 14, 30, 37, 39–43, 55, 73 f., 80, 126, 197, 202
Adusei-Poku, Nana 2, 19
Afropessimism 6, 18, 29, 56, 61, 65–66, 73–74, 88–89, 101, 102, 174, 192, 253
Afropessimism (the monograph) *see also* Wilderson, Frank B., III. 61
afterlife of property 9, 55, 91, 95 f.
afterlife of slavery 83, 92, 94, 95, 100, 237
Ahmed, Sara 19, 155
allegory 8, 105, 106, 117, 121, 154, 157, 161, 175, 248,
Allen, Paula Gunn 138–140
antiblackness 3, 58, 70, 74, 89 f., 95, 108, 176, 191, 251
anticipating generations 117, 175, 178, 184, 189
anticipatory wake 24, 54, 63, 79, 88, 91, 96–98, 249
anti-narrative *or* anti-narration 8, 24, 25, 62, 99, 104 f., 107 f., 216, 219, 224, 242, 244, 245, 250,
Applebaum, Barbara 19 f., 22
Athanasiou, Athēna 70 f.
Atlantic slavery 2, 4–6, 9, 20, 22, 34–37, 44, 69, 79, 83, 85, 88, 102, 117, 131, 137, 178, 185 f., 189, 191, 220–223, 243, 246–248, 251, 254

Bal, Mieke *see also* travelling concepts 21 f., 63, 99, 105, 110–113, 179, 253
Barbados 9 f., 77 f., 81 f., 103, 119 f., 123, 127, 132 f., 202, 225, 230–232, 238
belonging 70, 101, 117, 148 f., 166, 170, 189–195, 197, 201, 204, 208 f., 212, 216–219, 221
Beloved 1, 16, 25, 65, 100 f., 121, 138, 223 f., 226, 228, 230–235, 239, 249
Bennett, Joshua 55

Best, Stephen *see also* None Like Us; "On Failing to Make the Past Present" 7, 15–18, 46, 96, 100–103, 107 f., 120, 134, 166, 170, 193, 196, 202, 206, 214 f., 220, 223 f., 234, 237, 242, 247, 250
Bhandar, Brenna 3 f., 16, 55, 70, 86
Black feminism *or* Black feminist 4, 9, 13, 14, 18, 55, 59, 60, 63, 70, 71, 78, 83, 88, 94, 128, 177, 188, 193, 247, 251, 252
Black Optimism 89 f., 101
blacksmith, the 5 f., 12, 64, 69, 115 f., 156, 173, 184 f., 188, 194 f., 197, 200, 202–213, 215–219, 225, 232, 240
Brand, Dionne 175, 184
Broeck, Sabine 3, 19, 28, 32 f., 36–38, 42 f., 47 f., 55, 58, 66, 70 f., 76, 133, 158, 171–173, 230, 236, 253
Brooks, Peter 13, 17, 109, 196
Butler, Judith 13, 65, 70 f.

Campt, Tina 107, 184
capacity 28, 30, 33, 37, 44, 59, 68, 74, 77, 80, 82, 84, 104, 137, 142, 147, 151–153, 169 f., 174, 191, 200 f., 203, 235–237, 243, 247 f., 250
Carey, Brycchan 13, 47, 49–51
character 1, 4–8, 12–14, 24 f., 65, 68, 75–77, 83 f., 96 f., 99, 103–105, 108–122, 125 f., 128–131, 133–142, 144, 147–156, 161, 163, 169, 171, 174–178, 180 f., 184, 187, 189, 191–195, 203 f., 207–209, 212, 215–219, 221, 223–225, 230, 237, 244–246, 248–250, 253
– allegorical figures 6–8, 98, 104, 107–109, 111 f., 114
– bracketed character(s) 24, 99, 113, 114, 253
– fictional character 6 f., 20, 98 f., 104 f., 107–114, 253
– "insubstantial characters" 7, 104, 111, 246

Chesapeake 1, 9, 81, 123–125, 171, 225, 230, 247
Cho, Sumi *see also* post-racial 14, 176, 234
Christian, Barbara 79, 115, 134, 143, 155, 186, 227
Christiansë, Yvette 15 f., 248
– claim of right to property 26, 45, 155
coherence 44, 67, 72, 75, 105, 112–114, 168, 230, 233, 253
coming into being 28, 38, 119, 135
coverture 158, 160, 248
crawl space 214–216, 218, 228, 240–243
Critical legal studies 16, 55
Critical race studies 4, 17, 18, 28
critical vigilance 20, 22, 254

Davis, Adrienne D. *see also* sexual economy of slavery 34, 59, 78, 172, 221, 236, 238
Davis, Angela 59, 172
disnarration 113, 118, 122
dispossession 24, 36, 54, 63, 69–75, 98, 116, 137 f., 140, 142, 148, 151 f., 161, 183, 237, 249 f.
Douglass, Frederick 22, 65, 67 f., 70, 74, 88, 102, 201, 203, 253
Douglass, Patrice D. 22, 26, 54, 60, 65, 67 f., 70, 74, 88, 102, , 253
DuCille, Ann 13

empathy 60, 93
emplot-ability 7, 20, 113, 221, 251
Enlightenment 3, 18, 27, 29, 32, 37 f., 45, 51, 55, 104, 118, 247
Equiano, Olaudah 186, 224, 229–231, 233
equipment in human form 75
exile 84, 116, 137, 148 f., 153

fabulation 251 f.
– afro-fabulation 251
– critical fabulation 93, 95, 191
Fanon, Frantz 66
feudal rule 4, 36 f., 161, 174, 247 f.
flesh 2, 30, 50, 66 f., 73, 75 f., 82, 85, 112, 115, 238
Florens 5 f., 24, 63 f., 68 f., 81, 87 f., 96 f., 101 f., 106, 115–117, 119–121, 136 f., 141, 144, 148–156, 169, 172 f., 177, 179 f., 183, 185, 187, 189–221, 223, 225 f., 228, 230, 232–234, 236–244, 248, 250, 252
form 4, 6–8, 10, 12 f., 15–18, 22–24, 30, 40–42, 48, 60, 62 f., 65–68, 70, 79, 86, 90, 93, 95–98, 100–105, 107–109, 114 f., 126 f., 139, 149, 168, 192 f., 195, 197, 203, 213–216, 219 f., 222, 224, 227, 234 f., 237–239, 241–244, 246, 250, 252, 254
– and narrative 13, 24, 61 f., 97, 109, 219, 250, 254
– the agency of 100
Franklin, Benjamin 125
freedom 1–4, 6, 9, 11 f., 20, 23, 26–30, 33–35, 37–39, 44 f., 48 f., 51 f., 54–56, 59 f., 67, 69, 76, 80, 89, 96 f., 104, 112 f., 116–118, 122, 125, 129 f., 133–135, 141, 156, 174, 184, 189–191, 194, 201, 207–209, 211 f., 214–217, 227, 233, 244–248, 251
Fuentes, Marisa 59, 78, 127, 222
Fundamental Constitutions of Carolina, the 23, 26–28, 32, 43 f., 52
fungibility 24, 30, 43, 54, 60, 63, 69, 71–75, 88, 90, 98, 138, 140 f., 183, 197, 200–202, 249, 252

Generall Historie of Virginia 142
genocide 14, 68, 116, 137, 141, 142, 145–147, 149, 152–153, 183, 248
genre 60, 101, 184, 192, 195, 223, 226–228, 233, 235
Germantown Friends' Protest Against Slavery 23, 26–28, 47
Graeber, David 4, 27, 32, 132
grammar of property 64, 85, 97, 114, 117, 120, 137, 140, 149, 152, 153, 154, 155, 164, 167, 170–171, 175, 181, 188 f., 194, 206, 248

Hannah-Jones, Nikole 9, 247
Harris, Cheryl I. 4, 17, 27 f., 47 f., 51, 55, 59, 72, 86 f., 127
Hartman, Saidiya V. *see also* "The Belly of the World: A Note on Black Women's La-

bors"; *Lose Your Mother; Scenes of Subjection;* "Venus in Two Acts" 1–4, 19, 59–62, 65f., 69–79, 81, 83f., 88, 90–97, 104, 118, 129, 132, 138, 140, 175, 177, 184, 186f., 190f., 194, 198, 201, 203, 212, 214, 219f., 222f., 226, 234f., 237f., 243, 246, 249, 251f., 254
Hejinian, Lyn 105f., 114
Human 2–4, 6f., 12, 17–19, 23, 26, 33, 36–38, 40f., 47, 52, 54, 56–62, 66f., 70–74, 76f., 81f., 84–86, 92–94, 99, 102, 104f., 108, 112f., 115, 117–119, 125, 130, 142, 159, 161, 164, 169–171, 173–175, 180, 189–191, 193–195, 197–199, 202f., 207, 209, 213–222, 225f., 232, 234f., 238, 243–248, 250

impossible stories 20, 73, 243, 246
Incidents in the Life of a Slave Girl see also Jacobs, Harriet 214, 239
indentured servants 5f., 9–12, 115, 121, 157f., 164, 169, 209

Jackson, Zakiyyah Iman 67
Jacobs, Harriet 214f., 219, 224, 228, 233, 235, 239f., 242–244
Jacob Vaark 1, 4f., 24, 65, 68f., 76f., 82, 87, 103, 112, 115–122, 124f., 127, 129f., 135, 137, 139–141, 144, 146–149, 151–153, 155–157, 163f., 166f., 179, 181f., 184f., 188, 193, 195, 204, 207f., 212, 214, 216, 225f., 237f., 241, 250, 253
Jones-Rogers, Stephanie E. *see also They Were Her Property: White Women as Slave Owners in the American South* 159–161, 168f., 171, 174, 178, 182, 185, 238, 248

kinship 24, 54f., 59, 63, 73, 77–80, 84–87, 98, 117, 136, 149, 175, 178, 181–184, 188, 192, 206, 216, 221, 235, 241, 248f.
Kristeva, Julia 76

Laboring Women: Reproduction and Gender in New World Slavery see also Morgan, Jennifer L. 77

Laslett, Peter 23, 30f., 34, 38
Law and Literature 17, 109
liberal imagination 1f., 20, 23, 44, 48, 51, 54f., 59, 72, 96, 129, 133, 191, 246, 251
Lina 5, 24, 64f., 68f., 115f., 129f., 136–156, 164–169, 171, 175–177, 179–181, 183, 185, 187, 203–205, 211, 214, 216, 218, 232, 234
Lockean liberal subject *or* Lockean subjectivity 112, 122, 249
Locke, John 23, 26–28, 30–45, 48, 63, 118, 134, 207, 247
Lose Your Mother: A Journey Along the Atlantic Slave Route see also Hartman, Saidiya V. 91, 92
Lowe, Lisa 3, 45f., 55, 158, 161, 164

Macpherson, C. B. 27, 29, 32–34, 37
Martin, Valerie *see also Property* (novel) 15, 172f., 218, 236
Maryland *or* colonial Maryland 1, 5, 91, 125, 174, 225
Middle Passage 3, 29, 48, 56, 84, 88, 92f., 97, 100, 221, 223, 228–234, 238, 243, 248
Mills, Charles 55, 253
minha mãe, the 5, 24, 64, 81–83, 102, 115–117, 137, 172, 178, 202, 212, 217, 220–226, 228–230, 232–245, 248
Morgan, Edmund S. 8–11, 27, 55, 59f., 77–84, 87, 94, 127f., 149, 157–160, 182, 198, 220, 222f., 242
Morgan, Jennifer L. *see also Laboring Women: Reproduction and Gender in New World Slavery;* 8–11, 27, 55, 59f., 77–84, 87, 94, 127f., 149, 157–160, 182, 220, 222f., 242
Morrison, Toni *see also Beloved; Playing in the Dark; Sula* 1f., 12–16, 20, 24–28, 37, 60, 63, 65, 77f., 83f., 96, 99–101, 116, 118, 120–122, 124f., 128, 133, 135, 141, 153f., 161, 165–167, 176, 186, 192f., 197, 203, 210, 215, 217, 224, 226–228, 230f., 233f., 239, 246f., 249, 252, 254
Moten, Fred 89f., 107, 115, 184, 210, 217–219, 252

Multiculturalism   70, 178
myth   15, 103, 121, 123–126, 128, 139, 193

Narrative of the Life of Frederick Douglass, an American Slave, Written by Himself see also Douglass, Frederick   200
narratology   99, 105, 109f., 112f.
None Like Us see also Best, Stephen   15, 100
Nyong'o, Tavia   55, 79f., 87, 128, 149, 183, 251

Obama, Barack   14, 176, 234
object of property   75f., 86
"On Failing to Make the Past Present" see also Best, Stephen   15, 100
ownership   1, 3, 17, 20, 23f., 26f., 29, 33, 36f., 40, 43f., 46–48, 51f., 62, 69, 76f., 81, 99, 102f., 107, 114, 119, 121f., 132–134, 149, 159f., 167, 169, 174, 188–190, 192, 194, 197, 199–201, 208, 217, 221, 233, 246f.

Partus Sequitur Ventrem   9, 11, 77, 79f., 95, 97, 188, 242
Patterson, Orlando see also social death, *Slavery and Social Death: A Comparative Study*   43, 55, 66, 202
Paul, Heike   13, 15, 103, 123–126, 128, 134, 139f., 142f.
Playing in the Dark: Whiteness and the Literary Imagination see also Morrison, Toni   26, 118
Pocahontas   138–140, 142
positioning power   3, 8, 29, 65f., 76, 88, 92, 97, 137, 202, 249, 250
possessive investment   23, 52
post-racial see also Cho, Sumi   14, 15, 66, 173, 176, 234, 235
post-slavery   4, 6f., 18, 20–24, 29, 43, 48, 52–54, 56, 58, 62f., 65, 69f., 75, 88f., 97–99, 101f., 104f., 108, 113f., 141, 189, 191f., 196, 202f., 216, 249, 253
Prince, Gerald see also disnarration   62, 109, 111–113, 118, 219, 224, 233, 235f., 239, 241–244
Prince, Mary see also *The History of Mary Prince, a West Indian Slave. Related by Herself. with a Supplement by the Editor. to which is Added, the Narrative of Asa-Asa, a Captured African (1831)*   62, 109, 111–113, 118, 219, 224, 233, 235f., 239, 241–244
Property (novel) see also Martin, Valerie   172f., 236
– property/kinless constellation   85
property paradigm   4, 6–8, 14–16, 23f., 26, 30, 52, 54, 56, 69, 83, 96, 98f., 102–107, 112, 117, 120, 137, 169f., 173, 175, 177, 182–184, 187, 190, 192, 195f., 217, 219, 221, 239, 244, 247–249, 251
"pyrotechnics"   102, 254

Quaker   28, 47–52, 247
Queer studies   100

race/reproduction bind   78, 177
Randall, Alice   190
Rebekka Vaark   5f., 24, 68, 116, 137, 140, 147–150, 152f., 161, 174, 177f., 181f., 184, 187–190, 195, 199, 212, 216, 218, 228, 248
Red, White & Black: Cinema and the Structure of U.S. Antagonisms see also Wilderson, Frank B., III.   29, 66, 99, 138, 169
refusal *or* practicing refusal   1, 7, 8, 15, 24, 97, 98, 99, 101, 107–108, 114, 115, 135, 219, 237, 245, 246, 252, 254
relationality   6, 26, 29f., 44, 54, 70, 101, 171, 174, 218, 249
reproduction   4, 9, 19, 24, 54, 59, 63, 73, 77–81, 83, 85, 87, 94, 98, 131, 175, 177, 182, 188, 223, 226, 234f., 237, 242, 247, 249
reputation   35, 47–52, 70, 247
Roanoke   9, 122
Rolfe, John   138–140
[Routing the Argument]   22f., 26, 54, 99, 117, 137, 153, 175, 189, 220
rum   5, 77f., 103, 118f., 122, 127, 130, 132, 145
ruse   116f., 153, 164, 169, 182, 189f., 218

Said, Edward   21

Scenes of Subjection: Terror, Slavery, and Self-Making in Nineteenth-Century America see also Hartman, Saidiya V. 71, 190
Scully 5 f., 25, 115, 169, 185 f., 208 f.
self-making 2–4, 7 f., 20–24, 27, 52, 54, 69, 76, 88, 99, 102, 104 f., 114, 135–137, 152, 156, 174, 189, 194, 211 f., 217, 219, 221, 246, 249, 252
– self-possession 28, 37 f., 48, 59, 63, 74, 102, 198, 212
Senhor D'Ortega 77, 82, 115, 119, 130, 172, 208, 225 f., 237 f., 241
sentient being 26, 30, 56, 73, 75, 204, 209, 215
Serpell, Namwali 254
Sexton, Jared 18, 57 f., 61 f., 65 f., 70, 88–90, 176–178, 250
sexual economy of slavery *see also* Davis, Adrienne 77 f., 83, 117, 131, 221, 224, 243
Sharpe, Christina E. *see also In the Wake: On Blackness and Being* 9, 16, 18, 22, 48, 55 f., 60, 66, 86–91, 94–97, 117, 149, 172 f., 175, 178, 182, 188, 218, 220, 229, 236, 238, 250–252
Shilliam, Robbie 36, 39–42
shoes 15, 151, 180, 190, 193, 195, 204, 210–213, 216 f., 219, 225, 236, 241
Simone, Nina 210
slave narrative *or* neo-slave narrative 117, 192, 200, 214, 218, 220, 221–225, 227–228, 230, 232–235, 239–240, 242, 243, 244, 249,
– generic field of the 227
– literary tradition of the 200, 214, 218, 224, 242
– trope(s) of the 200, 224, 235
slave pass 200 f., 218, 228
Smith, John 13, 15, 124 f., 128, 139 f., 142, 227 f., 233
social death 7 f., 20, 24, 43 f., 55, 61 f., 67, 70 f., 89 f., 93 f., 97–99, 104 f., 108, 113 f., 117, 137, 189, 202, 214, 216 f., 223, 243 f., 247, 249–251, 253

"Social Death and Narrative Aporia in 12 Years a Slave" *see also* Wilderson, Frank B., III 61
social life 70, 75, 89 f., 114, 243, 251, 253
solidarity 60, 102, 116, 141, 153–157, 164–169, 174, 182, 190, 198, 250
Sorrow 5, 24, 64 f., 83, 85, 97, 99, 114–117, 128, 136 f., 141, 147–151, 153 f., 166, 168–170, 175–189, 203 f., 211, 213 f., 225 f., 251
Spillers, Hortense J. 2, 18, 55, 58 f., 66, 70, 73, 75, 79 f., 83–85, 128, 133, 136, 149, 177, 182, 188, 200, 220 f., 231, 238, 241, 250, 252
sugar 11, 40, 76–78, 103, 119 f., 122, 132, 134 f., 172, 205
Sula see also Morrison, Toni 65

telling 3, 20, 63 f., 88 f., 93, 96 f., 150, 152, 191, 194–196, 201–204, 211, 213–219, 221, 228, 232–234, 240 f., 243, 246, 248, 251 f.
tension 19, 24, 72 f., 107 f., 113, 193, 195, 217
"The Belly of the World: A Note on Black Women's Labors" *see also* Hartman, Saidiya V. 73, 77, 83, 235, 237
The History of Mary Prince, a West Indian Slave. Related by Herself. with a Supplement by the Editor. to which is Added, the Narrative of Asa-Asa, a Captured African (1831) see also Prince, Mary 239
They Were Her Property: White Women as Slave Owners in the American South see also Jones-Rogers, Stephanie E. 160, 238
travelling concepts *see also* Bal, Mieke 21, 253
True Relation of Virginia 142
Trump, Donald 86
Two Treatises of Government 23, 26 f., 30 f., 44

utopia 10, 117, 124, 175, 177 f., 184, 187, 189, 251

"Venus in Two Acts" *see also* Hartman, Saidiya V. 92, 93, 186, 191, 222, 234, 246,
violence   3f., 6f., 19, 24, 29f., 33, 36, 43, 54f., 60–69, 73, 75, 81, 88, 90f., 93, 98, 101, 113, 137, 141, 144f., 148, 151f., 161f., 172, 174f., 180f., 183, 185, 187, 191, 199, 202, 206, 219f., 222, 226, 228, 233, 237–239, 242f., 247, 249–251, 253
Virginia *or* colonial Virginia *or* North American beginnings 1, 2, 5, 7, 8f., 15, 20, 45, 64, 68, 75, 79, 82f., 91, 98, 103, 107, 115f., 122f., 125, 134, 138f., 153, 155, 156f., 161f., 174, 192f., 211, 230, 242, 247, 251
vocabulary   3, 6, 24, 84, 107f., 114, 187, 191, 205, 209, 216f., 219, 252

"wake work"   95
Walcott, Rinaldo *see also On Property*   55, 70
Warren, Calvin L.   14, 54, 63, 75, 201
Weheliye, Alexander   18, 56f.
Weinbaum, Alys E. 78, 177
whiteness   2, 4, 12, 14, 17, 19, 27f., 47f., 51f., 55, 59, 62, 68, 72, 85–88, 127
– and the liberal self   20, 62, 174, 196
– and reputation   48
– and value   72, 227
– the property of   17, 47f., 51, 115, 129, 216, 226, 244
– "Whiteness as Property"   86
white supremacy   76
Wilderson, Frank B. III. *see also Afropessimism* (the monograph); "Social Death and Narrative Aporia in 12 Years a Slave"; *Red, White, and Black*   7, 29, 50, 52, 54, 58f., 61f., 65–70, 74f., 99, 102, 104f., 113f., 138, 153, 169, 190f., 197, 251, 254
Willard Bond   5, 64, 115
Williams, Patricia J.   17, 55, 59, 66, 75, 78, 90
Wynter, Sylvia   2f., 55, 60, 93, 114, 127

Yancy, George   19f.

www.ingramcontent.com/pod-product-compliance
Lightning Source LLC
Chambersburg PA
CBHW020223170426
43201CB00007B/300